ANATOMY OF A SILICON COMPILER

THE KLUWER INTERNATIONAL SERIES IN ENGINEERING AND COMPUTER SCIENCE

VLSI, COMPUTER ARCHITECTURE AND DIGITAL SIGNAL PROCESSING
Consulting Editor
Jonathan Allen

Latest Titles

Symbolic Analysis for Automated Design of Analog Integrated Circuits,
G. Gielen, W. Sansen,
   ISBN: 0-7923-9161-6
High-Level VLSI Synthesis, R. Camposano, W. Wolf,
   ISBN: 0-7923-9159-4
Integrating Functional and Temporal Domains in Logic Design: The False Path Problem and its Implications, P. C. McGeer, R. K. Brayton,
   ISBN: 0-7923-9163-2
Neural Models and Algorithms for Digital Testing, S. T. Chakradhar,
V. D. Agrawal, M. L. Bushnell,
   ISBN: 0-7923-9165-9
Monte Carlo Device Simulation: Full Band and Beyond, Karl Hess, editor
   ISBN: 0-7923-9172-1
The Design of Communicating Systems: A System Engineering Approach,
C. J. Koomen
   ISBN: 0-7923-9203-5
Parallel Algorithms and Architectures for DSP Applications,
M. A. Bayoumi, editor
   ISBN: 0-7923-9209-4
Digital Speech Processing: Speech Coding, Synthesis and Recognition
A. Nejat Ince, editor
   ISBN: 0-7923-9220-5
Sequential Logic Synthesis, P. Ashar, S. Devadas, A. R. Newton
   ISBN: 0-7923-9187-X
Sequential Logic Testing and Verification, A. Ghosh, S. Devadas, A. R. Newton
   ISBN: 0-7923-9188-8
Introduction to the Design of Transconductor-Capacitor Filters,
J. E. Kardontchik
   ISBN: 0-7923-9195-0
The Synthesis Approach to Digital System Design, P. Michel, U. Lauther, P. Duzy
   ISBN: 0-7923-9199-3
Fault Covering Problems in Reconfigurable VLSI Systems, R.Libeskind-Hadas,
N. Hassan, J. Cong, P. McKinley, C. L. Liu
   ISBN: 0-7923-9231-0
High Level Synthesis of ASICs Under Timing and Synchronization Constraints
D.C. Ku, G. De Micheli
   ISBN: 0-7923-9244-2
The SECD Microprocessor, A Verification Case Study, B.T. Graham
   ISBN: 0-7923-9245-0
Field-Programmable Gate Arrays, S.D. Brown, R. J. Francis, J. Rose,
Z.G. Vranesic
   ISBN: 0-7923-9248-5

ANATOMY OF A SILICON COMPILER

edited by

Robert W. Brodersen
University of California
Berkeley

Kluwer Academic Publishers
Boston/Dordrecht/London

Distributors for North America:
Kluwer Academic Publishers
101 Philip Drive
Assinippi Park
Norwell, Massachusetts 02061 USA

Distributors for all other countries:
Kluwer Academic Publishers Group
Distribution Centre
Post Office Box 322
3300 AH Dordrecht, THE NETHERLANDS

Library of Congress Cataloging-in-Publication Data

Anatomy of a silicon compiler / edited by Robert W. Brodersen.
      p. cm. -- (The Kluwer international series in engineering and
computer science ; #181. VLSI, computer architecture, and digital
signal processing)
      Includes bibliographical references and index.
      ISBN 0-7923-9249-3 (acid free paper)
      1. Integrated circuits--Design and construction--Computer
programs. 2. Compilers (Computer programs)   I. Brodersen, Robert
W., 1945-   . II. Series: Kluwer international series in
engineering and computer science ; SECS 181.   III. Series: Kluwer
international series in engineering and computer science.   VLSI,
computer architecture, and digital signal processing.
TK7874.A59   1992
621.3815 ' 0285 ' 53--dc20
                                                          92-10809
                                                                  CIP

Copyright © 1992 by Kluwer Academic Publishers

All rights reserved. No part of this publication may be reproduced, stored in a retrieval system or transmitted in any form or by any means, mechanical, photo-copying, recording, or otherwise, without the prior written permission of the publisher, Kluwer Academic Publishers, 101 Philip Drive, Assinippi Park, Norwell, Massachusetts 02061.

Printed on acid-free paper.

Printed in the United States of America

TABLE OF CONTENTS

CONTRIBUTORS xi
ACKNOWLEDGEMENTS xiii

CHAPTER 1   Introduction and History   1
Robert W. Brodersen

1.1. WHAT IS LAGER 2
1.2. USERS VS. DEVELOPERS 3
1.3. HISTORY 4
1.4. ASSEMBLY VS. BEHAVIORAL SYNTHESIS. 5
1.5. BOOK ORGANIZATION 5

Part I   Framework and Design Entry

CHAPTER 2   The OCT Data Manager   11
Rick Spickelmier and Brian C. Richards

2.1. BASIC STRUCTURE. 11
2.2. POLICY VERSUS MECHANISM 13
2.3. THE OCT OBJECTS 14
2.4. THE OCT PROCEDURAL INTERFACE 16
2.5. OCT PHYSICAL POLICY. 20
2.6. OCT SYMBOLIC POLICY 22
2.7. SUMMARY 24

CHAPTER 3   Lager OCT Policy and the SDL Language   25
Brian C. Richards

3.1. LAGER POLICIES 26
3.2. THE STRUCTURE_MASTER VIEW AND THE SDL LANGUAGE . . . 26
3.3. THE STRUCTURE_INSTANCE VIEW 32
3.4. THE PHYSICAL VIEW. 32
3.5. SUMMARY 33

CHAPTER 4   Schematic Entry   35
Bob Reese

4.1. SCHEMATIC TOOL INTERFACE 35
4.2. DESIGN EXAMPLES. 38
4.3. SUMMARY 44

CHAPTER 5   Design Management   45
Brian C. Richards

5.1. PARAMETERIZATION AND LIBRARY SUPPORT 46
5.2. THE DESIGN FLOW STRATEGY 46
5.3. CONTROLLING THE DESIGN FLOW 49
5.4. THE DESIGN MANAGEMENT STRATEGY 53
5.5. SUMMARY 56

| CHAPTER 6 | Design Post-Processing | 57 |

Marcus Thaler and Brian C. Richards

- 6.1. CAPABILITIES 58
- 6.2. THE POST PROCESSING TOOLS 58
- 6.3. RUNTIME OPERATION 61
- 6.4. SUMMARY 63

Part II Silicon Assembly

| CHAPTER 7 | Hierarchical Tiling | 67 |

Jane Sun and Brian C. Richards

- 7.1. USER INTERFACE 68
- 7.2. MACROCELL DESIGN 68
- 7.3. TILER TECHNIQUES AND ALGORITHMS 78
- 7.4. CELL LIBRARY 81
- 7.5. SUMMARY 84

| CHAPTER 8 | Standard Cell Design | 87 |

Bob Reese and Barry Boes

- 8.1. SPECIFICATION 87
- 8.2. RANDOM LOGIC SYNTHESIS 89
- 8.3. EXAMPLE DESIGN 91
- 8.4. STANDARD CELL LIBRARY 93
- 8.5. LAYOUT GENERATION 95
- 8.6. EXTENDING THE LIBRARY 96
- 8.7. SUMMARY 100

| CHAPTER 9 | Interactive Floorplanning | 103 |

Seungjun Lee and Jan Rabaey

- 9.1. OVERALL FUNCTIONALITY 104
- 9.2. THE FLOORPLAN DESCRIPTION LANGUAGE 107
- 9.3. ALGORITHMS OF THE AUTOMATED PROCEDURES 108
- 9.4. EXAMPLES 124
- 9.5. SUMMARY 125

| CHAPTER 10 | Datapath Generation | 127 |

Mani Srivastava

- 10.1. THE BIT-SLICE DATAPATH MODEL 128
- 10.2. DESIGNING A DATAPATH 128
- 10.3. IMPLEMENTATION DETAILS 134
- 10.4. DATAPATH LIBRARY ORGANIZATION 138
- 10.5. LIMITATIONS 140
- 10.6. SUMMARY 140

CHAPTER 11   Pad Routing   141
Erik Lettang

- 11.1. ROUTING STRATEGY 141
- 11.2. USER INPUT 142
- 11.3. ROUTING ALGORITHM OVERVIEW 143
- 11.4. PLACEMENT ALGORITHM 144
- 11.5. ROUTING ALGORITHM 147
- 11.6. SUMMARY 154

Part III   Verification and Testing

CHAPTER 12   Design Verification   157
Wun-Tsin Jao and Rajeev Jain

- 12.1. VERIFICATION METHODOLOGY 157
- 12.2. NETLIST COMPARISON TECHNIQUES FOR VERIFICATION. . . . 159
- 12.3. ROUTING VERIFICATION 161
- 12.4. PERFORMANCE. 169
- 12.5. SUMMARY 170

CHAPTER 13   Behavior and Switch Level Simulation   173
Lars Svensson, Lars E. Thon and Seungjun Lee

- 13.1. THE THOR SIMULATOR 173
- 13.2. MODEL DESCRIPTION LANGUAGE 174
- 13.3. NETLIST LANGUAGE 176
- 13.4. BUILDING A SIMULATOR 177
- 13.5. DATABASE GENERATION 177
- 13.6. IMPROVED INTERFACE 180
- 13.7. SWITCH LEVEL SIMULATION 182
- 13.8. SUMMARY 185

CHAPTER 14   Chip and Board Testing   187
Kevin T. Kornegay

- 14.1. CHIP-LEVEL TESTING 188
- 14.2. BOARD-LEVEL TESTING 189
- 14.3. IMPLEMENTING SCAN PATH AND BOUNDARY SCAN 191
- 14.4. TEST CONTROLLER BOARD 193
- 14.5. TEST PATTERN GENERATION 195
- 14.6. SUMMARY 196

Part IV   Behavioral Synthesis

CHAPTER 15   DSP Specification Using the Silage Language   199
Paul Hilfinger and Jan Rabaey

- 15.1. BASIC CONCEPTS. 199
- 15.2. LANGUAGE FEATURES 202
- 15.3. EXAMPLES 211
- 15.4. SILAGE-BASED TOOLS AND ENVIRONMENTS 217
- 15.5. SUMMARY 219

CHAPTER 16 Synthesis of Datapath Architectures 221
Jan Rabaey, Chi-min Chu, Phu Hoang and Miodrag Potkonjak

- 16.1. HYPER OVERVIEW AND METHODOLOGY 223
- 16.2. BEHAVIORAL SPECIFICATION AND SIMULATION 225
- 16.3. MODULE SELECTION 229
- 16.4. ESTIMATION 232
- 16.5. EXPLORING THE DESIGN SPACE 235
- 16.6. TRANSFORMATIONS 238
- 16.7. SCHEDULING AND ASSIGNMENT 240
- 16.8. HARDWARE MAPPING 242
- 16.9. COMPARING IMPLEMENTATIONS OF THE IIR FILTER 246
- 16.10. SUMMARY 248

CHAPTER 17 From C to Silicon 251
Lars E. Thon, Ken Rimey and Lars Svensson

- 17.1. ARCHITECTURE EXAMPLES 252
- 17.2. THE RL LANGUAGE 254
- 17.3. FIR FILTER EXAMPLE 261
- 17.4. IMPLEMENTATION OF THE RL COMPILER 264
- 17.5. SUMMARY 267

CHAPTER 18 An FIR Filter Generator 269
Paul Yang and Rajeev Jain

- 18.1. ARCHITECTURAL REVIEW 270
- 18.2. FIRGEN FILTER DESIGN SYSTEM 272
- 18.3. ARCHITECTURE GENERATION 274
- 18.4. DESIGN EXAMPLE AND TEST CIRCUIT 279
- 18.5. SUMMARY 281

Part V Applications

CHAPTER 19 The PUMA Processor 285
Lars E. Thon

- 19.1. THE INVERSE POSITION PROBLEM 285
- 19.2. ALGORITHM SELECTION 288
- 19.3. ARCHITECTURE DESIGN AND EXPLORATION 290
- 19.4. FIXED POINT COMPUTATION 292
- 19.5. HIGH-LEVEL SIMULATION 295
- 19.6. CHIP DESIGN AND VERIFICATION 297
- 19.7. SUMMARY 297

CHAPTER 20 Radon Transform Using the PPPE 301
William B. Baringer

- 20.1. MACHINE VISION ALGORITHMS 302
- 20.2. A REAL-TIME RADON TRANSFORM ARCHITECTURE 304
- 20.3. THE RECONFIGURABLE RADON TRANSFORM SUBSYSTEM . . 306
- 20.4. THE RECONFIGURABLE RADON TRANSFORM ASIC 310
- 20.5. DESIGN PROCEDURE 314
- 20.6. SUMMARY 319

CHAPTER 21   Speech Recognition   321
    Anton Stölzle
    21.1. ALGORITHM AND ARCHITECTURE 322
    21.2. SYSTEM ARCHITECTURE 325
    21.3. CHIP ARCHITECTURES 330
    21.4. DESIGN PROCEDURE 334
    21.5. SUMMARY 338

CHAPTER 22   Conclusions and Future Work   339
    Robert W. Brodersen
    22.1. WHAT WORKED 339
    22.2. WHAT DIDN'T WORK 340
    22.3. FUTURE 341

APPENDIX A   Design Example   343
    Brian C. Richards
    A.1. RUNNING DMOCT TO GENERATE A DESIGN 343
    A.2. DESIGN POST-PROCESSING WITH DMPOST 346

APPENDIX B   Training and Distribution   353
    Bob Reese
    B.1. TRAINING 353
    B.2. LAGER DISTRIBUTION 354

INDEX 357

CONTRIBUTORS

William B. Baringer
Dept of EECS
U. C. Berkeley
Berkeley, CA 94720

Barry Boes
327 Arlington Circle
Ridgeland, MS 39157

Robert W. Brodersen
Dept of EECS
U. C. Berkeley
Berkeley, CA 94720

Chi-min Chu
Dept of EECS
U. C. Berkeley
Berkeley, CA 94720

Paul Hilfinger
Dept of EECS
U. C. Berkeley
Berkeley, CA 94720

Phu Hoang
Dept of EECS
U. C. Berkeley
Berkeley, CA 94720

Rajeev Jain
Department of Electrical Engineering
UCLA
Los Angeles, CA 90024

Wun-Tsin Jao
LSI Logic
Milpitas, CA

Kevin T. Kornegay
Dept of EECS
U. C. Berkeley
Berkeley, CA 94720

Seungjun Lee
Dept of EECS
U. C. Berkeley
Berkeley, CA 94720

Erik Lettang
Hewlett-Packard Company
San Diego Technical Graphics Division
San Diego, CA 92127-1899

Miodrag Potkonjak
NEC Research Institute
4 Independence Way
Princeton, NJ 08540

Jan Rabaey
Dept of EECS
U. C. Berkeley
Berkeley, CA 94720

Bob Reese
Electrical Engineering Department
Mississippi State University
P.O. Drawer EE
Miss State, MS 39762

Brian C. Richards
Engineering Research Laboratory
U. C. Berkeley
Berkeley, CA 94720

Ken Rimey
Department of Computer Science
Helsinki University of Technology
Espoo, Finland

Rick Spickelmier
Objectivity, Inc.
800 El Camino Real
Menlo Park, CA 94025

Mani Srivastava
Dept of EECS
U. C. Berkeley
Berkeley, CA 94720

Anton Stölzle
Dept of EECS
U. C. Berkeley
Berkeley, CA 94720

Jane Sun
Dept of EECS
U. C. Berkeley
Berkeley, CA 94720

Lars Svensson
IMEC labs
Kapeldreef 75
B-3001 Heverlee, Belgium

Marcus Thaler
Chalbisauweg 1
8816 Hirzel
Switzerland

Lars E. Thon
Engineering Research Laboratory
U. C. Berkeley
Berkeley, CA 94720

Paul Yang
Department of Electrical Engineering
UCLA
Los Angeles, CA 90024

ACKNOWLEDGEMENTS

The work described in this book took place since 1983 and was performed by a number of students and staff who contributed in a number of ways. Some contributed software, while others provided designs for the libraries and still others used the software and gave suggestions on improvements and extensions, not to mention finding and reporting the "occasional" bug. In most cases the authors of the various chapters were the developers of the tools that are described. However, this is not uniformly true and the contributions of those who were not able to participate in the writing of this book should be acknowledged, including Kahlid Azim, Steve Pope, Peter Ruetz, Gordon Jacobs, Teresa Meng, Shankar Narasaramy, Dev Chen, Mats Torkelson, Robert Yu, Phil Schrupp, Sue Mellers, Barbara Thaler, Paul Tjahjadi, and Etan Cohen. The Octtools developments led by Andrea Casotto and Rick Spickelmier and the Lightlisp interpreter written by Wendall Baker are basic components of the system. Particularly critical were the efforts of Rajeev Jain, Brian Richards and Bernard Shung who were responsible for the evolution of the basic framework and design management strategies. Brian also was responsible for "taming" Framemaker sufficiently to make the text and graphics formatting of the book possible.

In addition contributions from other Universities have been considerable. In particular, the research group of Rajeev Jain from UCLA has been a partner in the recent development of many parts of the system which are reported here. Similarly, Bob Reese and his group from Mississippi State University has been a considerable resource in not only adding functionality to the system but in organizing the distribution and training. The Institute for Technology Development, which is associated with MSU, provided the Standard Cell Library along with characterization and documentation. The testing software was provided by Mel Breuer's group at the University of Southern California and the simulators were from the Stanford University research groups of Tom Blank (THOR) and Mark Horowitz (IRSIM).

Of key importance was the support obtained from our Government and Industrial sponsors. The primary funding for the work described here was from the Information and Science Technology Office of DARPA. In particular, the program managers Paul Losleben and John Toole were strong supporters of this work, who continually provided the motivation for us to develop aggressive research goals and to attain them. Also, provided by DARPA support was access to the MOSIS fast turnaround implementation service of the Information Science Institute of the University of Southern California, which was under the leadership of George Lewicki. The ability to obtain fabricated chips within a couple of months of submission was critical in the development of this system. In addition, support from the Justice Department has been useful in providing feedback from practicing engineers. Additional support and feedback came from our industrial sponsors General Electric, IBM, Harris Semiconductor, LM Ericsson, TRW and LSI Logic.

ANATOMY OF A SILICON COMPILER

1

Introduction and History

Robert W. Brodersen

In the early 1980's, there were predictions of an integrated circuit design crisis, which was to occur when circuit complexities surpassed the hundred thousand transistor level. Commercial chips which have well over a million transistors are now available. What happened?

The answer lies in the widespread application of computer aided design tools which automated much of the low level record keeping and data base management needed for large designs as well as automatically performing design tasks such as routing and verification. These tools made it possible for large chips to be designed and verified with a reasonable time investment (10's of people--years).

The set of tools described here, however, are even more ambitious, since the goal is to produce a verified chip design within a few days. This coupled with a rapid fabrication capability, meant that a single designer could obtain a fabricated chip within 2 1/2 to 3 months.

To simplify the task in the initial stages of the tool development, designs were considered acceptable even if they were not highly optimized with respect to silicon area and performance. The strategy was to optimize at the architectural level, so that the system requirements could be met without requiring transistor level optimization. In fact, the lowest level units which the designer can specify are modules (adder, ram, etc.) and their interconnection. Even more desirable is to design at even higher level specifications, ranging up to hardware independent behavioral descriptions which further facilitates architectural exploration.

The basic approach is to use techniques that have become known as silicon assembly and compilation and the system which shall be described here in detail is called LAGER.

1.1. WHAT IS LAGER

LAGER is composed of design managers, libraries, design tools, test generators and simulators, which are interfaced to a common database. The basic approach is to use a set of libraries of hand designed cells that are configured so that they can be automatically assembled by layout generation programs. It is in the design of the libraries and the technology in which they are designed that the issues of performance and silicon area efficiency are most directly affected. Another major set of tools in LAGER involve using higher level descriptions of behavior to synthesize the structural description, which in turn is used to provide the necessary input data to the layout generators.

In order to minimize the size of the libraries, extensive use is made of parameterization. The various circuit blocks, such as adders or memory, are designed so that the size and other features can be varied through input parameters passed to the layout generators.

The extensive use of parameterization and the use of high level descriptions motivates the use of textual input as the primary entry mechanism. The description of a chip (or system) completely in text is a significant break from the conventional means of electronic design entry, which is typically based on schematics. In the approach used here, a variety of languages were used, including a language for describing the parameterized structure, a language for boolean expressions, a procedural language ("C") for behavioral description, and an applicative language (Silage) for behavioral description. A graphical input capability has also been integrated with enhancements to give it the flexibility of a language, but this has typically only been used for designs of low complexity.

The structural input description can be simulated, using the compiled simulator THOR. Each block in the cell library has a behavioral model written in "C" which allows a program to be generated that mimics the exact operation of the circuit which will be generated from that input description. It is at this level that the design is debugged, and compared to the input specification. Only after this task has been accomplished is the layout generation phase entered.

The input description is sufficient to automatically generate the complete physical description of all the geometries used to generate the masks for fabrication. From this physical description, the individual transistors and their sizes are extracted, as well as a netlist describing their interconnection. The extracted netlist is compared to the connections designated in the input description to check the connectivity. A switch level simulator, IRSIM, can also use this same data and be compared with the high level THOR simulation to determine if all timing constraints are met.

The large number of tools that are involved in this design system would be very time consuming to learn to use, necessitating the design management tools, DMoct and DMpost. DMoct simplifies the task of generating a design, by presenting a single user interface for the entire design process, providing error reporting and allowing conditional operation of the tools. A post processing design manager, DMpost, performs a similar function for the verification tools.

1.2. USERS VS. DEVELOPERS

There are two types of designers which must be supported with LAGER. At one level is the user who enters the structural and behavioral descriptions to produce a chip for a particular application. The other level is the developer who adds cells to the libraries or new tools to the system.

The user is primarily interested in the syntax and semantics of the various forms of entry, and primarily interacts only with the design management tools or with the interactive tools in the design process. This person is not required to be familiar with the underlying database or the techniques used in the automatic tools.

The developer, on the other hand, adds cells to the libraries and thus must be familiar with transistor level physical design. This requires not only the knowledge involved to implement a good design, but also the constraints that must be incorporated so that the designs are compatible with the automatic tools. In addition, if a new tool is to be incorporated then the developer must also be familiar with the policies involved in integrating with the underlying database, as well as possibly requiring extensions to the design management software and the descriptions to input the design.

1.3. HISTORY

In the first attempts at writing these tools, were single programs which performed the entire design task [Rabaey85, Ruetz86]. This, however, was only possible when the task was restricted to a limited range of architectures. The initial focus was on programmable microcoded multi-processors for digital signal processing [Rabaey85]. The degree of freedom included word-widths, the inclusion of major blocks such as the address arithmetic unit and decision making finite state machines and the number of processors and interconnection paths between them. The design of these circuits was from an assembly level language which was used to determine the parameters of the design to be generated as well as the microcode of the individual processors.

The use of this system for a variety of applications made it clear that more flexibility was desirable. For those applications which required high throughput, an optimized datapath was often required and microcoding was often not needed. This led to the development of a set of tools instead of one program, which perform tasks such as module generation and floorplanning. The data was transferred between these programs through the use of standard ascii files, without a centralized database. The flexibility of the circuits that could be generated was improved, but it was found difficult to add new tools as information about the design was distributed across a number of different files and formats. Also as the designs increased in size an increasing amount of time was involved in writing and parsing the design files.

A centralized database was then developed which was based on the object oriented programming package, Flavors, which ran under LISP [Shung88]. Integration of new tools and flexibility in their use was possible but the inefficiencies of LISP and the lack of a persistent database made it clear that this would not be adequate for the ever increasing size of the designs.

OCT, an object oriented database designed for VLSI applications, which had been under development for some time was then incorporated as the underlying database. This immediately gave access to a number of new tools which were already interfaced to OCT. In addition, the X windows interface was adopted and policies were adopted for the use of the OCT database so that data could be stored from high level flow graphs through to the actual physical design.

1.4. ASSEMBLY VS. BEHAVIORAL SYNTHESIS

The generation tools in LAGER can be divided into two basic categories which perform *behavioral synthesis* and *silicon assembly*. The *behavioral synthesis* components are exemplified by Hyper, Firgen and C-to-Silicon. These take a behavioral input, which is relatively hardware independent, that describes the specifications desired by the designer. They output a structural description (interconnection of circuit blocks and their parameters), which provides the input to the *silicon assembly* tools. These tools implement the actual physical design which can be used for fabrication, and include Flint, TimLager, dpp, Padroute, and Stdcell and are controlled by DMoct.

1.5. BOOK ORGANIZATION

The book is divided into five parts. Part I, which covers Chapters 2-6, describes the framework, including the underlying database, as well as the design management tools which control the design process. The data manager used is OCT [Harrison86], along with a specific set of policies that was developed for a library based, parameterized design methodology. A language was developed, SDL, which allows a text description that can be converted into the database by the design management tool DMoct. Also described is a schematic entry capability which allows direct graphical entry into the database. The post processing design manager, DMpost, is presented which provides a consistent user interface during the verification process.

The silicon assembly portion of LAGER is described in Chapters 6-10 which comprise Part II. First is TimLager, which is used to generate modules by tiling library cells which have been designed so that all connectivity is through abutment of adjacent signal lines (no routers are required). The sizes of these arrays are defined by parameters obtained from the user or higher level tools. Logic synthesis is provided by the script Bds2stdcell which converts a logic description written in the language BDS to a netlist. The actual generation of a standard cell module is accomplished by the script, Stdcell, which calls optimization tools for placement and routing. The placement and routing of modules is done by the interactive floorplanner, Flint. This tool is also used with TimLager to implement bit sliced datapaths, under control of the generator dpp, which also performs some modification of net lists to minimize the length of nets within the datapath. The last step in the silicon assembly process is done by Padroute, which performs the specialized routing task required to connect the signal, power and ground lines from the rectangular chip core to the bonding pad ring.

Testing and verification is described in Part III which covers Chapters 11-13. The on-chip test strategy makes use of scan paths and a limited level of automatic test pattern generation, provided by TGS. The direct verification of netlists is done by DMverify and verification through simulation by THOR and IRSIM.

The tools and representations used for the behavioral synthesis tools are described in Part IV in Chapters 14-17. The Silage applicative language, which is optimized for signal processing descriptions, provides a behavioral input description. The HYPER tools take this description and use transformations and scheduling algorithms to synthesize a datapath with the associated control that meets real time computation constraints with the minimum amount of hardware. C-to-Silicon is a synthesis tool which starts from a description using a subset of the procedural language "C" and a description of an architecture and produces the microcode and parameters necessary to generate a circuit. The third behavioral synthesis tool, Firgen, takes specifications in the frequency domain and generates high performance digital filter circuits.

Part V, Chapters 18-21, describe applications of the tools in a variety of areas. A chip for calculation of the inverse kinematics for a six degree of freedom robot (PUMA arm) is discussed which made use of the C-to-Silicon approach. A highly parallel architecture for calculation of the Radon transform, which is useful in machine vision, is described, followed by the design of a real-time large vocabulary speech recognition system.

It is suggested that the first chapters on the database and design management methodology be read since the following chapters assume familiarity with the concepts presented there. The subsequent chapters are more independent and can be read out of order.

REFERENCES:

[Harrison86] David S. Harrison, Peter Moore, Rick Spickelmier, A. Richard Newton, "Data Management and Graphics Editing in the Berkeley Design Environment," *The Proceedings of the IEEE International Conference on Computer-Aided Design*, November, 1986.

[Rabaey85] J. Rabaey, S. Pope and R. Brodersen, "An Integrated Automatic Layout Generation System for DSP Circuits", *IEEE Trans. Computer-Aided Design*, CAD-4, pp. 285-296, July 1985.

[Reutz86] P. Ruetz, S. Pope, and R. Brodersen, "Computer Generation of Digital Filter Banks," IEEE Trans. on CAD, Vol. 5, pp. 256-265, April 1986.

[Shung88] C. Shung, "An Integrated CAD System for algorithm-specific IC design," Ph. D. dissertation, University of California at Berkeley, June 1988.

Part I

Framework and Design Entry

2

The OCT Data Manager

Rick Spickelmier and Brian C. Richards

OCT is a data manager based on object-oriented principles for electronic CAD applications [Harrison86a, Harrison89b, Spickelmier90a]. OCT offers a simple interface for storing information about the various aspects of an evolving system design which can range from the lowest level of physical description of a chip to board level specification. The data manager provides a mechanism to store and retrieve information through a set of C language procedural calls that buffer the user from the actual data storage strategy. It is assumed that the underlying operating system is UNIX, as do most of the tools within LAGER, otherwise the data manager is relatively independent of the particular computing platform.

2.1. BASIC STRUCTURE

The basic unit in a design is the *cell*. This can be as small as a transistor or NAND gate, or as large as a complete board containing many components. A cell can consist of instances of other cells, such as a NAND gate consisting of several transistors or the floorplan consisting of an ALU, register file, *etc*. A cell can have many aspects or *views*, depending on what point in the design you are at and on what design style you use.

There can be a schematic view, showing in an abstract way what subcells the cell consists of and how they are connected. There can be the symbolic view, where additional information of rough relative placement, subcell size and shape, and initial implementation of interconnect might be kept. And even more refined, the physical view, where the implementation is fully defined with exact placement and specific geometry. In addition there can be quite different views, such as the simulation view, which might contain the description of the cell in a format that a particular simulator could understand. What views a cell may have, what is contained in those views, and how they are related are not issues addressed by OCT; these decisions, which are called the OCT policy, are specific to a given system, and are left to authors of the tools that use OCT. The particular policy used in the LAGER system is described in the next chapter.

Finally, there is the concept of *facets*. Cells are hierarchical, *i.e.* they contain instances of other cells which in turn may contain instances of other cells, *etc*. For various applications, the designer may wish to cut off this hierarchy; instead of continuing to traverse the hierarchy by processing the contents of a view, the cell might be represented by some simplified abstraction. For a graphics editor, the abstraction might be a bounding box, for routing the abstraction might be the terminals and routing regions of the cell, and for a netlist verifier, it might be just the terminals and nets on the boundary of the cell that need to be checked against those of neighboring cells (thus avoiding rechecking the interior of cell each time it is instantiated). We call these various abstractions *facets* of the view.

Each view has a facet named *"contents"* which contains the actual definition of the view, as well as various application-dependent interface facets. Again, OCT does not explicitly define what interface facets may exist, and what their relation to the contents facet might be. The only facet OCT has any explicit knowledge about is the *contents* facet, which is the facet that defines the name and number of external (formal) terminals that a view has. When an instance of a view or a new interface facet is created, the *contents* facet of that view is consulted to find out what terminals the instance or facet inherits from the view. Thus the *contents* facet must be created before any other facet is created and before any instance of that view is created.

The final specification for a facet is the version. The version allows there to be multiple time sequenced facets. Versions are generally ignored by most OCT applications and are only dealt with by data management systems [Silva89].

The facet is the unit that is edited in OCT. A particular facet of a view of a cell is opened and edited independent of the other facets of that view and the other views of that cell. The facet consists of a collection of objects that are related by *attaching* one to another (see Figure 2.1). A box is put on a particular layer by attaching it to that layer, a terminal is shown to be part of a net by attaching it to that net, a box is shown to implement a terminal by attaching it to that terminal. Objects can be attached to more than one object and can have more than one object attached to themselves. Being attached is not particularly strong: if you delete an object, you do not delete objects attached to it, you merely detach the connection. The structure created by attachments can be thought of as a directed graph of OCT objects

2.2. POLICY VERSUS MECHANISM

OCT provides a *mechanism* for representing data but places no meaning on the data. *Policy* is used for assigning meaning to the data represented using OCT. For example, OCT has objects that represent layers and geometry, but does not specify how the objects are related to give the *meaning* of a geometry implemented on a given layer. The *policy* states that a geometry that is contained by a

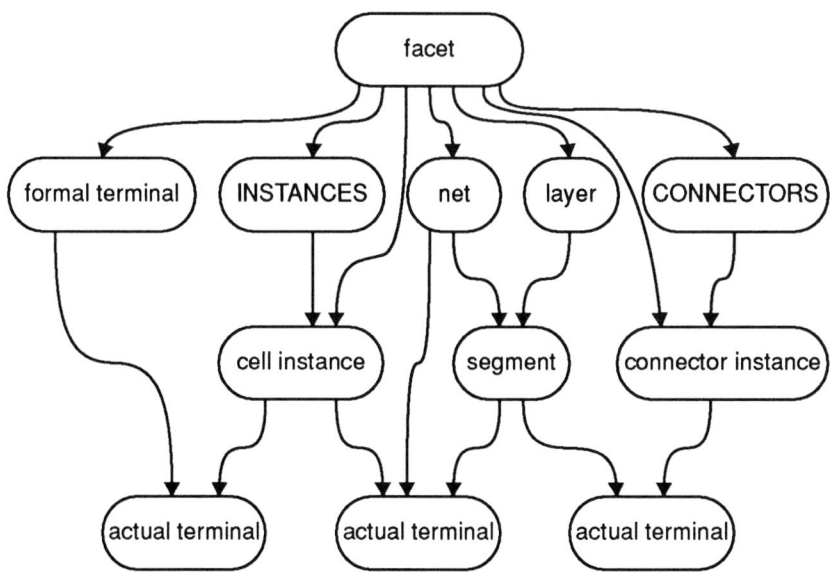

Figure 2.1: Sample Attachments which conform to OCT symbolic policy

layer is implemented on that layer. As another example, OCT has objects that represent nets and terminals, but does not specify how connectivity is represented. The *policy* describes how terminals and nets are used to represent connectivity. See [Spickelmier90b] for detailed information about specific OCT policies; the next chapter describes the extensions to those policies for LAGER. The next few sections describe the objects supported by OCT and the procedural interface that is used to create, modify, delete, and access those objects. The particular set of objects and procedures was chosen based on experience with previous systems. Similarly, the LAGER OCT policy evolved from an earlier version of LAGER which was based on LISP [Shung88].

2.3. THE OCT OBJECTS

OCT supports a set of primitive types chosen for their usefulness in VLSI CAD. All OCT objects have type and identification attributes. There are two identification attributes, **objectId** and **externalId**. An **objectId** is the per-session identifier for a specific object. Given the objectId, the procedure `octGetById` will return the corresponding object. This means that the **objectId** can be used as a form of pointer to the object, so the entire object need not be kept to reference it. An **objectId** is guaranteed to be unique across all objects in all open facets during an OCT session. When an object is deleted, the **objectId** is marked as invalid and may not be used in any further operations during the session. An **externalId** is an identifier that is scoped by the containing facet and exists across sessions. This is opposed to the *objectId*, which is unique across all objects in all opened facets. Combined with the facet that contains the object, the **externalId** is unique across all objects in all facets. OCT supports the following object types:

2.3.1. Design Objects

The basic design object in OCT is the facet, OCT_FACET. The facet can be read-only or modifiable, and can contain other objects, but may not be contained by any object.

2.3.2. Geometric Objects

OCT supports a wide range of geometric types: point, box, circle (containing support for donuts and arcs), path, polygon, edge, and label. OCT also provides a layer type for those wishing to implement a policy requiring geometry to be attached to a layer to specify that the geometry is on that layer (as is the case for all of the OCT policies used in LAGER).

2.3.3. Interconnection Objects

OCT supports two types of objects for the specification of interconnection, terminals and nets. Terminals can be associated with a view or an instance. A terminal of a view is called a formal terminal. A terminal of an instance is called an actual terminal. You cannot explicitly create an actual terminal. They are instead implicitly created when an instance is created, with an actual terminal being created and attached to the instance for each formal terminal in the view itself. Formal terminals can only be created in the *contents* facet of a view; all other facets may only reference them. A net can be used to represent a logical connection between terminals by attaching the terminals to the net. Terminals and nets have a width attribute that can be used for representing vectors of terminals and nets.

2.3.4. Hierarchy Objects

OCT supports the building of hierarchical designs with the instance object, OCT_INSTANCE. An instance is a reference to a facet, called the master of the instance. Instances have an origin and a transformation. When instances are created, an actual terminal for each formal terminal of the facet is created and attached to the instance.

2.3.5. Annotation Objects

OCT supports properties and bags as a means of specifying extra information not directly supported by OCT. A property contains annotation that can be attached to any object (including another property). A property consists of a name, a type, and a value of that type. The supported types are: integer, floating point number, string, collection of bytes, integer array, and floating point array. The bag object, OCT_BAG, is used for collecting other objects. Typical policy applications include grouping OCT_INSTANCE objects into a category, or containing a list of design parameters. Using bags and properties, arbitrarily complex structures (composite objects) can be built up.

2.3.6. Change List Objects

OCT supports the ability to monitor changes in a design. The change list and change record objects are used for monitoring changes in an OCT facet. For each set of operations and objects that is to be monitored, a change list is created and change records are created and attached to the change list whenever a monitored change takes place.

2.4. THE OCT PROCEDURAL INTERFACE

The procedural interface, along with the basic concepts, make up the heart of OCT. The particular set of objects can change, but the procedural interface defines the way users think of OCT. Since CAD tools tend to navigate through designs, rather than make complex ad-hoc queries (as in rational database systems), OCT has a number of routines for traversing a design. OCT also has a limited number of simple query routines for those cases where appropriate (accessing objects by name). For a complete list and description of the OCT procedural interface, see [Spickelmier90c]. In order to ease the use of the OCT procedures (and based on feedback on the first version of OCT), the procedures signal the majority of errors via raising an exception (using the *errtrap* facility [Spickelmier90c]. This eliminates the necessity for programmers having to surround all OCT procedures with a check of the return value of the procedure.

2.4.1. Starting a Session

In order to access the database, initiate and terminate the interaction with the database, OCT provides the procedures `octBegin` and `octEnd`. The former will flush out any existing design facets from the database manager, and the latter will force existing data to be purged if new facets are opened.

After `octBegin`, the facets containing objects must be opened before the objects themselves can be viewed. There are three routines for opening facets. `octOpenFacet` is the primary method for opening the top level of a design or creating a new facet. `octOpenRelative` is a version of `octOpenFacet` that opens up a facet with a location relative to some other OCT facet. This could be used for opening up cells in the same library or in the same workspace or configuration. `octOpenMaster` opens up the master (describer) of an instance. `octOpenFacet` and `octOpenRelative` can be used for creating new facets, opening up old ones (for read or modify access), and for purging the modifications to the current version and bringing in a fresh copy (revert).

Once a facet is opened, all objects contained in the facet are available to the user. Whether all of the objects are read into memory when the facet is opened or paged in and out as needed is an implementation detail, and both forms have been implemented.

When there is no longer a need to have the facet open (i.e. objects contained in the facet available), the facet can be closed, via `octCloseFacet`, with all changes committed to persistent storage, or freed, with all changes thrown away (aborted). Pairs of opens and closes/frees are reference counted so that the objects are still available until the last close/free is called. This allows programming libraries to open and close/free facets without affecting the open status of the facets being used by the program that uses the programming library. A facet can also be committed to persistent store and have the objects available after the commit (also known as commit and hold). This can be accomplished by `octFlushFacet`. The OCT procedural interface also provides a procedure for committing the facets to an alternative name (`octWriteFacet`), and for copying a facet (`octCopyFacet`).

2.4.2. Access to Objects

The OCT procedures do not give the user access to the actual object that was made available (referred to as 'OCT space') due to an `octOpenFacet`, `octOpenMaster`, or `octOpenRelative` call, but instead they copy the user-visible portion of the OCT object into a user-supplied structure. This provides a layer of safety: the user can modify and destroy his copy of the object without corrupting the data in OCT space. A commit operation (such as `octModify`) is necessary to update OCT space.[1]

All objects have two methods for directly accessing them and some have a third method. There are two identifiers associated with each object. The first is a per-session object identifier (objectId) that is unique across all facets that are currently available (contained in open facets). The procedure `octGetById` will retrieve an object using an objectId. The second identifier is called an external identifier (xid). This identifier is unique inside of a facet, and combined with the facet name is unique across all facets (even those that have not been opened). This identifier persists across sessions and can be used for relating objects across facets. The xid of an object is returned by the `octExternalId` procedure. The procedure `octGetByExternalId` is used to retrieve an object based on its xid.

The third method of retrieving an object based on an identifier is by name. Some object types have names associated with them (properties, bags, instances,

---

[1.] The only exception to this is in the case of strings. The char pointers for strings (bag names, property values, etc.) are pointers into the data in Oct space. Thus directly modifying or freeing the string is not allowed and will corrupt 'Oct space'. Strings of copied out objects should have been put in static buffers that are only valid until the next Oct call.

layers, terminals, nets, and labels). The call `octGetByName` takes a name, type, and container and returns the object of the given name and type contained by the container. If multiple objects of the same type and container have the same name, the particular object that is returned is not defined; OCT does not try to force uniqueness of names, this is considered a policy.

2.4.3. Navigation of Designs

There are two ways to navigate in OCT. One is by direct reference as described previously (`octGetById`, `octGetByName`, *etc.*). The other is by the use of `generators`. Generators are used for sequencing through objects of a given type or set of types that are attached to an object (such as all terminals attached to a net). Generators must be *initialized* by a call to `octInitGenContents` or `octInitGenContainers`, and objects are returned in sequence by subsequent calls to `octGenerate`. If objects that would occur in the generation sequence are modified such that they will not be in the generation sequence (deleted, detached from the container), they will be skipped. If new objects appear (creation or attachment to the container) they will be seen by the generator. OCT generators are also safe, meaning that if the object that is being generated over is deleted, the generation sequence terminates cleanly, or if the object that appears next in the sequence is deleted, the generator continues on to the next object in the sequence.

2.4.4. Creation, Deletion, and Modification of Objects

OCT supports the creation of objects via the `octCreate` procedure. Objects in OCT are always created relative (attached) to some other object. Objects can be deleted with the `octDelete` procedure. Objects are modified by taking the original object (returned by any of the OCT retrieval procedures), modifying the fields that are to be changed, and then calling `octModify` to tell OCT about the changes. In addition to altering the attachment structure of a facet by using the creation and deletion procedures, there are two procedures for directly changing attachments, `octAttach` and `octDetach`.

2.4.5. Support for Hierarchy

OCT allows the specification of hierarchy by providing instance objects. Instance objects are used for referring to OCT facets. The OCT facet referred to by the instance is known as the *master* of the instance. There are two ways to reference facets using instances: *relative* and *absolute*. Relative means that the context (location) of the facet that contains the instance is used to find the master of the instance. Relative allows the entire design to be moved without modifying the instance objects. This method would be used for the cells specific to a particular design. Absolute means that the context is not used to find the master of the instance; all the necessary information is contained in the instance object. This method would be used for accessing cell libraries; the location of the cell library does not change as the design moves.

When an instance of a master is created, a copy of the formal terminal of the master, called an actual terminal is created for each formal terminal in the master (and attached to the instance).

2.4.6. Change Propagation and Dealing with Inconsistencies

One difference between a static data format (*i.e.* EDIF [EIA87]) and a database and interface (*i.e.* OCT) is how changes and inconsistencies are dealt with. Static data formats are snapshots of a design at a particular point in the design process, they do not change. Databases are dynamic and thus constantly changing, therefore techniques for dealing with these changes must be developed.

There are two types of changes that are automatically propagated by OCT: changes to the bounding box of a master and changes to the formal terminals of a contents facet. If the bounding box of the master changes, the bounding box information associated with the instances of that master are updated. If the name of a formal terminal in the contents facet of a view changes, the name is changed in all other facets of the view and all instances of all facets of the view. If a new formal terminal is created in the contents facet, the terminal is added to all other facets of the view and to all instances of all facets of the view. If a formal terminal is deleted from the contents facet, the terminal is deleted from all other facets of the view and all instances of all facets of the view. These operations are performed transparently by OCT.

Under certain circumstances, there is one type of change that will not be done transparently. When a formal terminal is deleted, if the equivalent formal terminal in another facet is contained by something other than the facet or contains objects itself, it will not be deleted and instead will be marked as *inconsistent*. Once the attachments to the formal terminal are removed, the next time the facet is opened, the formal terminal will be deleted. This procedure also applies to actual terminals; if the actual terminal that corresponds to a deleted formal terminal is contained by something other than the instance or contains objects itself, it will not be deleted and instead with be marked as *inconsistent*. Once the attachments to the actual terminal are removed, the next time the facet is opened, the actual terminal will be deleted.

Besides the inconsistencies that can be encountered when terminals change, inconsistencies can also occur when the masters of instances disappear. Masters can disappear for three reasons: the master has been removed, the master has been moved, or the facet that references the master has been moved. The later case is only a problem if the master is referenced via relative paths; if the master is referenced via an absolute path, the facet that contains an instance of the master can be moved anywhere where the absolute path can be reached. When opening up a facet that contains instances, if the masters can not be found, the instances are marked as inconsistent.

There are many instances in CAD tools where finding out what has changed in a design is important. The region searching package supplied with the OCT Tools [Spickelmier90a] uses change lists to update internal data structures when changes occur in the data base. The graphics editor VEM, which is used for viewing physical layouts, uses change lists for undo, for determining when new layers have been added, and for determining what regions of the screen need to be redrawn after a command [Harrison89a, Harrison89b]. OCT supports change monitoring via change lists and change records.

A change list object has two fields, one for the types of operations to be monitored and one for the types of objects that operations of the selected type should be recorded. Multiple change lists can be created, each looking for any combination of operation/object pairs. The change record contains information about the type of the operation, the xid of the object changed, and the xid of the container of the object changed. Performing an `octBB` on the change record will return the bounding box of the object before the operation. In the case of a change record that represents a deleted object, generating the contents of the change record will return the deleted object.

2.5. OCT PHYSICAL POLICY

The OCT "physical" view of a design is the simplest form of design representation, storing the design in terms of geometrical shapes[Spickelmier90b]. The OCT_FACET object, representing the contents of a "physical" view, contains primarily OCT_LAYER, OCT_TERM, and OCT_INSTANCE objects. Each OCT_LAYER typically corresponds to a photographic mask used for designing a custom chip or printed circuit board, and can contain any number of OCT_BOX, OCT_CIRCLE, and other geometrical objects. Figure 2.2 illustrates how objects are connected to represent some simple geometrical shapes.

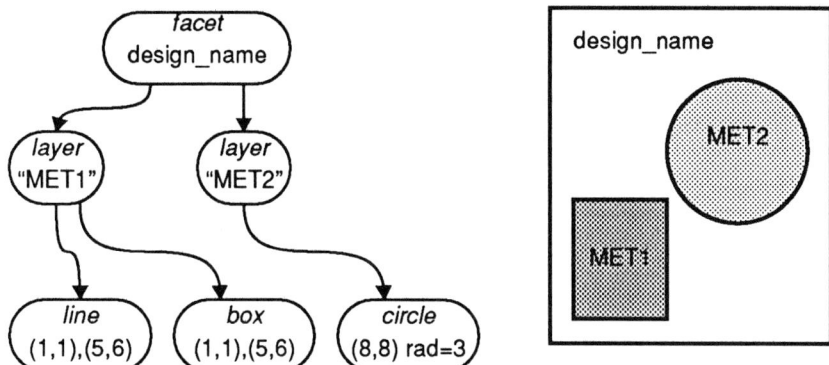

Figure 2.2: Representing physical geometry in OCT

The OCT_TERM objects define terminals, which serve as ports for passing information between the different levels of design hierarchy. A terminal will usually contain one or more geometrical object, also contained by an OCT_LAYER, to define the implementation. For a VLSI macrocell, a terminal is typically implemented as a piece of metal or polysilicon near the edge of the corresponding circuitry. For a printed circuit board design, each pin on a package or connector might have an associated terminal. These OCT_TERM objects contained by the OCT_FACET are called "formal" terminals.

The OCT_INSTANCE objects refer to instances of smaller, self-contained designs, which most commonly have "symbolic" or "physical" views. The referenced subcircuit is referred to as the master of the given instance. The OCT_INSTANCE objects automatically contain copies of the OCT_TERM objects defined in the master. These copies are called "actual" terminals.

For the most part, the remainder of the OCT Physical policy defines annotation of parts of the facet. Several OCT_PROP objects define which policy and technology is used for the design. Also, OCT_INSTANCE objects are generally grouped together under an OCT_BAG named the "INSTANCES" bag. This bag is used by tools to categorize different types of instances; for most "physical" views, the same instances are grouped under the "INSTANCES" bag as they are

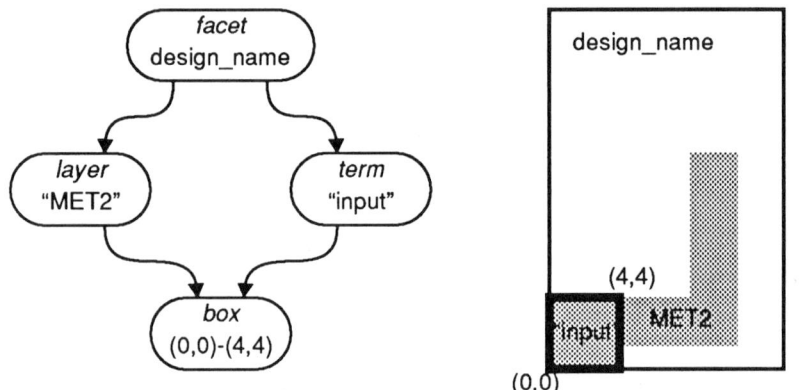

Figure 2.3: Representing a terminal in OCT

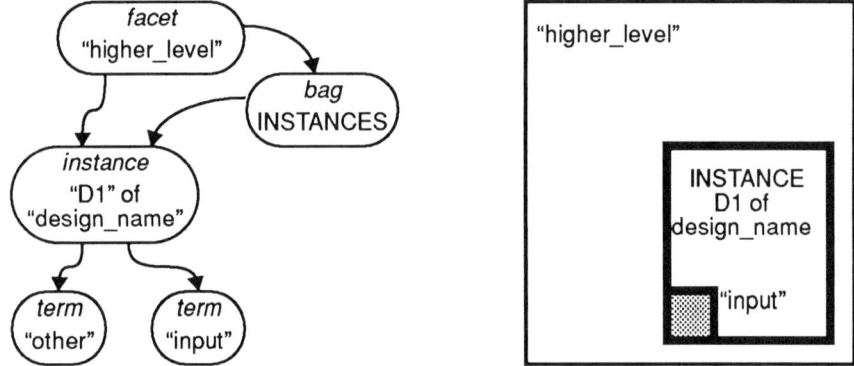

Figure 2.4: Instantiation of a design in OCT

attached to the facet. The need for this distinction is more apparent in the OCT Symbolic policy which follows.

2.6. OCT SYMBOLIC POLICY

The OCT Symbolic policy extends the Physical policy, adding interconnectivity information. Figure 2.5 illustrates how a net object is used to represent connections between one or more terminal. An OCT_NET object, contained by the OCT_FACET, can contain one or more OCT_TERM, where both formal and actual instance terminals are allowed. A net may refer to anything from a terminal to an

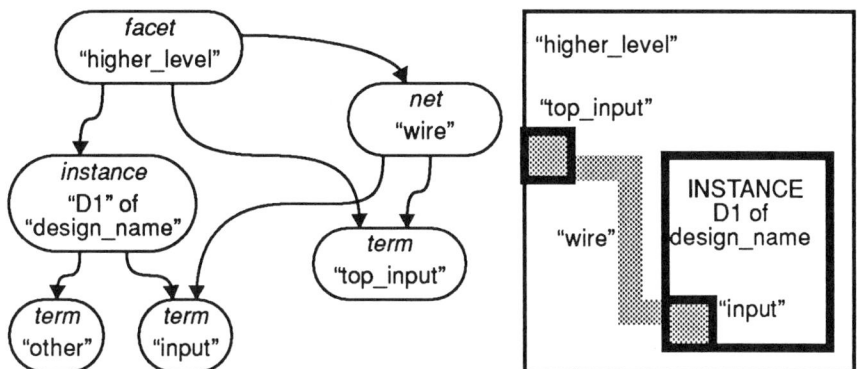

Figure 2.5: Interconnection with nets in OCT

abstract data type which is passed between the contained terminals. For the most part, only the OCT_SEGMENT object is used to represent geometry in a symbolic view. A design will consist of instances of physical or symbolic masters, with segments interconnecting the terminals. The OCT_SEGMENT object defines two points and a width, and may be diagonal. If more than one segment is needed to define a path between terminals, instances called *connectors* are placed at the corners between adjacent segments. Connector instances can define connections between layers, so that the segments of a given net need not be defined on the same layer. These instances are identical to the other OCT_INSTANCE objects, with the exception that they are contained by a bag named "CONNECTORS". In

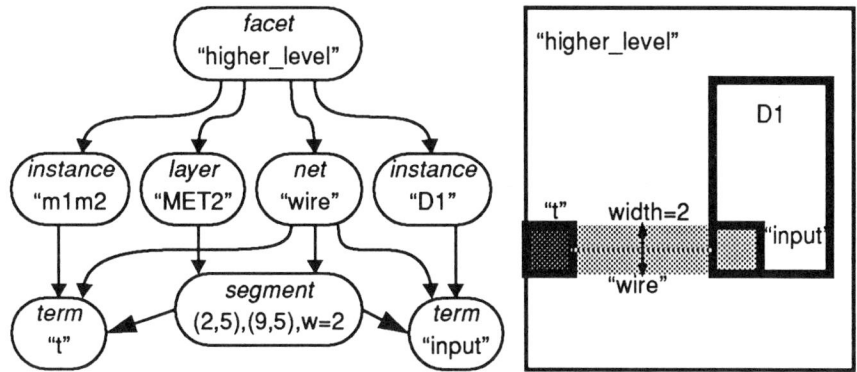

Figure 2.6: Connectors and segments on a net.

general, all connector instances and segments could be deleted without losing any interconnectivity information, and hence they are grouped separately from the instances under the "INSTANCES" bag, described earlier.

2.7. SUMMARY

A key advantage of OCT is that by adhering to a consistent policy, tools from a variety of sources can be made to work together. In addition, the adherence to object-oriented principles considerably simplified the task of interfacing new tools.

REFERENCES:

[Harrison86a] David S. Harrison, Peter Moore, Rick Spickelmier, A. Richard Newton, "Data Management and Graphics Editing in the Berkeley Design Environment," *The Proceedings of the IEEE International Conference on Computer-Aided Design*, November, 1986.

[Harrison86b] David S. Harrison "VEM: Interactive Graphics for Oct," *Master's Thesis*, The Department of Electrical Engineering and Computer Sciences, U. C. Berkeley, Berkeley California, 1989.

[Silva89] Mario Silva, David Gedye, Randy H. Katz, and A. Richard Newton, "Protection and Versioning in Oct," *The Proceedings of the ACM/IEEE Design Automation Conference*, pages 264-269, Las Vegas, Nevada, June 1989.

[Spickelmier90a] Rick L. Spickelmier, editor, *Oct Tools Distribution 4.0*, U. C. Berkeley Electronics Research Laboratory, Berkeley, California, March 1990.

[Spickelmier90b] Rick L. Spickelmier, Jeff L. Burns, and A. Richard Newton, "Policy Guides for OCT," *Technical Report UCB/ERL M90*, U. C. Berkeley Electronics Research Laboratory, Berkeley, California, January 1990

[Spickelmier90c] Rick L. Spickelmier, Peter Moore, and A. Richard Newton, "A Programmers Guide to Oct," *Technical Report UCB/ERL M90*, U. C. Berkeley Electronics Research Laboratory, Berkeley, California, January 1990

[Shung88] C. Shung, "An Integrated CAD System for algorithm-specific IC design," *Ph.D dissertation*, University of California at Berkeley, June 1988

[EIA87] Electronic Industries Association, "EDIF - Electronic Design Interchange Format Version 2.0.0," Electronic Industries Association, 1987.

3

Lager OCT Policy and the SDL Language

Brian C. Richards

The OCT database, described in the preceding chapter, provides a mechanism for storing system design information, without imposing rigid rules for specifying a design. Instead, guidelines or policies are offered, recommending certain conventions for representing different types of design information. Designs containing structural information, including interconnections or netlist specifications, are typically represented by following the OCT Symbolic Policy, detailing how OCT objects are used to represent the structure. The OCT Physical Policy is used to represent structure defined by geometrical shapes. Within LAGER, several policies are defined, for representing a system at different stages of the design process.

The majority of tools that use the OCT database use either the "symbolic" or "physical" policies for representing both the input description driving the tool, and for saving the results. It follows that LAGER uses these policies as well for passing design information from one tool to another. Since the basic "physical" and "symbolic" policies do not have parameterization and bus representation capability, the structure_master and structure_instance policies have been defined to provide this capability.

3.1. LAGER POLICIES

Most tools tied into the OCT database follow either the physical or symbolic policies described previously. Within LAGER, there are three common policies for describing different stages of the design flow. These views are generally extensions to the OCT "physical" or "symbolic" policies described in the preceding chapter.

The user interfaces to OCT in two ways: 1) textually, through the Structure Description Language (SDL), and 2) graphically using a commercial tool (see Chapter 4).

3.2. THE STRUCTURE_MASTER VIEW AND THE SDL LANGUAGE

A parameterized design can be represented using the "structure_master" view, which can be generated from either graphical or textual design specifications. SDL is a textual description, which uses a LISP-like syntax to describe the design, and is converted by the Design Manager, discussed in the following chapter, into the structure_master view. The relationship between the structure_master view and the SDL file is one to one; all design information in the SDL file is transferred to the structure_master view, with the exception of comments. Hence, an SDL file can be produced from a structure_master view, with the same information as the original.

The first non-comment line in an SDL file begins with the *parent-cell* declaration, defining the OCT cell name:

```
(parent-cell mycell)
```

In the resulting structure_master view, the OCT_FACET object will be named:

"mycell:structure_master:contents"

The SDL file must be named "mycell.sdl", corresponding to the name in the parent-cell declaration for consistency checking by the design manager, DMoct.

Several aspects of a design can be parameterized. Arrays of nets, called busses, can be defined, where the width of the bus is derived from a parameter.

Chapter 3   Lager OCT Policy and the SDL Language   27

Arrays of instances can also be defined, allowing a single description to describe a large number of possible circuits. Also, parts of the structure can be generated conditionally, including both nets and subcell instances.

A subset of Common LISP, called LightLisp [Baker90], is used to evaluate parameters. The structure_master view is used in conjunction with user-defined parameters by the LAGER design manager, DMoct, to produce a unique version of the desired design that uses the given parameters. This "instance" of the parameterized design is represented in a "structure_instance" view, described in the next section.

Aside from structure, the parameters also serve to pass information about a circuit to tools, eliminating the need for user interaction in most cases. In the case of a ROM circuit, the number of words and the word size would be specified, as well as the contents of the ROM locations.

Parameters are defined in an OCT_BAG named FORMAL_PARAMETERS, with one OCT_PROP for each parameter (See Figure 3.1). Each property has a parameter for its name, and either has a string value containing a LightLisp expression, or has no value at all. If an expression is given, it will be used as a default value if the parameter named in the property is not provided by the user or from a higher level of the design hierarchy.

The FORMAL_PARAMETERS are defined in the SDL file with the *parameters* declaration. The parameters defined in Figure 3.1 can be entered as follows:

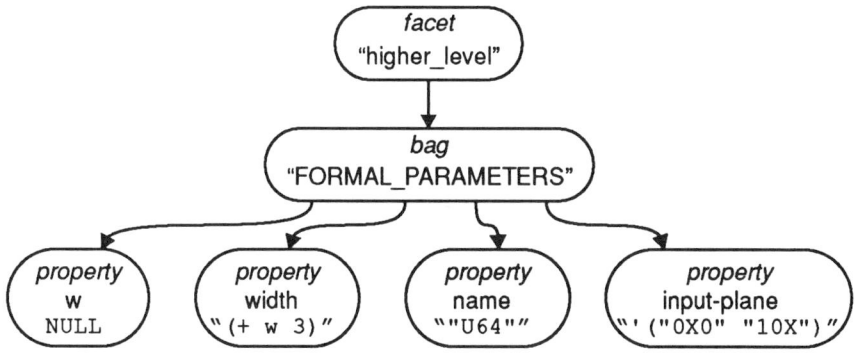

Figure 3.1: Formal Parameter specification in the structure_master view.

```
(parameters
    w
    (width (+ w 3))
    (name "U64")
    (input-plane '("0X0" "10X"))
)
```

The parameter *w* is not given a default value, and thus the corresponding property in the structure_master view has an OCT_NULL value, rather than an OCT_STRING value.

To pass parameter values through a design hierarchy, instance-specific parameters are defined as properties in the ACTUAL_PARAMETERS bag under each instance. Parameter values are passed explicitly; the string value of each property will be evaluated in LightLisp, and passed to the subcell, to be used as a value for the FORMAL_PARAMETERS property by the same name.

The SDL file defines the hierarchical structure of a design using the *subcells* declaration. To declare the instance in Figure 2.4, the following text may be used:

```
(subcells
    (design_name D1 ((w w)))
)
```

In this case, the parameter w is passed to the subcell, so that its value may be used for parameter evaluation in the cell design_name. The values are assigned explicitly; the first w is the name of the subcell's parameter, and the second, which may be a general LightLisp expression, is evaluated using the current set of parameters.

The structure_master view defines interconnections between instance and facet terminals with the same policy as is used in the *symbolic* view (Figure 2.1), although extensions have been included to describe busses of signals in a parameterized fashion. These connections may be made using either the *instance* or the *net* declaration syntax. The difference is whether connections are grouped by instance or net. Using instance declarations, the connections in Figure 2.5 can be declared as follows:

Chapter 3   Lager OCT Policy and the SDL Language

```
(instance parent (
        (top_input wire)
        ; Other terminal - net pairs on the parent.
))
(instance D1 (
        (input wire)
        ; Other terminal - net pairs on D1.
))
```

Alternatively, the connections could be described using the *net* syntax:

```
(net wire (
        (parent top_input)
        (D1 input)
))
```

The choice of which syntax to use is left to user preference.

In the OCT database, the instance object can store two-dimensional transformation information, which can be used for providing placement hints to VLSI and printed circuit board CAD tools. This information can be described in the SDL file using pre-defined instance properties, as in the following example:

(top_input wire)

; Other terminal - net pairs on the parent.uinformation can be described in the SDL file using pre-defined instance properties, as in the following example:

```
(instance D1 (X 1) (Y 2) (T MXR90) (
        ; terminal - net pairs in D1.
))
```

These properties are evaluated in LightLisp, and are included in the resulting instance object for the x translation, y translation, and manhattan rotation fields. Tools such as `oct2rinf`, an interface to a commercial PCB router, allow the designer to define parameterized placement hints.

The structure of a design can be controlled by parameters. In the simplest case, an instance or net can be included conditionally into a design by placing a CONDITIONAL property with a LightLisp value under the OCT object in question. For the above net in an SDL file, this might appear as follows:

```
(net wire (CONDITIONAL (> w 1)) (
        (parent top_input) (D1 input)
)
```

A zero or nil value will cause the object to be eliminated when a design instance is produced. Although not detailed here, arrays of instances can be defined as well, so that a variable number of instances can be described.

Busses can be represented in the structure_master view by an OCT_NET object containing a NETWIDTH property. The width will be evaluated, and the single net in the structure_master view will be expanded into the specified number of distinct nets in the structure_instance view, with integer indices appended, ranging from 0 to NETWIDTH-1 by default.

To define more elaborate connectivity, a net and terminal can contain one or more MAP bag. A given MAP bag is contained by exactly one OCT_TERM and one OCT_NET object, defining a detailed relationship between the terminal and net. The MAP contains one or more properties defining how terminal and net indices are numbered, or in the extreme, completely renamed. Also, a given terminal-net pair may share several MAP bags.

The MAP bag contains one or more of several properties containing Light-Lisp expressions. To define busses, two properties can be used. The MAP bag may contain a property named WIDTH, defining the number of indexed terminal and net pairs that should be connected. If the NETWIDTH property is given under the OCT_NET object, the integer value of the given LightLisp expression will be used as the default value for the MAP width. To give maximum flexibility to the designer for naming and indexing bus nets, the LightLisp variable _i can be used in expressions to define the net and terminal indices. If only the WIDTH is given, then the terminals and nets will be enumerated from 0 to WIDTH-1. The

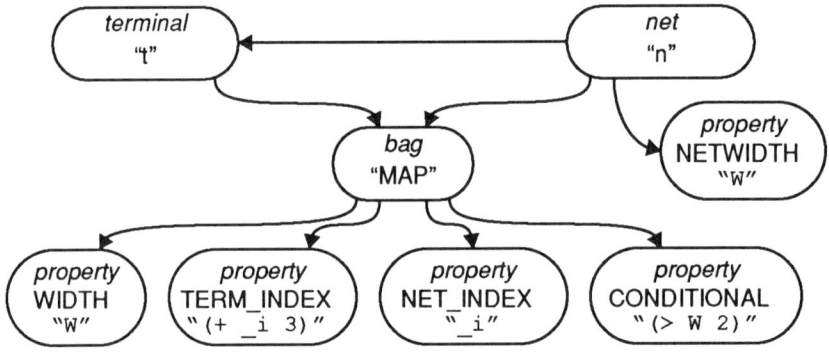

Figure 3.2: Example of terminal to net bus mapping.

following SDL lines illustrate how terminal and net indices may be controlled explicitly:

```
(instance something (
        (t n (width W) (term-index (+ _i 3))(net-index _i)
        (CONDITIONAL (> W 2))
))
```

Figure 3.2 shows the resulting OCT representation in the structure_master view. Each terminal-net connection may be conditional, according to the value of the CONDITIONAL property, which is re-evaluated for each connection. Using these and additional MAP features, complex terminal to net assignments for busses can be achieved.

Many tools expect that formal terminals (facet terminals) have certain properties to give hints about placement, power requirements, etc. These can be defined with the *terminal* declaration:

```
(terminal top-input (TERM_EDGE LEFT)
        (TERM_RELATIVE_POSITION 0.65))
```

In the above case, the OCT_TERM 'top-input' will have an OCT_PROP named "TERM_EDGE" with the string value "LEFT". The "TERM_RELATIVE_POSITION" property will be real-valued.

In addition to parameters and busses, the "structure_master" view also defines the tools that are to be executed to generate an instance of a design. Two properties have been defined to specify tools; the STRUCTURE-PROCESSOR and LAYOUT-GENERATOR properties. The STRUCTURE-PROCESSOR defines an optional tool that will modify the structure (netlists and hierarchy) of a design, such as a datapath compiler or a logic synthesis program. The LAYOUT-GENERATOR is a program that produces a physical design from netlist information and parameters. In an SDL file, these properties are declared with the *structure-processor* and *layout-generator* declarations:

```
(structure-processor dpp)
(layout-generator "Flint -a")
```

If arguments are given to the tool, then the command should be placed in quotes.

The SDL syntax includes several other constructs which are detailed in the **sdl(5)** manual pages [Lager91]. Other features include arrays of instances, hierarchy flattening, and a general syntax for placing nested properties and bags on OCT facets, instances, nets and terminals.

3.3. THE STRUCTURE_INSTANCE VIEW

The "structure_instance" view is essentially an OCT "symbolic" policy view, although there is usually no physical information (geometry) stored. Only structural information, with unique instances and individual nets is required. As an extension to the "symbolic" view, the FORMAL_PARAMETERS bag is copied from the structure_master view, with all parameters evaluated, and stored as constants in properties with the corresponding names. Figure 3.3 shows how the parameters defined in Figure 3.1 would be evaluated and represented in OCT. Notice that a list of parameters in LightLisp is expanded into several properties with the same name as the parameter, with enumerated indices. Integer, real and string data types can be represented. Notice that the property w now has an integer value; the user provided this value, since there was no default.

The names of tools are carried over from the structure_master view, included as properties under the design facet. The structure_instance view should contain all of the information that is required to generate the layout of the defined structure, including tool names and options.

3.4. THE PHYSICAL VIEW

The "physical" view is used for the primitive design description, and is produced by all layout generation tools. In general, either the OCT "physical" or OCT "symbolic" policy can be generated by each tool. By convention, LAY-

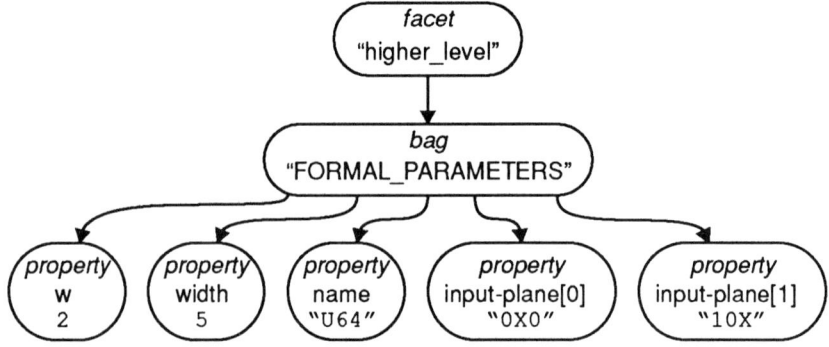

Figure 3.3: Formal Parameter specification in the structure_instance view.

Chapter 3   Lager OCT Policy and the SDL Language   33

OUT_GENERATOR tools generally produce files for the `Magic` layout editor [Scott85] as well, which are used for design post-processing, including extraction and design rule checking.

3.5. SUMMARY

The SDL language is a powerful textual representation for design structure which can offer parameterization of nearly all aspects of a design. The flexibility of the SDL text description is also fully represented in the OCT database using the structure_master policy, allowing alternative front-ends to be used without sacrificing any of the parameterized features of the SDL language. By placing this information in the OCT database, the user can thus choose a preferred style of user interface.

REFERENCES:

[Baker90] W. Baker, "Volume 6: Light/Oct/Vem/Lisp," *Oct Tools Distribution 4.0*, U. C. Berkeley Electronics Research Laboratory, Berkeley, California, March 1990

[Lager91] "Volume 2: Lager Tool Set," U. C. Berkeley, U. C. Los Angeles, Mississippi State University, Institute for Technology Development, June 1991.

{Scott85} Walter S. Scott, Robert N. Mayo, Gordon Hamachi, and John Ousterhout, editors, "1986 VLSI Tools: Still More Works by the Original Authors," *Technical Report No. UCB/CSD 86/272*, December, 1985

Schematic Entry

Bob Reese

The LAGER system consists of tools, libraries, and policy. It is the LAGER policy which allows the user to incorporate new tools into the design methodology. The SDL file, the structure_master view, the structure_instance view and the physical view represent the data formats which tools can use as entry points into the LAGER design pipeline. The structure-processor and layout-generator specifications are the links through which DMoct can be instructed to invoke a tool. These links can be used to interface various commercial tools to LAGER. This section details a schematic capture/simulation system which has been integrated into the LAGER system.

4.1. SCHEMATIC TOOL INTERFACE

While SDL provides for text entry, the schematic drawing tools and VHDL simulator from Viewlogic Systems, Inc. provides a means for graphical entry of design information into OCT using the LAGER policy described in Chapter 3. This allows the user to create parameterized schematics for his design, using a state-of-the art commercial interface. There is nothing unique about the Viewlogic schematic capture/simulation package that particularly lends itself to interfacing with LAGER - the interfaces described in this chapter could be duplicated for most other commercial schematic capture/simulation packages.

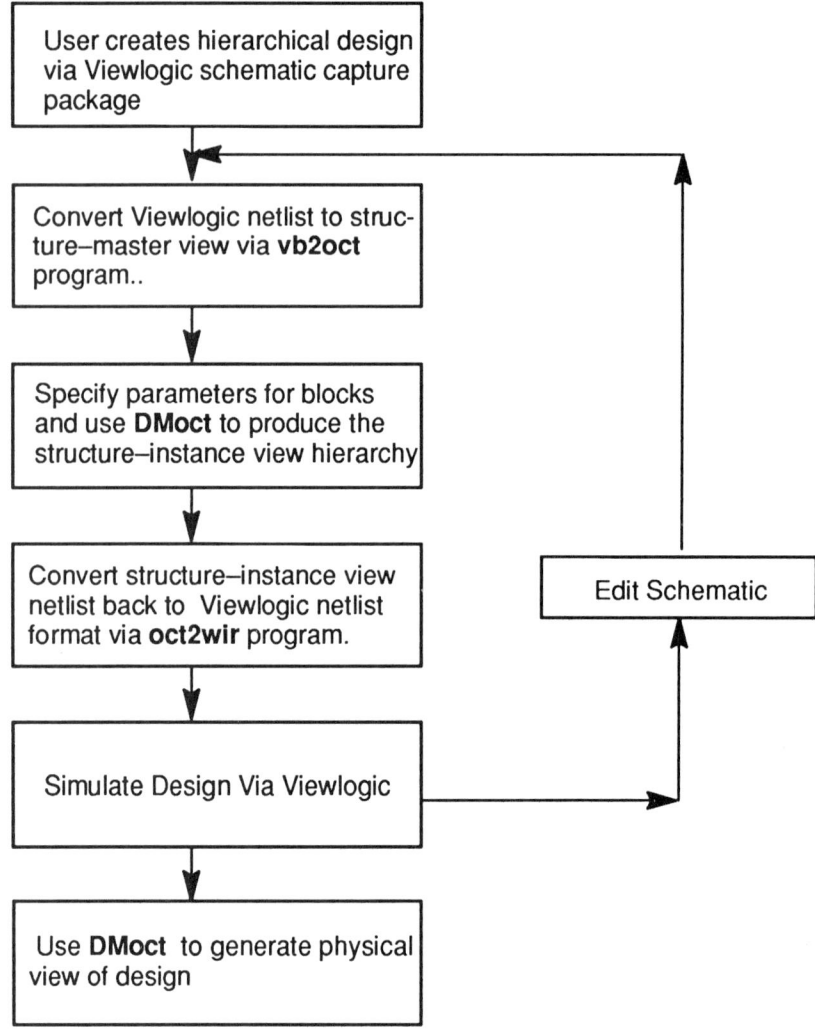

Figure 4.1: Schematic capture/simulation interface for LAGER

The operation of the schematic tool interface is shown in Figure 4.1. Four of the basic file types in the schematic tool are:

1. sch: schematic placement information.
2. wir: netlist (bare netlist, no schematic placement information).
3. sym: symbol (icon form of a schematic).
4. vsm: simulation netlist (created from individual wir files, flattened view for simulator).

The conversion of Viewlogic-Viewbase schematics to a LAGER view is done by `vb2oct`. The `vb2oct` program converts each schematic to a structure--master view. Full timestamp checking is done so that only those schematics which are younger than the corresponding structure_master view are converted. It is important to realize that the schematic represents the structure_master view and *not* the structure_instance view. For designs with no parameters (designs with constant bus sizes, non-parameterized components such as standard cell leafcells) these views are essentially the same. `Viewlogic` simulation can be performed on unparameterized designs without creating any LAGER views because the schematic represents the exact entity to be simulated. However, simulation of a parameterized design requires that you:

1. generate the structure_master view hierarchy using `vb2oct`.
2. generate the structure_instance view hierarchy using `DMoct`.
3. generate the `Viewlogic` wir files for each block using the `oct2wir` program.
4. generate the `Viewlogic` simulation file using the `siv2vsm` program.

The first two steps are available from `Viewlogic` menus customized for LAGER. The last two steps can be automated as a side effect of the structure_instance view generation by specifying the `oct2wir` and `siv2vsm` programs as structure-processors in the appropriate schematics. This is the desired method as new wire files and simulation netlist files are only generated if the schematic has been updated. If the schematic has changed then `vb2oct` will create a new structure_master view; this will cause `DMoct` to regenerate the structure_instance view the next time it is invoked; the creation of the structure_instance view involves running user-specified structure-processors such as `oct2wir` and `siv2vsm`. Special components and attributes added to the schematic by the user are used to define such LAGER constructs as structure-processors and layout generators. Once a design is simulated, the layout can be generated in the normal manner using `DMoct`. This requires that `vb2oct` had been run previously to generate the structure_master view hierarchy.

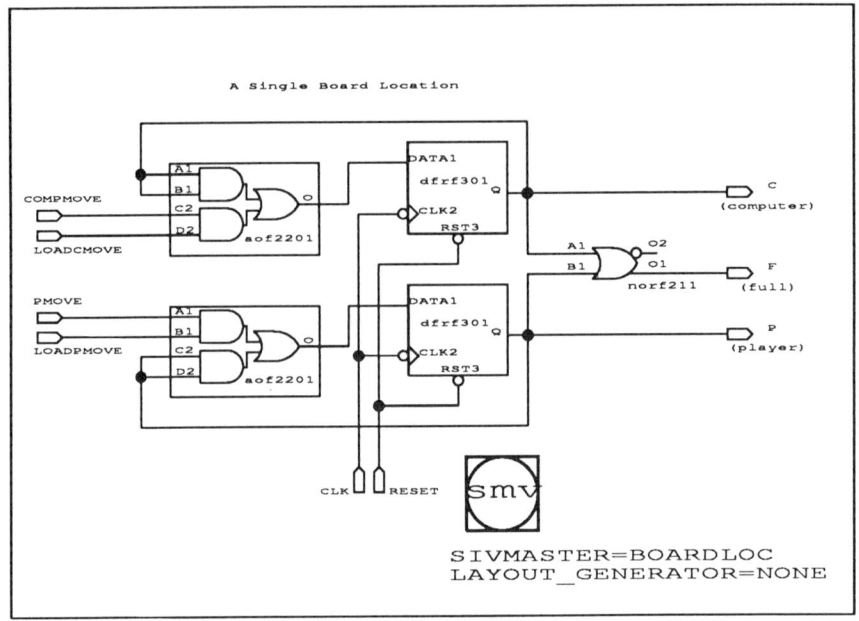

Figure 4.2: Logic schematic without parameters

4.2. DESIGN EXAMPLES

Figure 4.2 shows some random standard cell logic entered as a schematic. The schematic can be thought of as the graphical view of the SDL file. Items which are entered as statements in the SDL file are now entered as either components or attributes in the schematic. The special SMV component is used as a place for putting LAGER specific attributes such as SIVMASTER, DEPENDS-ON, STRUCTURE-PROCESSOR, etc. This schematic has no parameters and thus can simulated without invoking `vb2oct`.

Figure 4.3 is a schematic of a parameterized component called "MOVE" which is used as part of a chip that plays TicTacToe. Note that this schematic only contains formal terminal definitions; its internal logic is synthesized by the `Bds2stdcell` structure-processor from a BDS description (see Figure 4.3). The bag component labeled SP is used to create an OCT bag named SP in the structure_master view. Attributes attached to a bag component get translated to OCT string properties attached to the OCT bag object. The SP bag contains a list of

Chapter 4 Schematic Entry 39

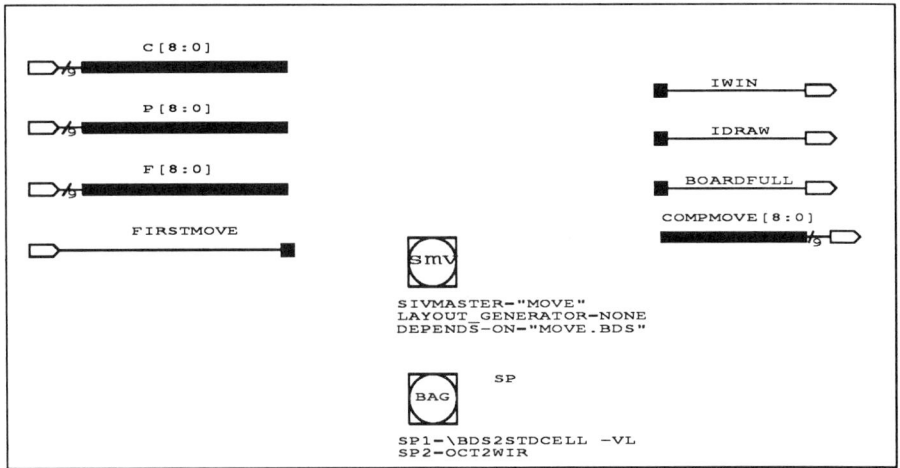

Figure 4.3: BDS block schematic

structure-processors to run on the structure_instance view. The SP1 structure processor is executed first, the SP2 structure_processor next, and so forth. Note that the SP1 structure processor is Bds2stdcell, the standard LAGER structure processor used for logic synthesis (see Standard Cell chapter). The SP2 structure processor is the oct2wir program which will create the wir file for the block from the structure_instance view produced by the Bds2stdcell structure-processor. The oct2wir structure processor is necessary because the wir file is needed in order to simulate this block with the VHDL simulator. The oct2wir program is run after the Bds2stdcell script; this order is required in order to generate the wirelist of the synthesized logic. The ability to specify multiple structure processors or layout generators for a block allows us to make the standard LAGER scripts very generic. Users can combine the standard LAGER processors and vendor-specific processors in any manner they choose.

Figure 4.4 shows the top-level schematic of the TicTacToe chip; the move block is just one component of the top-level schematic. The TERM_EDGE attribute on each terminal is used by the standard cell layout generator to control terminal placement on the generated layout. Note that the "-flatten" option is passed to the Stdcell layout generator; this causes the Stdcell script to flatten the standard cell hierarchy before attempting layout generation.

Figure 4.5 shows a pad frame created with the schematic interface for the TicTacToe chip. The FACET:PAD parameter of each pad and the pads parameter

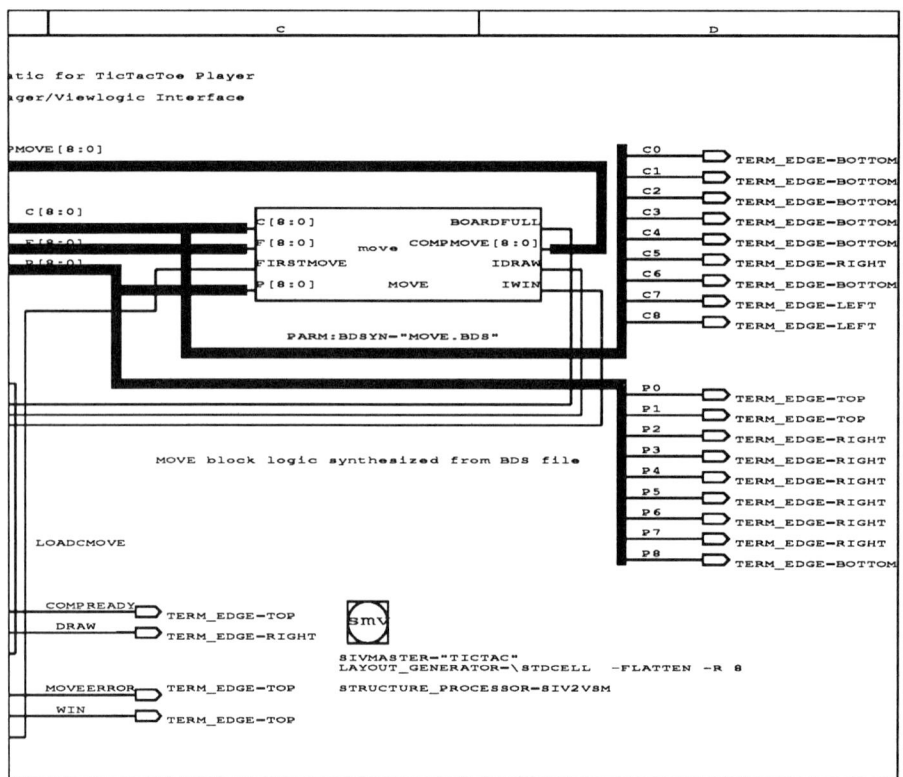

Figure 4.4: Part of a top-level standard cell design schematic

on the top level schematic are used by the Padgroup structure processor in creating the padframe. The FACET: prefix to a symbol attribute. For example, FACET:PAD=10 attaches that attribute as a property to the instance in the structure master view which corresponds to this symbol. The Padroute layout generator is used to construct the physical view of the padframe and to perform the pad to core routing.

4.2.1. Parameterized Designs

Figure 4.6 shows a datapath design using elements from the datapath compiler library for dpp (See Chapter 10). In this schematic, the datapath nets are parameterized through the attribute NETWIDTH=\N. The value for N is passed through the formal parameter N defined in the FORMAL_PARAMETERS component bag (this is equivalent to the parameters statement in SDL). Note that the

Chapter 4　　Schematic Entry　　41

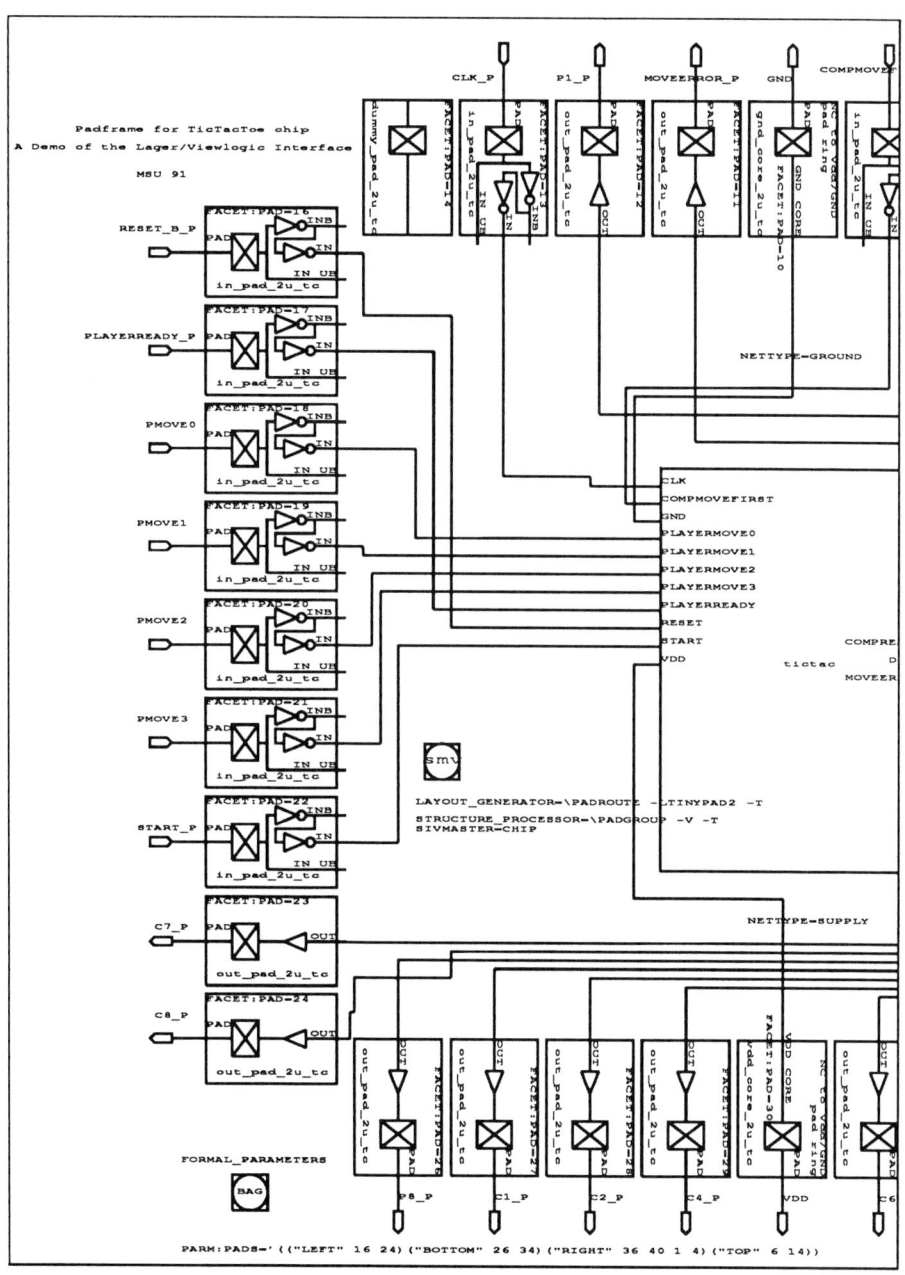

Figure 4.5: Part of a pad frame schematic

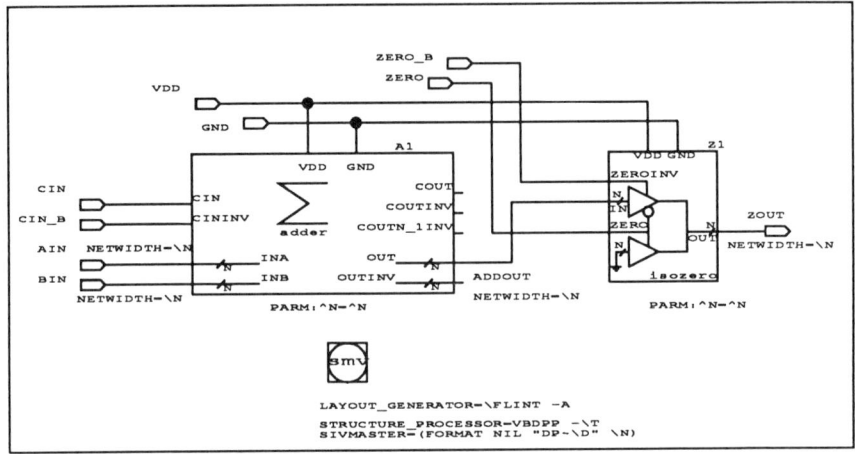

Figure 4.6: Datapath schematic

parameter N is also passed to each of the datapath components through the PARM:^N=^N attribute on each component. The '^' and '\' symbols are case modifiers to compensate for the lack of case sensitivity in the schematic tool. '^' makes all characters lower case while '\' only lower cases the next character. The structure-processor used is vbdpp which is a special version of the normal dpp structure processor. vbdpp invokes oct2wir and the VHDL compiler whenever the structure instance view is generated so that the simulator can be run. Because the schematic is parameterized and cannot be simulated "as is", vb2oct must first be run to produce the structure-master view hierarchy, and then DMoct must be run to produce a structure-instance view hierarchy with N bound to some fixed number, such as N=16. Only then can the resulting wirelist be simulated from within the schematic tool.

Figure 4.7 is part of a schematic of a parameterized standard-cell multiplier which illustrates some additional features of parameterized nets. The parameterized busses are drawn as single nets with the NETWIDTH attribute defining the width of the bus. The value of the NETWIDTH attribute can be an arbitrary Lisp expression such as (* 2(+ N 1)). In this case, N is a parameter defined in this schematic. The sc_register cell is a parameterized N-bit register. The parameter N specifies the width of the register. This parameter is represented by the PARM:N attribute on the sc_register symbol. The PARM: notation is used to distinguish this attribute from other attributes which are non-LAGER specific. In this exam-

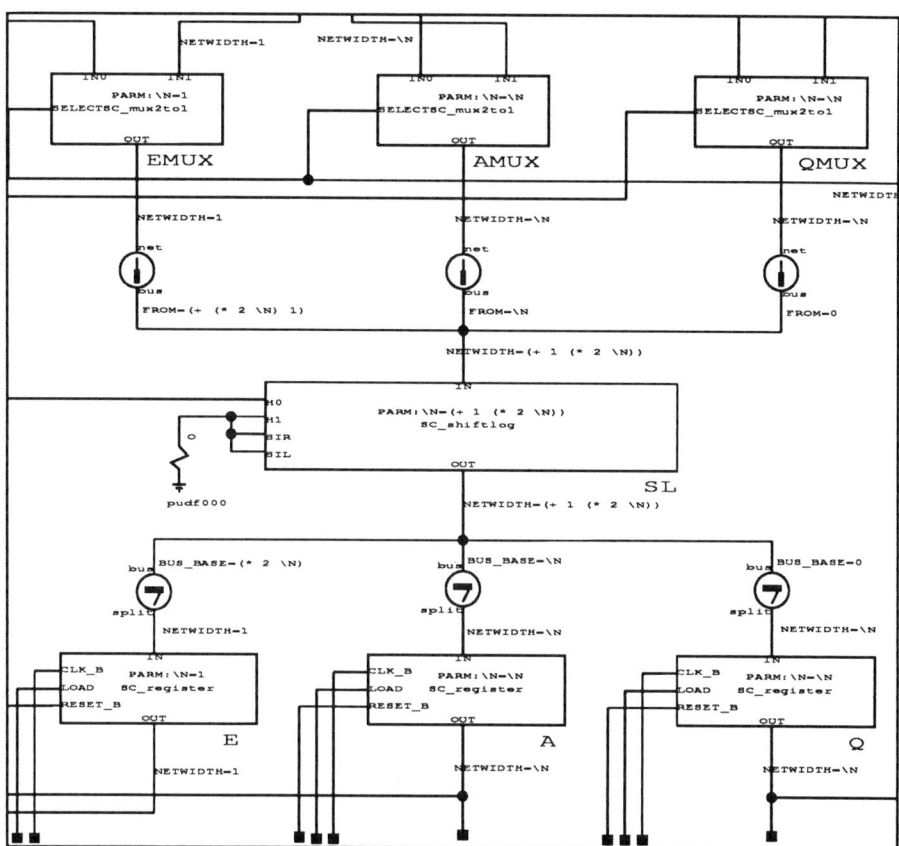

Figure 4.7: Part of a parameterized multiplier schematic

ple, PARM: specifies that this attribute should be attached as property to the actual parameters bag for this instance.

The BUS_SPLIT component is used to split off a portion of a parameterized bus to form another bus of smaller size (the split bus). The BUS terminal in an input and connects to the larger bus and the SPLIT terminal connects to the split bus. The BUS_BASE attribute is used to define the index at which the split bus will be taken from the larger bus (other attributes such as BUS_INCR and SPLIT_INCR give more control over which bus lines will be taken from the larger bus and placed in the split bus). Again, the values of these attributes can be Lisp expressions. The BUS_MERGE component is to use to merge a smaller bus into a larger bus. The FROM attribute defines the starting index in the larger bus for connecting the smaller bus.

4.3. SUMMARY

The Viewlogic schematic capture tool, which is interfaced to the LAGER tool set, provides the user with an alternative to textual design entry. By defining a strategy for representing the structure_master view policy using a schematic capture systems normal capabilities, nearly all parameterization features offered by DMoct can be supported in a graphical environment. Since this capability is not restricted to a particular tool, the designer can choose text or graphical entry for any of the LAGER tools.

5

Design Management

Brian C. Richards

`DMoct` is the central design manager for the LAGER system design tool set and libraries. As a design manager, it automates the generation of a hierarchical system which may consist of pieces created by a large variety of tools. The user has access through the design manager to a large and ever growing library of designs, from VLSI leafcells to large subsystems with a wide range of custom and commercial parts. A designer can tailor a system to specific needs by taking advantage of several parameterized subsystems. With this design manager, the user is not required to be an expert at the large number of underlying tools that make up LAGER.

Where the software developer might use the UNIX "make" utility, the hardware system developer can use `DMoct`. This offers file time-stamp checking to identify parts of a design that must be regenerated. Unlike the "make" utility, `DMoct` is also capable of dealing efficiently with hierarchical designs, where each level of the design has a set of commands for regenerating that part of the design. Each subsection of a design is treated as a black box when used within a larger system. This modular design style allows designers to focus on system architectures without large levels of effort at the lower levels of circuit design.

5.1. PARAMETERIZATION AND LIBRARY SUPPORT

Modules or subsystems are generally parameterized so that no more circuitry is generated than is needed. If the designer wants a memory with 13 bits of data and 21 locations, then the memory type is chosen (Dynamic, Static, Read-Only), and the size information is passed as design-specific information (width = 13, size = 21). Pre-designed circuits can thus be reused with maximum flexibility. Parameterized interconnections, such as variable-width groups of signals, or busses, can be described so that the number of bits in such a bus can be evaluated according to system parameters.

An important feature of the design manager is the ability to maintain libraries of pre-designed and tested subsystems. VLSI and common board level systems can be made available to many users as black-box systems which have specified connections and accept design-specific parameters. In some cases, there may be more than one design for a subsystem (such as a RAM or multiplier), and the designer can try the different versions by merely selecting a different module from the library to try meeting size, shape or speed requirements.

5.2. THE DESIGN FLOW STRATEGY

DMoct is, as the name suggests, based on the OCT object-oriented database. The OCT database supports a hierarchical design style, where a given level of the hierarchy may be some actual circuitry (a VLSI layout or a TTL part for example), or an interconnection of these circuits. If several circuits are interconnected within one level of a design hierarchy, they may be called subcells, modules, or macrocells of that design, and each reference to a given subcell is called an instance of that subcell. A library cell is usually treated as a subcell in a larger design.

The user can enter designs by using the LISP-like Structural Description Language (SDL) files described in the previous chapter. DMoct will then read the SDL files to produce the structure_master view described previously. Once the parameterized structure_master view has been created for all parts of a design, the parameters are entered from a parameter file, on the DMoct command line, or by responding to interactive requests for parameter values from DMoct. Then, a second OCT view, the "structure_instance" view is created. In the structure_instance view, all parameters requested by the structure_master view are evaluated by calling to the built-in LightLisp interpreter, subcells are instantiated (referenced as instances of a subcell), and all interconnection busses are expanded into individ-

Chapter 5   Design Management   47

ual signals or nets. The structure_instance view is the generic structure description which is recognized by all of the OCT-based tools. For most tools, the structure_instance view contains all information required to generate a design, although references to other technology and design files may be included.

Once the structure of a design has been produced in the structure_instance view, some modifications to parts of the structure may be required, to incorporate additional information into the design, or to improve the area or speed of a design. Tools that modify the structure of a design are called structure-processors as described in Chapter 3, which read one or more structure_instance views and produce one or more modified versions. As an example, the designer might have a behavioral description of part of a design along with a structure_instance view describing the boundary connections or terminals of the design. The structure-processor could then synthesize the subcell structure_instance views for this circuit (See Figure 5.1). An example of this for logic synthesis will be given in Chapter 8.

A datapath compiler is another example of a structure-processor [Srivastava87]. Given a set of subcircuits that are to be assembled, the structure of a

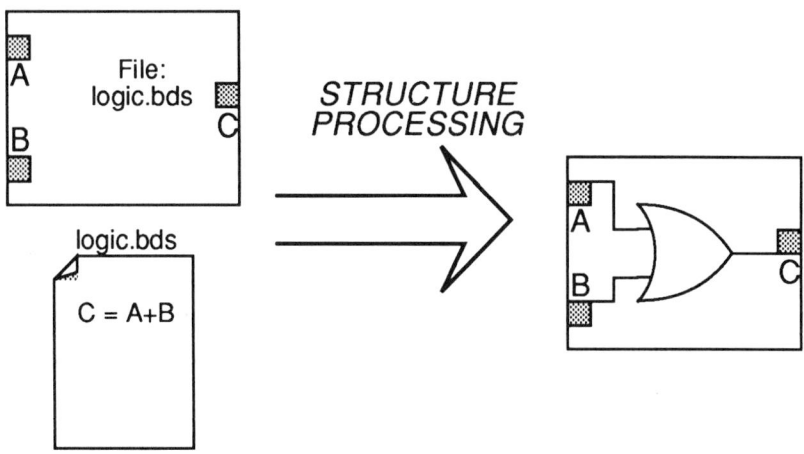

Figure 5.1: Structure Processing with Behavioral Descriptions

datapath slice can be replicated, and size and placement hints can be provided to lower level tools to produce an efficient design, as in Figure 5.2.

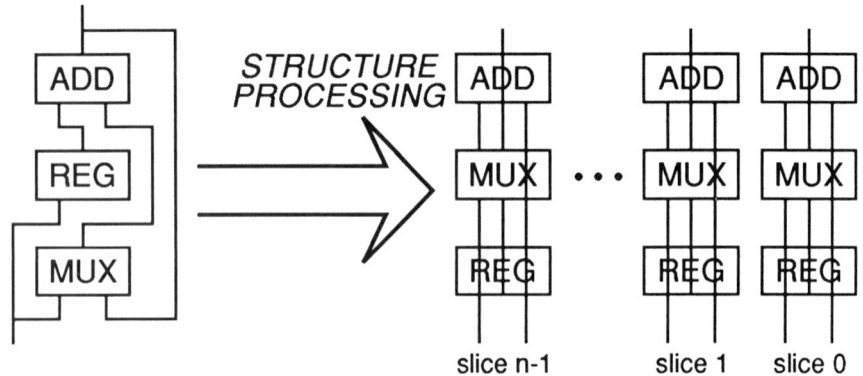

Figure 5.2: Structure Processing a Datapath Structure

Once the structure of the design is available, functional simulation can be performed. The THOR simulator described in Chapter 13 is used for this purpose. Many problems can be identified by simulating a design at this point, before actually generating the final design.

With a complete structural description of the design, DMoct will then run tools to generate the physical design descriptions, such as in Figure 5.3. Each structure_instance view designates which "layout generator" tool is to be executed to produce the "physical" view. These layout generators generally accept a structure_instance view description, and will produce a physical description of the design in OCT. In the case of VLSI design, files are also optionally produced for the Magic layout editor [Scott85].

Several post-processing operations can then be done on the resulting physical database, such as design extraction, switch-level simulation, design rule verification and actually preparing the design for the fabrication vendor. This is done outside of the DMoct design manager in an application called DMpost that will be discussed in the next chapter.

Chapter 5 — Design Management

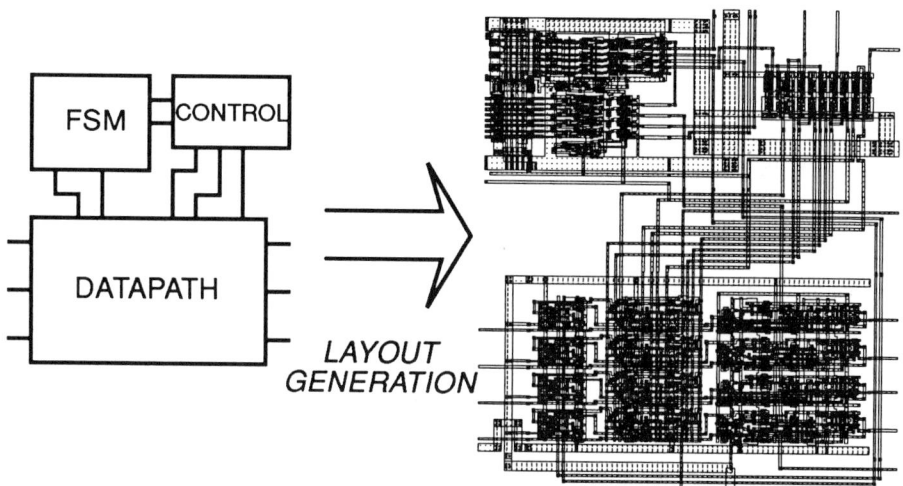

SIV with connectivity and modules   Final layout, ready for fabrication

Figure 5.3: Layout Generation from structure_instance views

5.3. CONTROLLING THE DESIGN FLOW

The user has a great deal of flexibility with DMoct to manage the evolution of a design. Given a design description, there are three distinct stages of design generation, discussed in the previous section, each of which can be individually controlled, or all can be executed with a single command. An important feature of DMoct is that design parameters may come from several sources. Also, hierarchical designs can be regenerated in a controlled fashion, selecting parts of a design to be regenerated unconditionally, as well as sections to ignore, to improve performance while experimenting with design iterations.

5.3.1. Design flow from description to layout

The three distinct tasks for DMoct to manage can be individually controlled by DMoct. The tasks are as follows:

1. Read the SDL files to produce parameterized structure_master views
2. Substitute parameters, and expand instances and nets to produce the structure_instance view netlist
3. Run designated tools on the structure_instance views to produce the physical layout in OCT.

Typically, DMoct is run in one step to produce the layout given one or more SDL files. If an SDL file or design parameter has been revised and DMoct is executed, the file modification time stamps are inspected to determine which parts of a design should be regenerated, and only those sections that are inconsistent will be regenerated.

In many cases, the designer does not want to run all steps of a design. For instance, if a high level simulation is to be performed on the netlist, then there is no need to spend the time creating the physical layout. Alternatively, if no changes are made to SDL files, but new parameters are to be given, then there is no need to check the SDL files. The actions taken can be explicitly controlled by DMoct command line options to perform only select design operations. In addition, the time-stamps can be ignored entirely, and all of the design files can be regenerated unconditionally (as long as they are writable). If a design has been generated incrementally, and the designer has manually prevented any consistency checks, the full design should be regenerated from the beginning to make certain that there are no hidden inconsistencies in the design.

5.3.2. Specifying design parameter values

There are many ways that parameters can be provided for controlling the generation of a design instance. Default values can be provided in the database, and parameters can be entered interactively, from files, or on the DMoct command line.

Default values can be entered in the SDL files when parameters are defined, and are saved in the structure_master view database. The default values are defined using arbitrary LightLisp s-expressions, and thus may be constant numbers or strings, or may depend on other parameters, according to the order that the parameters are defined in the SDL file.

If no value is provided for a parameter, the user will be prompted by DMoct to enter a LightLisp expression for the value. If the designer knows in advance which parameters are needed, such as the width of a datapath, these values can be passed on the DMoct command line. For small designs, this is within reason, but for more than a handful of parameters, this is extremely cumbersome, especially if the design will be regenerated.

For most significant designs, the parameters are entered using a parameter value file, the name of which is specified on the DMoct command line, or

Chapter 5   Design Management   51

requested interactively. This file contains one or more parameter name and value pairs, each enclosed in parentheses. In cases where the design is synthesized from a higher level description, this parameter file may be generated automatically.

5.3.3. Selective generation of hierarchical designs

DMoct will be routinely run more than once on a given design, as simulations and revisions are followed up by final design generation. If only minor changes are made to a design, the regeneration time can still be significant, since all database files must be read in to verify design consistency. To speed up this process, DMoct allows the user to designate parts of a design to regenerate, or parts to ignore.

Sometimes, a minor revision is made to a subcell in a large design, leaving most of a circuit unaffected. In this case, the circuit or circuits affected may be named according to their design instance names on the DMoct command line. This will cause all subcells of those designs to be checked for consistency, and regeneration will be performed as needed. Then, all cells containing that design will be regenerated, without checking other subcells. This often reduces the regeneration time significantly, so that the new circuit may be simulated promptly.

Alternatively, the designer may want to specify one or more subcells that are *not* to be checked. An example of this is the task of attaching pads to the core of a VLSI design; if the core is believed to be correct and consistent, then the design could be quickly assembled without searching through the design hierarchy.

The capability to selectively disable time-stamp checking has proven valuable in many large designs, but must be used with care; the burden of checking the design for consistency has been entrusted to the designer. Generally, a design that was made through several iterations of DMoct should be regenerated from the beginning to check the consistency of the full design before considering the design finished.

5.3.4. Support for PCB and larger system designs

When designing systems beyond the scope of VLSI, some additional design management issues arise. With VLSI design, the system is implemented as a single end-product; a chip. When designing printed circuit board systems using custom VLSI parts, the structure of the design will describe a system which has an electrical hierarchy differing from the physical implementation hierarchy. To a circuit simulator, the system is a single entity; to the physical CAD tools, the system is a collection of independently manufactured components, boards and backplanes. Also, different CAD tools will view a single design in differing ways. For instance, a PCB design tool considers a custom VLSI part to be a primitive package with specified pin locations, whereas a VLSI tool expects to find pads and circuitry.

To support the description of physical hierarchies, DMoct can selectively treat a complex circuit design as a component with no subcircuits, or as a complete electrical design. The notion of PACKAGECLASS has been adopted to control the treatment of a circuit. From the command line, the designer can designate how designs with specific PACKAGECLASS properties are to be treated. Examples of systems where the physical and electrical descriptions differ include VLSI components, programmable logic devices, and multiple component modules.

5.3.5. Controlling CAD tools

DMoct reads the structure_instance views of a design to determine which tools are to be executed. As discussed previously, there are two types of tools; *structure-processors*, which modify netlists and/or hierarchy, and *layout-generators*, which produce physical designs given parameters and netlists. DMoct can modify the parameter lists passed to these tools, and can avoid running them entirely.

Often, the designer would like to pass additional instructions to tools without having to modify SDL files or structure_master views in OCT. DMoct allows additional parameters to be passed to tools, identified by the name of the tool. A typical use of this is to request that a layout generator run in an automatic batch mode, or to control the effort or technology used by a placement tool.

To debug new designs or new tools, DMoct can avoid running the structure-processors or layout-generators entirely. Since structure-processor tools actually replace the original structure_instance view netlist with a new netlist, possibly with a different design hierarchy, the original structure_instance view is usually modified in-place, losing the original information from DMoct. This information can be saved by disabling structure-processing, so that the original database can be investigated.

5.4. THE DESIGN MANAGEMENT STRATEGY

Several underlying mechanisms are involved in managing the design flow. Library support, parameter evaluation, and consistency checks are among the tasks which must be performed by DMoct. Also, clear and unambiguous user feedback is a crucial part of DMoct to guide the designer through a project, and to help solve problems.

5.4.1. Libraries and Customization: The path-search mechanism

To support libraries, DMoct uses a start-up file, generally referred to as a *lager* file. The file follows a LISP-like syntax, and is essentially a set of parenthesis-enclosed lists. The first element of each list is a key, and the remaining elements are directory names in the UNIX filesystem. The key, which can be an arbitrary string, provides a mechanism to categorize the search space. A full search is specified by the file name and a key, and the file is searched for in each of the directories listed under that key. The key names themselves are part of the policies associated with the various tools.

A library of routines is available to read and search for files according to the *lager* file. These routines are used by many of the LAGER tools. A path search mechanism, Getpath, can be used by shell scripts to access the *lager* file database. This mechanism is the core of the library support within LAGER. When tools (including DMoct) start up, they usually read one or more *lager* files, which include a common file maintained for all users and customized lager files for an individual user or even for a particular design. If a given key appears in more than one file, then each list of paths is checked in turn.

Following is an excerpt from a *lager* file:

```
(DMoct.sdl
        ./
        ~lager/cellib/actel
)
(bin
        ~/bin
        $LAGER/bin
)
```

In this example, there are two search paths with keys *DMoct.sdl* and *bin* respectively. DMoct itself uses the list of directories specified under the key *DMoct.sdl* to search for SDL files. A default directory list for this purpose is maintained in the common lager file. Users can have private libraries referenced in the *lager* file in the current or home directory, although it is strongly recommended that the common *lager* file should be updated and the private library installed, so that others can use the new cells, and avoid duplication of effort.

5.4.2. LightLisp

LightLisp [Baker90] is an interpreter for a subset of Common LISP. It has been chosen to provide parameter evaluation capability to DMoct, since it is much smaller and more portable than commercially available Common LISP implementations.

DMoct is actually linked together with LightLisp libraries, and can be used as an interactive LightLisp interpreter to debug LightLisp parameter evaluation problems. It is possible to add user-defined functions to DMoct, by simple additions to a LightLisp start-up file (DMoct.11). Some of the functions included in the standard file are:

```
log2 integer-length bitsize if1 stringify termpos termsize
mkpinstring mkpinlist getpin pinlist pin getval
```

These functions help to simplify common parameterized property declarations.

Each level of a design hierarchy uses a separate set (or *closure*) of variables for parameter evaluation within that part of the design. For the most part, all variables are passed explicitly through a design hierarchy, to avoid ambiguity for similar subcells. DMoct allows the user to define global variables that are accessible to all LightLisp expressions, either by defining them in the DMoct.11 file, or by defining them on the DMoct command line.

5.4.3. UNIX File Time-stamp Checking

In addition to providing the flexibility of LightLisp to develop parameterized designs, DMoct also identifies those parts of a design that are out of date, and can selectively regenerate only those parts that have been changed, in analogy to the UNIX "make" utility. The basic rules for determining if a design will be regenerated are listed below.

Structure_master view

A structure_master view will be regenerated if there is a corresponding SDL file that is more recent, or if any of the subcells of this design are more recent.

Structure_instance view

A structure_instance view will be regenerated if the structure_master view is newer, or if any of the subcell structure_instance views are newer. Also, if any of the parameters have changed, the structure_instance view will be recreated. In addition, the structure_instance view may depend on changes in foreign files (other than SDL and OCT views). Any foreign files that affect a given structure_instance view should be named in a "depends-on" property under the structure_master view facet. Many structure-processor tools require foreign files (e.g.: BDS files in Chapter 8).

Physical view and Magic files

With the exception of leafcells (CELLCLASS LEAF), DMoct does not generate any physical views on its own. DMoct calls layout-generator tools named in each structure_instance view if necessary. This is done if the physical design is older than the corresponding structure_instance view, or if any of the physical subcells are newer. The time-stamps are checked on magic files as well, to determine if a design should be regenerated.

The user can instruct DMoct to ignore all file time-stamps, and regenerate everything. Any time a tool is changed, a design should be regenerated, even if DMoct thinks that it is up to date. Options exist to cause all views to be regenerated, starting from the SDL files, if they exist, or only regeneration of specific views, either structure_master view, structure_instance view, or physical, if desired.

The selective design generation capabilities discussed in Section 5.3.3. essentially disable time-stamp checking on select parts of a design. As a result, a large circuit can be regenerated more quickly than if all parts of a design are checked. This capability is intended to speed up design iteration, but the designer should regenerate the complete design from scratch before submitting a design for fabrication, to verify the consistency of the design hierarchy.

5.4.4. Error Diagnostics

DMoct collects all the error diagnostics into a single readable file. DMoct passes the name of this diagnostic log file to the tools that it executes. The log entries from DMoct are formatted to cut long lines, and to indent according to the active part of the design hierarchy. As DMoct runs, diagnostics are output to indicate what DMoct is doing. These diagnostics and more detailed information are saved in the log file, which can be referred to in the event that problems should occur.

Typical errors are often indications that one or more of the *lager* files are out of date, or the desired SDL file is not installed in the correct place. `DMoct` gives the user an opportunity to solve the problem without having to restart the design process.

5.5. SUMMARY

`DMoct` is the design manager for the tools used in LAGER. With the Light-Lisp interpreter built in, `DMoct` has powerful parameter evaluation capabilities. Time stamp checking and design flow control over user-designated tools are offered, along with clear feedback to the user. Also, library support makes a wealth of existing designs available to users, to avoid duplication of effort.

REFERENCES:

[Baker90] W. Baker, "Volume 6: Light/Oct/Vem/Lisp," *Oct Tools Distribution 4.0*, U. C. Berkeley Electronics Research Laboratory, Berkeley, California, March 1990

{Scott85} Walter S. Scott, Robert N. Mayo, Gordon Hamachi, and John Ousterhout, editors, "1986 VLSI Tools: Still More Works by the Original Authors," *Technical Report No. UCB/CSD 86/272*, December, 1985

[Srivastava87] M. B. Srivastava, "Automatic Generation of CMOS datapaths in the LAGER Framework," Masters Thesis, University of California at Berkeley, May 1987.

6

Design Post-Processing

Marcus Thaler and Brian C. Richards

Once a design has been generated by DMoct, it has to be verified and tested before the final data for fabrication can be created and shipped to the manufacturer. Simulation data must be extracted from a structural or physical representation, design rule checking (DRC) performed, CIF (Caltech Intermediate Form) layer representation generated, and the verified design sent to MOSIS for fabrication. DMpost supports a set of tools that perform these tasks, while hiding all the details and command line options of the tools from the user. It essentially provides the link between the OCT design database and the post-processing tools by generating the required input files.

All the necessary technology parameters involved in a design are derived either from command line options or read in interactively and are passed to the tools called by DMpost, thus minimizing the possibility of erroneous usage. DMpost itself is a UNIX C-shell script that can easily be updated and extended as LAGER is updated with minimal portability problems. Many of the tools used by DMpost are based on design files for the Magic layout editor.

6.1. CAPABILITIES

Figure 6.1 shows the role of DMpost in the design process. For functional simulation using THOR, DMpost extracts behavioral models of cells and a corresponding net-list from the structure_instance view. For logic level simulation using IRSIM, circuit information is extracted from the physical view. To check for design rule violations, generate the CIF layer representation, and edit the layout, DMpost invokes the layout editor Magic. DMpost collects all the files required by the simulators, including technology dependent files or files generated by Magic, into appropriate directories so the designer has easy access to them for further processing.

6.2. THE POST PROCESSING TOOLS

The main tools involved in design post-processing or verification include the Magic layout editor and the simulators, THOR and IRSIM.

6.2.1. The Magic layout program

Magic [Scott85] is an interactive tool for creating and modifying the geometric layers of a VLSI circuit. All cell libraries are designed with Magic and then converted to the OCT physical representation. In contrast to many other layout editors, Magic has a wide range of capabilities beyond color painting. For example Magic performs on-line design rule checking during editing as well as in batch mode on complete designs, based on given layout rules for a particular technology. One particularly important feature is that the MOSIS implementation service provides complete design rules in the Magic technology file format. Magic also contains a hierarchical circuit extractor which provides netlists for the circuitry, useful for logic level simulation or for Spice circuit simulation.

Magic is a symbolic layout editor that uses simplified design rules and circuit geometrical structures, making layout editing easier. This style of layout depends on a uniformly spaced grid in both X and Y directions where the grid size represents the minimum feature size. Designs in Magic use lambda based rules. Magic is restricted to Manhattan (non-diagonal) designs, but this has rarely been found to be a limiting factor, and has considerable advantages. In LAGER, the layout editing capability of Magic is *only* used for library development. It is not used for layout generation once the design with DMoct is under way.

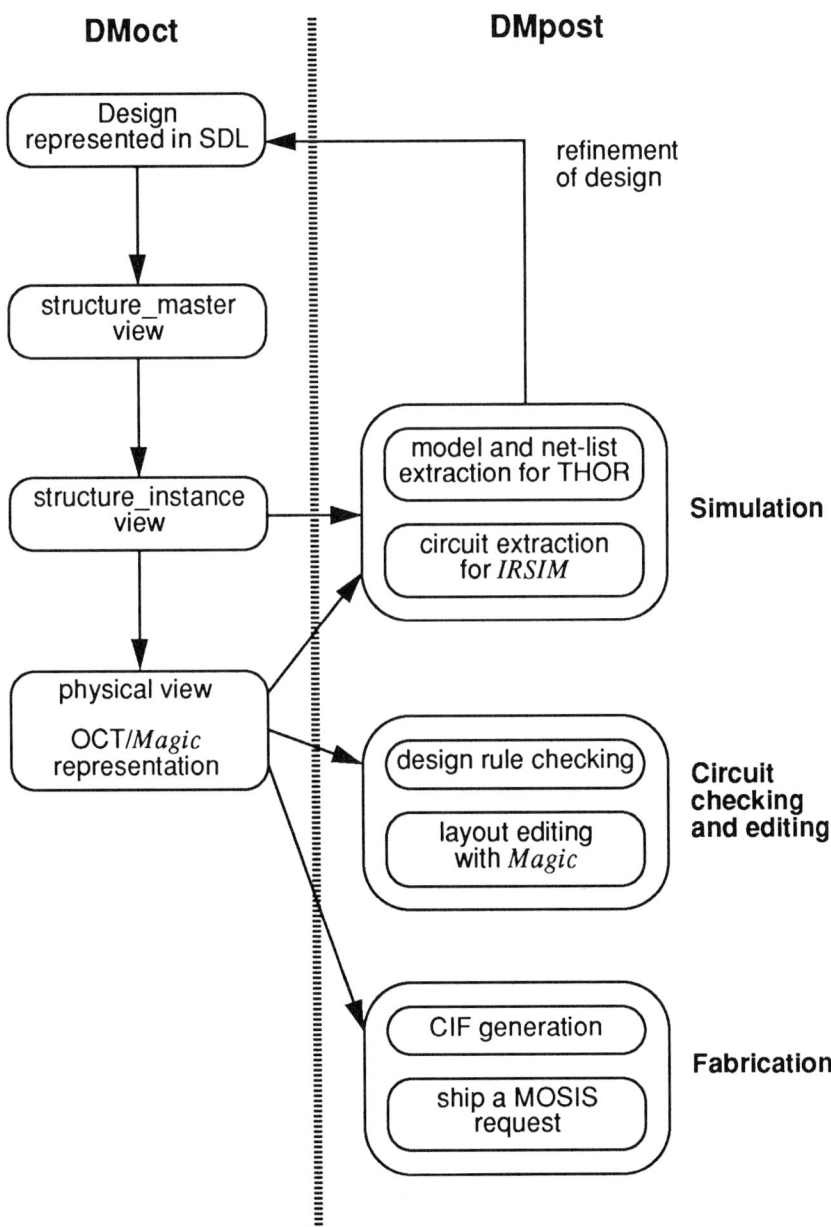

Figure 6.1: The Role of DMpost in the design process

6.2.2. THOR

THOR [Alverson88] is a functional simulator based on hierarchical behavior models, including register-transfer level down to the gate level models. In LAGER, THOR models accompany all library cells which are part of the library installation, such as PLA's, RAM's, adders, latches, logic gates, *etc.* When appropriate or necessary, the designer describes higher levels in the design hierarchy with more complex models. Since THOR is a compiled simulator that models behavior above the transistor level, THOR simulations run much faster than a corresponding switch-level simulation using IRSIM, described in the next section.

THOR is equipped with a graphical user interface, the analyzer, that interactively presents the simulation results. The analyzer reports results in the form of timing diagrams as they would appear from an actual logic analyzer. Signals are displayed and updated as the simulation proceeds, and the user can pan and zoom through the data. Signals can be viewed individually or as busses, and their order of appearance can be rearranged interactively.

In LAGER, the behavioral models are parameterized and stored as string properties attached to the OCT facets of the cells. They can be extracted from the structure_instance view with MakeThorSim, along with a netlist of all the models used in the entire design. In addition, MakeThorSim generates a call to the analyzer for displaying all nets to which formal terminals are attached.

6.2.3. IRSIM

In contrast with the THOR functional simulator, IRSIM [Salz89] is an event driven logic or switch level simulator for nMOS and pMOS circuits, and models first order timing behavior. Two transistor models are supported by IRSIM. The first is a switch model where each transistor is represented by a voltage controlled switch. The second is a linear model where each transistor is modeled by a voltage controlled switch in series with a resistor, and each node in the circuit has a capacitance attached to it.

The stimuli and user commands can be passed to the simulator from either a start-up command file or interactively by the user. The first mode is typically used for initialization. The simulation results are displayed through an interface similar to the analyzer graphical interface used by THOR.

6.3. RUNTIME OPERATION

`DMpost` must be run from the directory containing the top level or working directory of the design. `DMpost` will create a subdirectory in the working directory, if it does not exist, to store the `DMpost` and simulation files for `THOR` and `IRSIM`.

As mentioned earlier, `DMpost` is a UNIX C-shell script and is controlled either by command line options or interactively asks for required parameters. As with most LAGER tools, `DMpost` records all runtime information produced by `DMpost` itself or by any tool called by `DMpost` and saves the information in a log file. In addition, `DMpost` informs the user about each task as it is executing, which is helpful in monitoring progress on large, time-consuming designs.

To locate all of the technology and configuration files used by the different tools, `DMpost` uses the `GetPath` mechanism provided by LAGER.

6.3.1. Organization

The input to `DMpost` can either be a structure_instance view or a physical view of a design.

Figure 6.2 shows an overview on the organization of `DMpost` for a design example named MyDesign. Given the structure_instance view, `DMpost`, calls `MakeThorSim` to generate the MyDesign.csl file which contains the net-list and the `THOR` model files. Given the physical view, `DMpost` uses `Magic` for further processing. If the physical representation exists only in OCT, `DMpost` converts the physical view to a CIF representation which is readable by `Magic`. On the other hand, if the `Magic` files already exist, `DMpost` will copy all `Magic` files into the layout directory. This forces `Magic` to regenerate all extraction files, making sure they use identical technology parameters. Otherwise, `Magic` would use already created extraction files. Once the extraction files are generated, `Magic` will flatten the design into one single simulation file `MyDesign.sim` in the layout directory.

`DMpost` is also capable of passing commands to `Magic` to run a design rule check (DRC), generate a CIF file according to the chosen technology and process parameters, or start `Magic` interactively, eventually resulting in an updated set of `Magic` files.

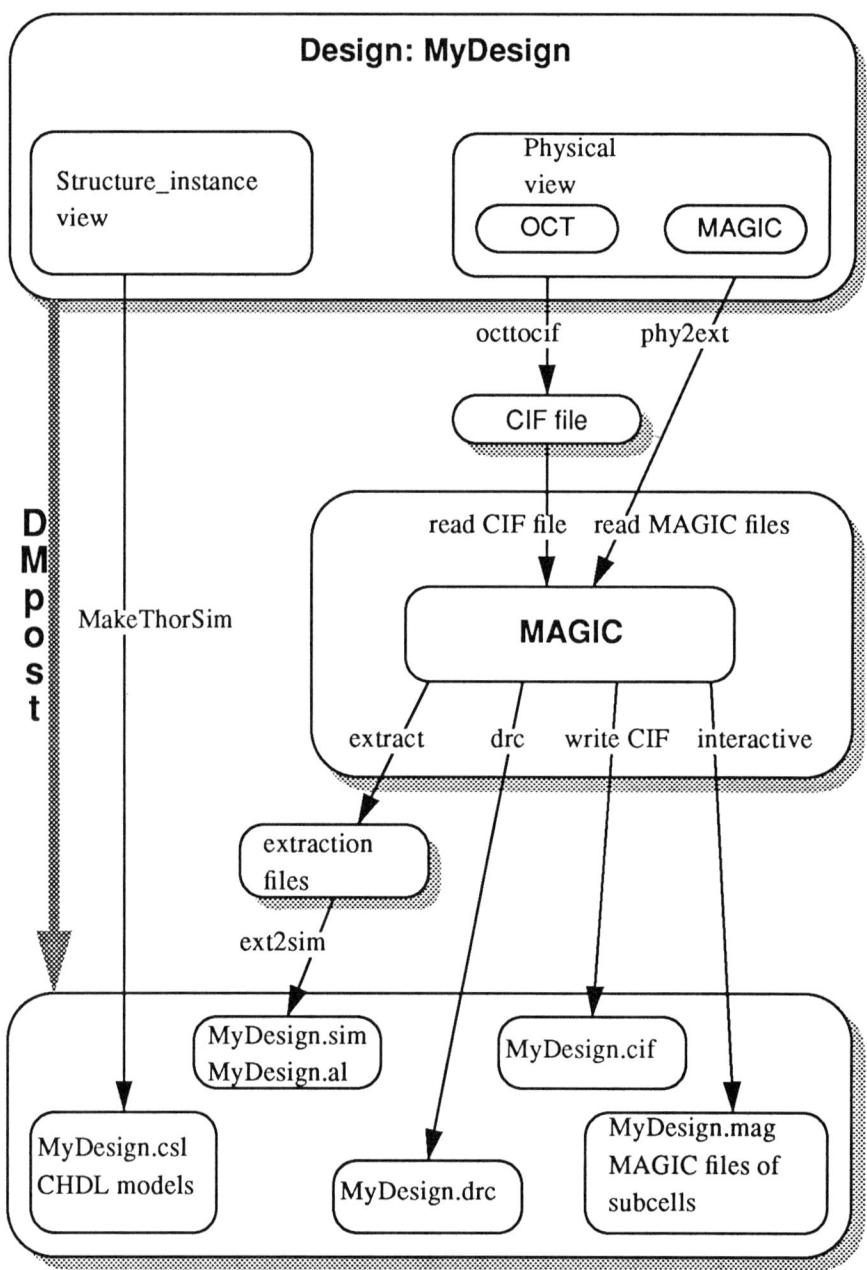

Figure 6.2: Organization of DMpost

6.3.2. User Control

The commands that the user can give to `DMpost` can be divided into four categories.

- Options for specifying the desired post-processing operations.
- Commands for generating `THOR` simulation files, or extracting electrical circuit information for `IRSIM`.
- Commands for enabling design rule check to verify that a design can be fabricated.
- Commands for preparing a CIF file, and format a request to MOSIS to fabricate a chip.

With the exception of running `Magic` and preparing for MOSIS, the options may be used in any combination or order. If no options are given then `DMpost` will extract the circuit, perform a design rule check, and create a CIF file.

`DMpost` requires the user to specify the technology used for post-processing. The minimum feature size must be given in microns per lambda, where 2 lambda corresponds to the minimum feature width. Presently, 0.6, 0.8, 1.0, and 1.5 microns per lambda are supported. The p-well, n-well or generic process well type must also be defined, where the latter allows the fabrication vendor to decide which well type should be used.

Several options specify where the files generated by `DMpost` are placed. The destinations of layout files, simulation netlist files, and run-time diagnostic log files can be changed. `DMpost` can be run in verbose mode, to provide diagnostics to the user, or the operations can be previewed without actually running any tools. Also, to save space, files can be used directly from libraries where appropriate, and post-processing files can be deleted from previous post-processing attempts. Other options affect `THOR` model generation, such as disabling automatic generation of the analyzer. This is useful if the user wants to explicitly control the generation of the analyzer or an alternative user interface.

6.4. SUMMARY

`DMpost` supports designers during the verification phase. From behavioral simulation through submitting a fabrication request, `DMpost` offers a consistent interface to each application.

REFERENCES:

[Alverson88] R. Alverson, T. Blank, K. Choi, A. Salz, L. Soule, and T. Rokicki, "THOR user's manual," Technical Report CSL-TR-88-348 and 349, Stanford University, January 1988.

[Salz89] A. Salz, M. Horowitz, "*IRSIM*: An incremental MOS switch-level simulator," *Proceedings of the 26th ACM/IEEE Design Automation Conference*, June 1989, pp. 173-178.

{Scott85} Walter S. Scott, Robert N. Mayo, Gordon Hamachi, and John Ousterhout, editors, "1986 VLSI Tools: Still More Works by the Original Authors," *Technical Report No. UCB/CSD 86/272*, December, 1985.

Part II

Silicon Assembly

7

Hierarchical Tiling

Jane Sun and Brian C. Richards

TimLager is a hierarchical tiler which exploits the feature for which VLSI technology is well suited - the implementation of regular arrays of subcells. It is a general purpose, hierarchical macrocell layout generator that is used to generate parameterized bit-slice modules such as adders, registers, multiplexers and parameterized array-based modules such as RAMs, PLAs and ROMs. TimLager assembles the layout by abutting instances of leafcells, which are the lowest level entities seen by TimLager and represented by OCT physical views. TimLager allows high level control over construction of macrocells through its tiling specification language.

A key characteristic of a TimLager macrocell is that all the interconnection between leafcells is made through abutment of the leafcells. The leafcell layout can be either manually designed or generated from the macrocell place and route tool, Flint. Since TimLager macrocells are assembled by placement, leafcells must be pitch-matched and no routing is involved in the macrocell generation. Hence, the area and performance of the macrocell can easily be predicted based on the leafcell characteristics.

The macrocells produced by `TimLager` can be connected by `Flint` to other datapath blocks, standard cells, or `TimLager` macrocells to generate a macrocell for the next higher level in the design hierarchy.

7.1. USER INTERFACE

To generate a tiled macrocell with `TimLager`, two distinct sets of inputs must be provided. The first is the leafcell layouts and the (parametrized) tiling specification for the generic macrocell. The second is the specific macrocell name and parameter values for a macrocell instance. Usually, a `TimLager` *user* will only need to provide the second set, because the leafcell layouts and tiling specification for the generic macrocell are already designed by a *developer* and installed into the `TimLager` cell library.

An example of a `TimLager` macrocell is an N-bit latch. The general block diagram and the actual layout for a 4-bit macrocell is shown in Figure 7.1, while the SDL description of the black-box view is given in Figure 7.5. The latch leafcells are the memory cell, *latchcell*, and the control signal buffers in the cell, *latchctl*. A tiling specification describes how the leafcells are abutted to form the N-bit latch. A strength of `TimLager` is its ability to allow the tiling specification to be dependent on parameters. In this example, the word width, *Nbit*, is the only parameter, as shown in the SDL description. To have `TimLager` generate a 4-bit macrocell, the user needs only to provide the specific instance name, such as "my_latch", and the *Nbit* parameter value, 4.

A PLA (programmable logic array) is another example of a `TimLager` macrocell. The leafcells for this macrocell are the AND-OR plane cells and the input and output signal drivers. The tiling specification describes how the leafcells are abutted to form the AND-OR plane circuit and the driver circuits that buffer the input and output signals to and from the AND-OR plane. The parameters are the bit patterns of the AND and OR planes, as well as the input word and output word width.

7.2. MACROCELL DESIGN

7.2.1. Design Flow

The design flow for a `TimLager` macrocell is shown in Figure 7.2. The developer needs to determine the black-box view of the macrocell (i.e., what are the I/O terminals), what the parameters are, what set of leafcells are needed, and

Chapter 7 Hierarchical Tiling

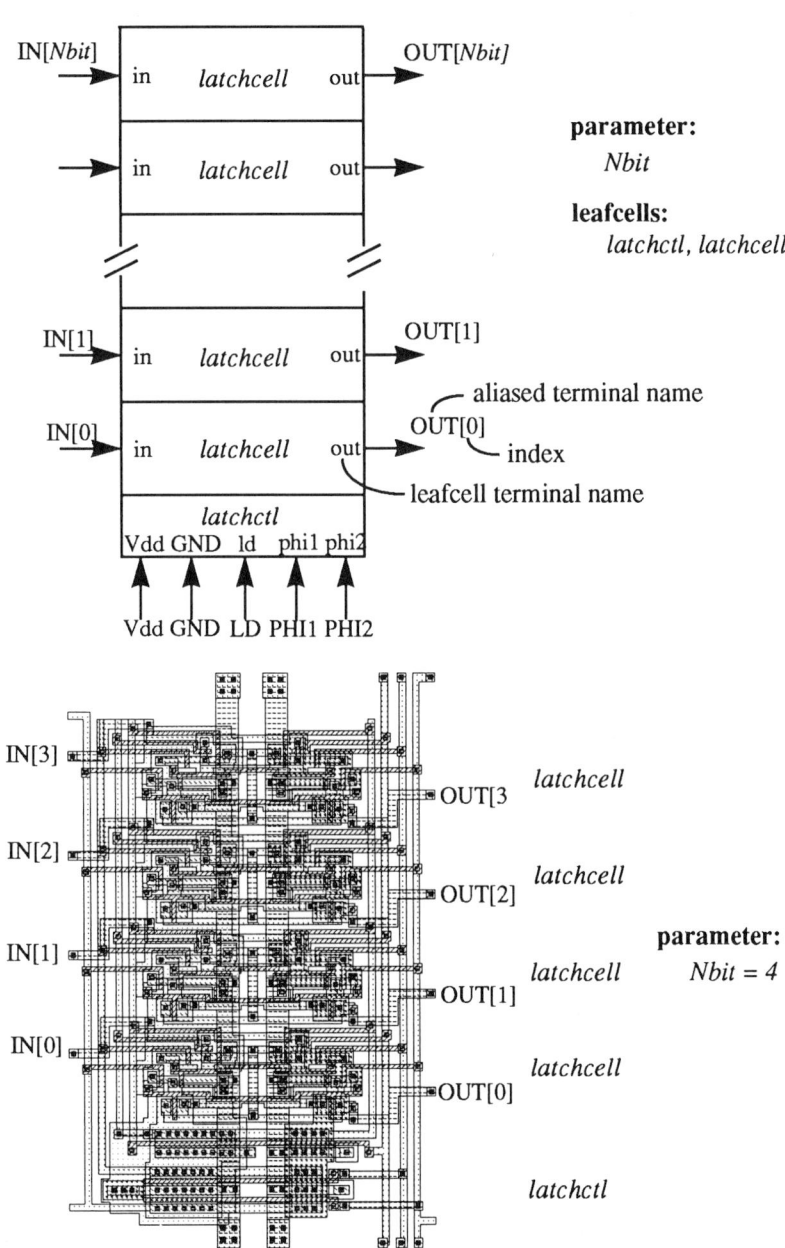

Figure 7.1: Latch Macrocell: block diagram and layout

how instances of the leafcells are placed to compose the macrocell. The developer then uses this specification to create a set of leafcell layouts, an SDL file and a tiling procedure. The tiling procedure describes how the macrocell is constructed from the leafcell instances, as a function of the parameters.

Once the basic macrocell design is complete, TimLager can generate various layout instances, given the instance name and parameter values from the user. The complete set of inputs to TimLager, one from the developer and the other from the user, is shown in Figure 7.2.

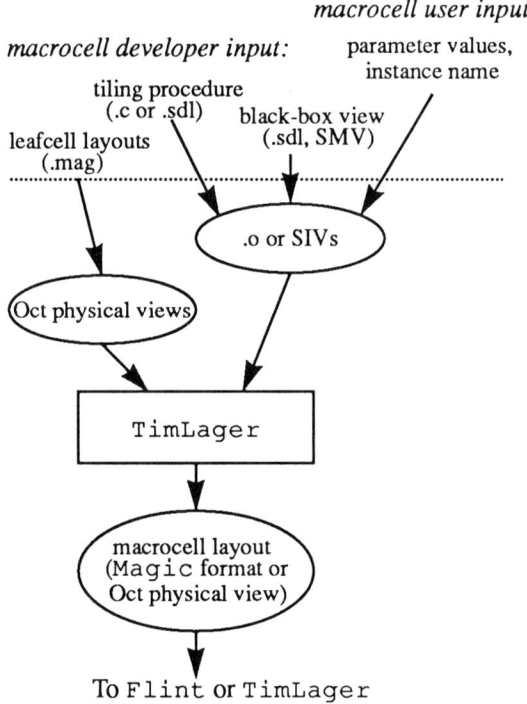

Figure 7.2: TimLager design flow

To actually generate a macrocell layout, TimLager reads the leafcell layouts in the OCT physical format, and loads the tiling instructions either from the structure instance view or the .o object file compiled from the C tiling description. The macrocell instance name and parameter values are contained in the structure instance view. `TimLager` can be invoked manually or it can be run automatically through `DMoct`.

`TimLager` produces the macrocell and its subcell layouts in two possible formats: the OCT physical view or the `Magic` file format. The `Magic` layout files are deposited in a directory called "layout" residing in the current working directory.

The rest of this section elaborates on the various phases of the design flow.

7.2.2. Leafcell Layout

The main goal at the leafcell level is to design the minimum number of leafcells needed to form the macrocell. Typically, the leafcell layout is manually designed with the layout editor `Magic`. Instances of the leafcells can be overlapped, rotated, mirrored, and then abutted to form the macrocell array(s). When array dimensions are large, signal lines formed by connecting common signals among leafcells will have significant load capacitances, and this must be given consideration in the circuit design process. For some applications, the rows and columns of the array need individual control circuits, or they may need circuits that interface them to other (external) circuits. In all these cases, the developer also needs to design peripheral leafcells such as driver and latch circuits. Though not mandatory, an efficient design style is to constrain control, clock, data, and power signal lines to run straight through the entire leafcell so that they connect upon leafcell abutment to form the global line. Metal 1 and metal 2 lines should run mutually perpendicularly and should be used for global signals; polysilicon lines should be used only for connections local to the leafcell. The total area of a leafcell layout is heavily influenced by the number of vias in it, so that number should be minimized.

The leafcells in the current `TimLager` cell library have been designed using the MOSIS scalable design rules and are compatible with minimum channel lengths of 0.8, 1.2, 1.6, 2.0 and 3.0 microns (lambda = 0.4, 0.6, 0.8, 1.0, 1.5).

7.2.3. Tiling Procedures

The tiling procedure instructs `TimLager` how to generate the macrocell layout by placing and abutting (tiling) leafcell instances. The procedure can be written in the C programming language or directly in SDL. In either case, the user can express the desired design hierarchy, leafcell tiling dependency on parameters, and placement transformations. Both specification methods are described

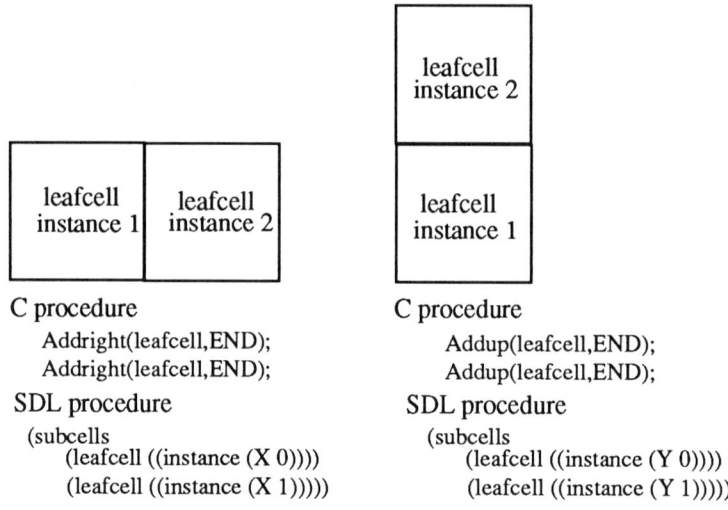

Figure 7.3: Placement functions (in C) or properties (in SDL)

here. There are other macrocell generators such as *Mocha Chip*[Mayo86] which allow graphical specification of the macrocell structure in terms of leafcell layouts, but they do not support hierarchy, which is essential for use within LAGER.

C Tiling Method

To support the C tiling specification, `TimLager` provides a library of placement functions used for procedural tiling of leafcell instances, as well as parameter access functions used to read parameter values from the structure instance view.

The two placement functions Addright() and Addup() are illustrated in Figure 7.3. `TimLager` tiles the array in the upward direction or toward the right of the current array of leafcells already tiled. An Addup() performed on a leafcell will place an instance (of the leafcell) by aligning the bottom left corner of the *instance* with the top left corner of the *current array*. Performing an Addright() on a leafcell will place an instance by aligning the bottom left corner of the *instance* with the bottom right corner of the *current array*. In Figure 7.1, the latch is a one-dimensional vertical array, so only Addup() is used. Both the Addup() and Addright() functions use the same arguments. The syntax of the two functions are:

```
Addup(leafcell_name,[optional arguments], END)
Addright(leafcell_name,[optional arguments], END)
```

The optional arguments OVERLAP, MX, MY, R90, R180, R270, [OFFSETX, integer], [OFFSETY, integer], specify geometric transformations in which the leafcell instance is overlapped, mirrored, rotated, or offset when it is placed. Optional arguments of the form [TD,old_terminal_name,ALIAS,new_terminal_name, INDEX, integer] allow the instance terminal names to be aliased and indexed at the parent level, as illustrated by the latch example. Thus the terminal names on the leafcells can be changed arbitrarily to names more pertinent to the design as seen at the top level. The optional arguments NONE, LEFT, RIGHT, TOP, BOTTOM, indicate which sides of the leafcells have terminals that are to be promoted to parent terminals of the macrocell.

The C tiling procedure for the latch example of Figure 7.1 is shown in Figure 7.4. Even numbered *latchcell* leafcell instances are placed without geometric transformation, and odd-numbered *latchcell* instances are mirrored about the X-axis as specified with the argument **MX**. All the *latchcell* instances have input and output data terminals that become parent terminals under aliased and indexed names, IN and OUT, respectively.

Addup() and Addright() also allow optional arguments that provide a stretching mechanism to pitch-match one macrocell to another, and a mechanism for adding metal feed-through lines between cells. These options are useful for optimization of higher level cells that are constructed from multiple TimLager cells. In particular, these features are used in the datapath optimization tool, dpp, as discussed in Chapter 10.

The Addright() and Addup() functions specify how cells are placed relative to cells that have already been placed. Hence the macrocell structure depends on both the order and the arguments of the function calls. Addup() and Addright() functions can be embedded in for-loops to form parameterized array structures. They can also be embedded in if-else statements to make conditional choices of which cell to place or the particular geometric transformation to apply. This is shown in the latch tiling procedure, where the parameter is *Nbit*. The for- loop is controlled by *Nbit* so that instances of the leafcell *latchcell* are tiled *Nbit* times. Inside the loop, the if-else statements select the MX geometric transformation for odd-numbered *latchcells*, and no mirroring for even-numbered *latchcells*.

```
/*
 * latch.c: tiling procedure for an N-bit latch.
 */
#include "TimLager.h"
latch()
{
   int i,Nbit;
   Nbit=Getparam("Nbit");
   Open_newcell(Read("name"));
/*
 * tiling procedure
 */
/* addup control leafcell */
   Addup("latchctl",BOTTOM,TD, "ld",ALIAS, "LD",
                       TD, "phi1",ALIAS, "PHI1",
                       TD, "phi2",ALIAS, "PHI2",END);
/* addup one-bit latch leafcells */
   for(i=0;i<Nbit;i++){
     if ((i%2) == 0)
       Addup("latchcell",RIGHT|LEFT,TD,"in",ALIAS,"IN",
              INDEX,i,TD,"out",ALIAS,"OUT",INDEX,i,END);
     else
       Addup("latchcell",RIGHT|LEFT,MX,TD,"in",ALIAS,
          "IN",INDEX,i,TD,"out",ALIAS,"OUT",INDEX,i,END);
   }
   Close_newcell();
}
```

Figure 7.4: C tiling procedure for latch macrocell

For complex macrocells, it may be desirable to introduce additional levels of hierarchy in the tiling procedure. This means that the subblocks are created which in turn are tiled to form the top level layout. To express hierarchical layout, the developer writes the tiling procedure for each individual block layout as a C function, and calls them in such an order so that the subblocks have been generated before we use them at a higher level.

Each of the subblock tiling procedures must call the Open_newcell (*block_name*) library function for TimLager to start placement at the lower-left corner (0,0) for the block named, *block_name*. The end of the block tiling proce-

Chapter 7   Hierarchical Tiling

```
; latch.sdl : top-level SDL file for latch.
(parent-cell latch)
(parameters Nbit)
(layout-generator TimLager)
(net in    (width Nbit) ((parent IN)))
(net out   (width Nbit) ((parent OUT)))
(net ld    ((parent LD)))
(net phi1  ((parent PHI1)))
(net phi2  ((parent PHI2)))
(net Vdd   ((parent Vdd)))
(net GND   ((parent GND)))
(end-sdl)
```

Figure 7.5: Top-level SDL for latch macrocell

dure must call the Close_newcell() library function which marks the end of the block tiling procedure. The latch tiling procedure shows the use of these two necessary library functions.

For `DMoct` to integrate a `TimLager` into a larger design, the developer must create an SDL file which specifies the black-box I/O-terminal view of the macrocell and the macrocell parameters. Hierarchical tiling procedures need only be accompanied by a single SDL file for the top-level. The top-level SDL file for the latch example is shown in Figure 7.5.

The tiling procedure for each macrocell is compiled and stored in its own directory of the cell library, along with associated OCT physical view of the leafcells. At run time `TimLager` dynamically links the required procedure and executes it. The parameter values are read from the structure_instance view and a physical view is generated for the macrocell layout.

SDL Tiling Method

The macrocell developer can also describe the complete tiling procedure in with the SDL language. In this specification method, the SDL for the top-level macrocell gives the black-box view, and it also describes how lower level blocks or leafcell instances are geometrically transformed and placed to form an array. The SDL file for any lower level block describes in turn its black-box view and tiling procedure. Finally, the there must be an SDL file *for each individual leafcell type*, showing the black-box view of that leafcell. Hierarchical description and parameter access are natural features of SDL, as is promoting a leafcell terminal to a parent terminal of the macrocell. Also, loop and conditional procedure control is provided by SDL LightLisp expressions. In the top level SDL tile, a "-s" flag must be specified for the layout-generator `TimLager`. The flag indicates that the tiling procedure information is given in the structure_instance view of the macrocell.

The SDL tiling procedure for the latch example appears in Figure 7.6. The individual leafcell SDLs are not shown. To describe the equivalent of Addup() and Addright() placement, the user attaches X and Y properties to each leafcell instance, as illustrated in Figure 7.6. The X or Y property takes on an integer value or an expression that evaluates to an integer. Note that the X and Y properties specify the absolute position of the instance, as opposed Addright() and Addup(), which place the instance relative to the instances that have already been placed. The top and block level SDLs are most conveniently written in the pinlist style (as opposed to the netlist style), since placement properties can then be conveniently attached to each instance as it occurs.

For the latch example, no X property is required for the *latchcell* instances, since they are tiled in the vertical direction only. Also note that the order of declaration of instances does not affect the placement. SDL is a non-procedural language, and all the entire text of the SDL file is read before any action is taken. To apply geometric transformations, the user can attach T properties to any leafcell instance. A T property can take one of the values: "NONE", "MX", "MY", "R90", "R180", "R270", "MXR90", "MYR90". Also, the developer can rename terminals by placing an ALIAS property on the terminal. The new terminal name and must then be used to refer to that terminal at the next level of the SDL hierarchy.

C Tiling versus SDL Tiling

Originally, `TimLager` supported only C language tiling procedures. The C language offers the macrocell developer powerful constructs that can be used to describe complex and detailed tiling procedures. However, this method did not support timestamp checking of individual blocks and leafcells in a hierarchical macrocell. Since the use of SDL already had automatic timestamp checking and naturally provided all the non-placement and transformation features that the C description provided, SDL was enhanced with the X,Y, and T property options to support tiling tasks.

```
;latch.sdl: top-level SDL and tiling procedure for latch.
;
(parent-cell latch)
(layout-generator "TimLager -s")
; go through structure_instance view to get tiling info!
(parameters Nbit);
; ----- top-level view of latch macrocell -----------
(instance parent (
    (IN IN (width Nbit))
    (OUT OUT (width Nbit))
    (LD LD)
    (PHI1 PHI1)
    (PHI2 PHI2)
    (Vdd Vdd)
    (GND GND)     ))
; ------ pinlist and tiling procedure ------------
; addup control leafcell
(subcell
    (latchctl ((CTRL (Y 0)) ) ) )
(instance CTRL (
    (ld LD)
    (Vdd Vdd)
    (GND GND)
    (phi1 PHI1)
    (phi2 PHI2)
))
; addup one-bit latch leafcells
(dotimes  (i Nbit)
    (subcell (latchcell ((LATCH (cond (evenp i) )
        (Y (+ i 1)) ))))
    (subcell (latchcell ((LATCH (cond (oddp i))
        (Y (+ i 1)) (T "MX")) )))
    (instance LATCH   (
        (in IN (term-base i) (net-base i))
        (out OUT (term-base i) (net-base i)))
)
```

Figure 7.6: SDL tiling procedure for the latch macrocell

Besides automatically gaining the timestamp checking capability, the macrocell developer can easily instruct `TimLager` to tile leafcells that are actually macrocells produced by `Flint`. In this case, the developer can succinctly write the black-box view of a macrocell and a simple tiling procedure all in one file.

The cost of using the SDL for tiling procedure specification is in layout generation speed. The extra expense over using C tiling procedures is incurred during generation of the macrocell structure_instance view from the structure_master view. Using a C tiling procedure, `TimLager` dynamically loads the .o tiling procedure, and then runs the tiling. When SDL tiling procedure are used, structure_instance views are generated for each leafcell type (recall that each leafcell type has an SDL) before `TimLager` can actually perform tiling. If a macrocell design includes many leafcell types or the tiling procedure includes many control loops, the structure_instance view generation time is significant compared to the actual `TimLager` run time.

For simple macrocell array structures, with few leafcell types, it is often easier to use the SDL tiling method. For large hierarchical macrocell designs, it can be better to use the C tiling method or a mix of SDL and C tiling. For example, the developer could write a C function for complex and detailed tiling of lower level blocks from leafcells. Then, the placement of the blocks for the overall macrocell could be described one top level SDL file, along with the macrocell black-box view.

7.3. TILER TECHNIQUES AND ALGORITHMS

As discussed in the previous sections, the heart of `TimLager` is a small set of C functions that give relative placement instructions for tiling leafcells and subblocks, both collectively referred to as subcells in this section. The routines Open_newcell(), Addright(), Addup(), and Close_newcell() are called to produce the layout of the design. The same routines are called when using SDL tiling, but in this case they are called from inside TimLager itself instead of explicitly in a developer-supplied C function.

Several characteristics are required of subcells that are to be tiled. The cells are tiled so that adjacent cells automatically abut, without requiring specific translation information. This tessellation of cells is guided by the tiling boxes of each cell. By default, the tiling box is the bounding box of all the layout geometries. In many cases, however, there are geometries extending beyond the desired tiling box, so the developer may give a specific tiling box by placing the label "obox" in

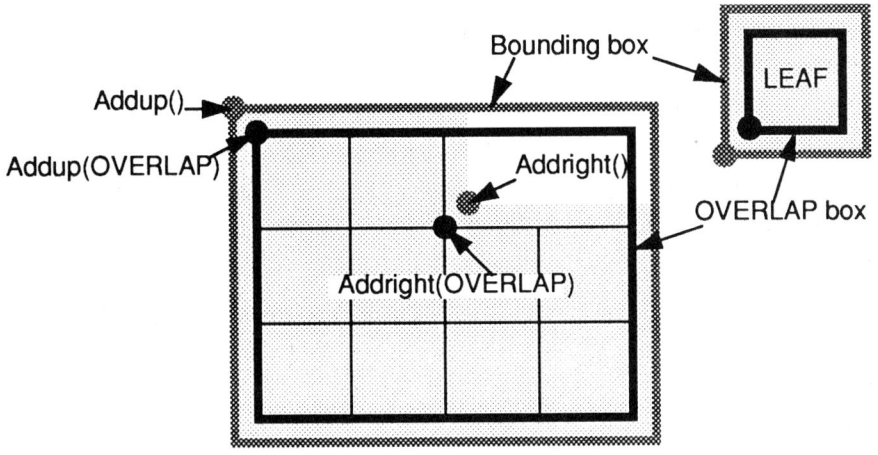

Figure 7.7: The two tiling boxes maintained by TimLager

a Magic design file, or the "OVERLAP" layer in an OCT physical view. This is often useful in CMOS designs, where space can be saved by sharing wells and contacts on the boundary of adjacent subcells.

The developer may specify that either the bounding boxes or the OVERLAP boxes of adjacent cells should abut. To support both alternatives, TimLager maintains two current tiling location coordinates, one for the OVERLAP box and another for the bounding box. The locations designate where the lower left-hand coordinate of the next subcell can be placed. If the developer specifies the OVERLAP option in the Addup() or Addright() function, the OVERLAP coordinates are used, otherwise the bounding boxes are aligned (default).

TimLager maintains simultaneously a bounding box and an OVERLAP box for the cell under construction, both of which contain all of the respective bounding and OVERLAP boxes of subcells (see Figure 7.7). These boxes are updated following each Addup() or Addright() call, along with the current tiling location pointers. Addup() or Addright() are identical, except that the Addup() function resets the next OVERLAP box tiling point to the top-left corner of the current OVERLAP box, and similarly updates the bounding box tiling point. Addup() and Addright() both call a more general routine that performs the actual tiling.

When a subcell is tiled, the tiling box of the subcell is transformed according to Addup() or Addright() command arguments. The resulting tiling box is then translated so that the lower left-hand corner of the tiling box lies on the appropriate current tiling location.

The terminals of a subcell can be promoted to the next level of hierarchy if (and only if) they are on the boundary (bounding box or OVERLAP box) of the cell. To be promoted, they must also be declared as parent terminals in the SDL file.

If a C routine is used for the tiling, each call to Addup() or Addright() must specify which terminals (if any) should be promoted. Each call has the form Addup (subcellname, termsidemask, <other arguments>, END). The termsidemask is an integer which specifies the sides of the subcell from which terminals should be promoted. The macro NONE or any bitwise-OR combination of the macros LEFT, RIGHT, TOP, BOTTOM can be used It is also possible to specify by name exactly which terminals to promote (on any side)

Addup() and Addright() can also take TD (terminal data) arguments specifying that any given terminal be renamed (aliased) and possibly indexed when promoted. The alias specified must correspond to a terminal declared in the SDL file. The format of a TD specification is Addup(..., TD, old_terminal_name, ALIAS, new_terminal_name, INDEX, integer,...). The INDEX argument, if any, will be appended to the new terminal name inside square brackets following the terminal name. This conforms with LAGER bus naming conventions.

If the SDL tiling method is used, the terminals of a subcell instance are promoted to the tiled design only if they connect to a formal or "parent" terminal in the structure_instance view, through a net. The parent terminal name in the structure_instance view is given to the terminal in the layout. This has the advantage of unambiguously defining terminal names, without regard to the position of the terminal in the subcell. Also, the structure_instance view may be generated by a tool other than DMoct as long as it conforms to the LAGER policy, so that tiling commands can be created or modified by structure-processors.

When tiling according to a C routine, TimLager uses the Getpath file search mechanism to find the files containing the tiling routine and the leafcells. The Getpath keyword TimLager.o in the lager file delimits a list of directories to search for TimLager tiling routines. This list is usually machine-dependent, since the ".o" files are created for each machine.

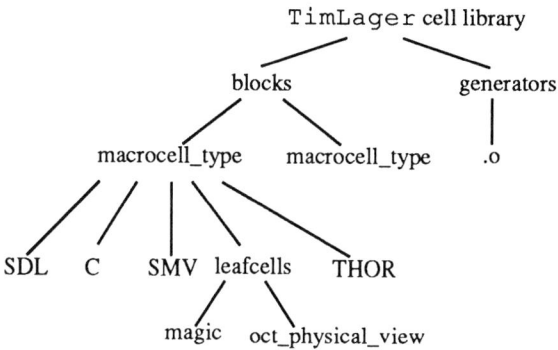

Figure 7.8: `TimLager` cell library organization

7.4. CELL LIBRARY

7.4.1. Library Organization

Figure 7.8 shows the directory organization of the main `TimLager` cell library. The library consists of individual directories for each macrocell, located under the *blocks* directory. Every macrocell is represented in the library by a top-level SDL and a C tiling procedure, or by one or more SDL files that also contain the tiling procedure. Each SDL file has a corresponding structure master view. The *leafcell* directory contains the physical leafcell layouts in Magic format and OCT physical format. The THOR directory contains THOR behavioral models for functional simulation. Current LAGER cell library policy places the machine independent files and OCT views in the *blocks* directory, while the machine-dependent .o files reside in a separate *generators* directory.

The organizational structure of files for a `TimLager` macrocell allows a "family" of macrocells to be conveniently placed in one macrocell directory. Macrocells are considered a family if they serve the same type of function and share leafcells. For example, a PLA macrocell and FSM macrocell can be implemented with the same input and output plane leafcells, but need different I/O leafcells on the border of the main array (the FSM will need latches). In one generic directory, the individual SDL, structure_master view, and C descriptions reside along with a common leafcell directory that contains leafcells that are shared by both types of macrocells, and along with other leafcell directories that contain leafcells specific to one of the macrocell types.

Table 7.1 lists the macrocells currently installed in the main TimLager cell library. All but one are parameterized array-type cells that serve PLA, memory, or I/O pad functions. *Clock* is just a single leafcell but still is closer to the TimLager design paradigm than to the standard cell or datapath libraries. TimLager cells designed for datapaths are stored in a separate dpp library, as discussed in Chapter 10. The TimLager.kappa library contains special purpose macrocells for the kappa processor. These three libraries represent the TimLager macrocells provided as part of the standard LAGER distribution. Users can create their own TimLager libraries to complement the standard libraries.

7.4.2. Extending The Library

In the case that the macrocells provided by the TimLager cell library do not meet the particular functional or performance requirements of a user's application, the LAGER library management philosophy allows the user to design and install their own macrocells. The guidelines presented here apply specifically to the main TimLager library cells. Guidelines for datapath cells appear in Chapter 10.

The user may extend the TimLager library by putting the new macrocell design in either the main TimLager library directory or in a directory outside of the LAGER directories. Either way, the user should follow the directory organization presented in the previous section. If the user or developer prefers, all the files and views including the .o files for a macrocell may reside in one directory. The user's *lager* file should be updated to contain the names of any such additional directories.

The leafcell layout files are called *leafname.mag*. The C tiling procedure must be named *macrocell.c*, where *macrocell* is the name of the new macrocell. To use the TimLager C function library, it is only necessary to include a header file TimLager.h at the beginning of the *macrocell.c* file.

After installing all Magic, SDL and C files, the user must create the structure_master view (by running DMoct), the tiling .o file (by running the C compiler) and the OCT physical views (by running mag2oct on each Magic leafcell file). At this point, the user can run DMoct and TimLager to generate instances of the new macrocell. To complete the installation, the developer should

macrocell	purpose/parameters
pla	Programmable logic array. Implements the input and output planes of a truth table. Macrocell parameters are the bit-pattern of the planes.
fsm	This is the pla macrocell with input and output latches. Can be used as a finite state machine.
ram3T	Dynamic RAM with optional ROM locations. Macrocell parameters are the word size, memory capacity, and ROM addresses and contents.
dpram	Dual-port RAM. Macrocell parameters are the word size and memory capacity.
fifo	Static first-in-first-out stack. Macrocell parameters are the word size, memory capacity and flag option.
latch	Clocked semi-dynamic latch with load control. Macrocell parameter is the word size. Used only for non-datapath designs.
clock	Two-phase non-overlapping clock generator. No parameters.
scpads	I/O pads for CMOS technology. The cell library provides a set of pads for 3micron, 2micron, 1.6micron and 1.2micron feature sizes. Each set consists of a variety of leafcell pads including input, output, Vdd, GND, Corner, ... type pads. To generate the chip pad frame, appropriate pad leafcells are tiled in a row by TimLager to form a side of the frame. Scpads library does not provide a top-level SDL. The user writes a SDL tiling procedure specifically for each application.

Table 7.1: Example Macrocells in `TimLager` Cell Library

create THOR functional models, as well as documentation in textual and schematic form. For example, the documentation for the latch example is shown in Figure 7.9.

> **MACROCELL-NAME** latch
>
> **SYNOPSIS**
>
> The latch macrocell is an n-bit clocked semi-dynamic register.
>
> **INTERFACE**
>
> Input Terminals: IN[i], i=0,1,...,*width*-1.
>
> Output Terminals: OUT[i], i=0,1,...,*width*-1.
>
> Control Terminals: LD. LD=load.
>
> Clocks: PHI1 and PHI2 for two phase non-overlapping clock signal.
>
> **PARAMETERS**
>
> *width* - number of bits stored in latch
>
> **DESCRIPTION**
>
> The n-bit width data is written into the latch at the IN[i] terminals and is read out at the OUT[i] terminals. To write, the LD control signal should be 1; during read, the LD signal should be 0. Each storage cell (contained in the leafcell *latchcell*) in the n-bit latch is clocked by PHI1, PHI2, and their complements, and each cell is controlled by LD and its complement. The complements are generated locally in a separate cell (contained in the leafcell *latchctl*).
>
> **SIZE**
>
> leafcell: W = 164 lambda, H = 49 lambda
>
> macrocell: W = 164 lambda, H = (49* *width* + 64)
>
> **PERFORMANCE**
>
> A 10-bit latch has been fabricated on a 3 micron run, and its functionality and performance verified at 20MHz.
>
> **LAYOUT-GENERATOR** TimLager
>
> **FILES** latch.sdl, latch.c

Figure 7.9: Documentation for Latch Macrocell

7.5. SUMMARY

Among the LAGER ensemble of layout generation tools, `TimLager` supports the tiled layout design style. Starting from a parameterized tiling procedure and associated leafcell layouts (from the cell library), and from the user-specified parameter values, the tool constructs bit-sliced or arrayed macrocells by leafcell *abutment*. The tiling procedure is a structural specification of the macrocell in terms of placement of leafcell instances, and can be described either in C or SDL. Since there is no routing involved in the macrocell generation, the user can easily and accurately predict macrocell area and performance from knowledge of the pre-designed leafcells, and use the predictions to make higher level design decisions.

Typical circuit layouts generated by TimLager are datapath stages in the bit-slice design style, rows of I/O pads, static or dynamic RAMs, ROMs, PLAs, and crossbar switches. For macrocells that have relatively simple structure and parameters, such as the memory circuits and I/O pads, TimLager is self-sufficient. However, some macrocells such as datapaths (complex structure) and PLAs (complex parameters) are generated with the assistance of *structure processors*. LAGER provides the structure processors dpp for datapaths and plagen for PLAs. dpp takes care of the routing between stages of the datapath, whereas TimLager itself only tiles each separate datapath stage. plagen allows the specification of PLAs at a high level and translates the specification into the low-level inplane and outplane parameters used by TimLager.

REFERENCES:

[Mayo86] R. Mayo. *Mocha Chip: A System for the Graphical Design of VLSI Module Generators*. IEEE International Conference on Computer Aided Design, 1986.

8

Standard Cell Design

Bob Reese and Barry Boes

One of the most efficient ways to implement random logic is through the use of standard cells. For this purpose, LAGER contains a cell library of standard cells along with logic synthesis and place-and-route tools. The standard cells can be used to create blocks of random logic to be used in conjunction with other LAGER library blocks (datapath blocks from the `dpp` library and/or special-purpose tiled macro cells from the `TimLager` library), or the design can consist solely of standard cells. When creating a standard cell design, the user can explicitly specify the standard cell netlist, or synthesize the standard cell netlist from a high-level description.

8.1. SPECIFICATION

One method for specifying a standard cell design in LAGER is explicitly to specify the netlist as an SDL file. Figure 8.1 shows a schematic of a simple decoder built using standard cells. The SDL file which describes the decoder is also included in Figure 8.1. The components invf101 and nanf211 are cells from the standard cell library. The statement `(layout-generator Stdcell)` in the SDL file causes DMoct to use the `Stdcell` script to produce the layout for this design.

88   Silicon Assembly   Part II

When using an SDL file for a standard cell design, the user can control two characteristics of the layout: The terminal edge placement and the number of rows. A TERM_EDGE property defined for a terminal instructs the Stdcell program to bring that terminal to the specified edge of the layout. If the TER-

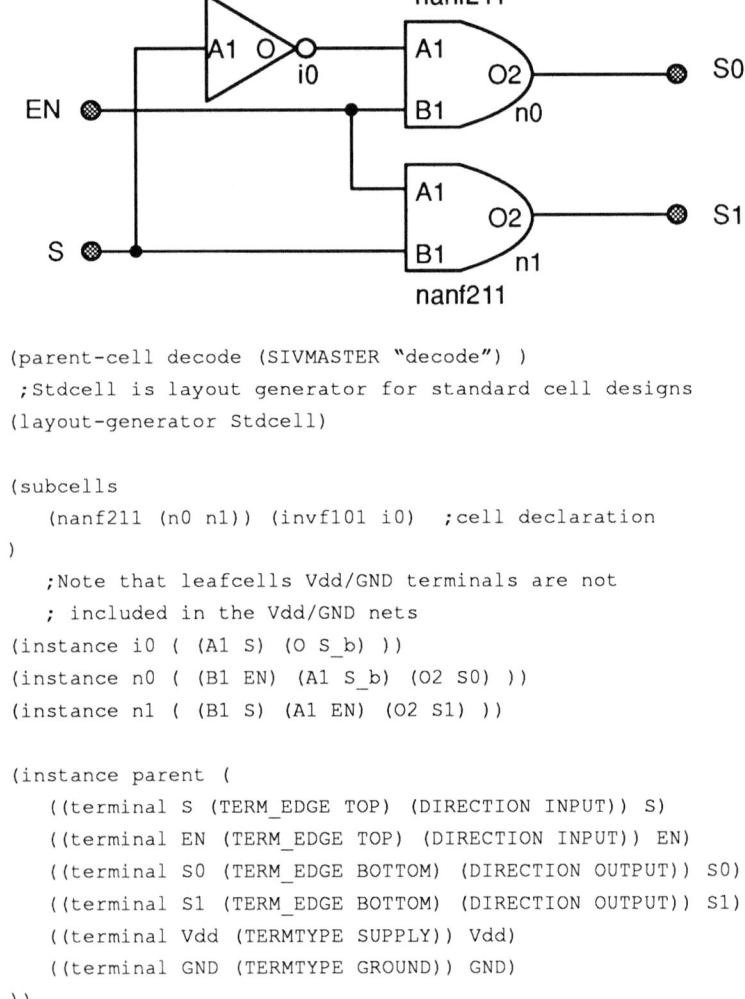

```
(parent-cell decode (SIVMASTER "decode") )
;Stdcell is layout generator for standard cell designs
(layout-generator Stdcell)

(subcells
    (nanf211 (n0 n1)) (invf101 i0)   ;cell declaration
)
    ;Note that leafcells Vdd/GND terminals are not
    ; included in the Vdd/GND nets
(instance i0 ( (A1 S) (O S_b) ))
(instance n0 ( (B1 EN) (A1 S_b) (O2 S0) ))
(instance n1 ( (B1 S) (A1 EN) (O2 S1) ))

(instance parent (
    ((terminal S  (TERM_EDGE TOP) (DIRECTION INPUT)) S)
    ((terminal EN (TERM_EDGE TOP) (DIRECTION INPUT)) EN)
    ((terminal S0 (TERM_EDGE BOTTOM) (DIRECTION OUTPUT)) S0)
    ((terminal S1 (TERM_EDGE BOTTOM) (DIRECTION OUTPUT)) S1)
    ((terminal Vdd (TERMTYPE SUPPLY)) Vdd)
    ((terminal GND (TERMTYPE GROUND)) GND)
))
```

Figure 8.1: Standard Cell example with schematic and associated SDL file.

M_EDGE property is not specified, then the terminal is routed to a randomly chosen edge. The number of rows can also be specified.

When writing SDL files for a standard cell design, two special rules must be followed. The first rule concerns hierarchical design. The `Stdcell` layout generator expects a flattened netlist of standard cells. The user may create a hierarchy of SDL files to describe a standard cell design, but if so must also direct the DMoct program (in the top level SDL file) to flatten the hierarchy before passing the structure_instance view to the `Stdcell` program for layout generation.

The second rule concerns the Vdd and GND terminals. Note that in the decoder example, the Vdd and GND terminals of the individual cells are not included in the Vdd and GND net statements. This is because the Vdd and GND terminals of the individual cells are automatically connected via abutment whenever a row of standard cells are generated. Thus there is no need to include these Vdd/GND terminals in explicit nets. In fact, specifying these terminals in Vdd/GND nets will cause the layout generator to attempt to route these nets and will cause an erroneous layout to be generated. It is necessary to specify parent Vdd and GND terminals for the entire design, hence two TERMINAL statements defining Vdd and GND terminals are included in the SDL file.

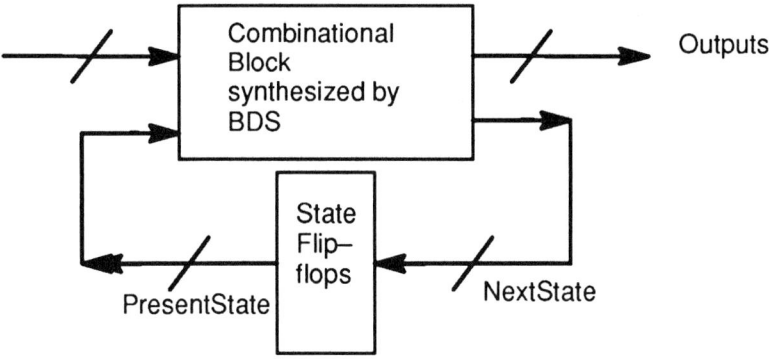

Figure 8.2: Finite State Machine Example

8.2. RANDOM LOGIC SYNTHESIS

Standard cell logic can alternatively be synthesized from a high-level specification in the BDS language [Segal90]. The BDS file is translated by the

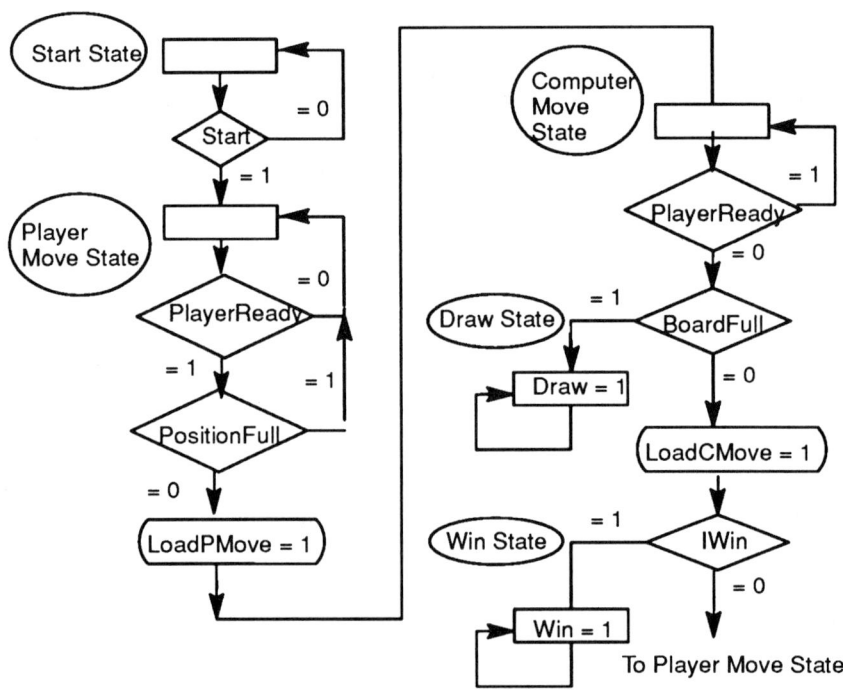

```
Start state entered on Reset; Win, Draw states exited only on
Reset
Inputs: Start - starts game
        PlayerReady - indicates when player move is ready
        Position Full - selected position by player is
        already full
        Board Full - no empty spaces remain on board
Outputs: LoadPMove - write player move to board
        LoadCMove - write computer move to board
        Win - indicates that the computer has won
        Draw - indicates that the computer has drawn
Always results in a win or draw for the computer
```

Figure 8.3: State Diagram for Tic-Tac-Toe Player

BDSYN program into BLIF format (Berkeley Logic Interchange Format) which is then handed to the misII program for logic optimization and technology mapping onto the standard cell library [Brayton87]. This sequence of steps is packaged into a single shell script called Bds2stdcell, which the user can specify as a *structure-processor* in the SDL file.

Chapter 8 Standard Cell Design 91

One use for the logic synthesis capability is to synthesize finite state machines. Figure 8.3 shows a generic finite state machine. A BDS description can be used to synthesize the combinational logic block. SDL files are then used to define the formal terminals for the logic block and connect the terminals to the state flip-flops. Figure 8.3 shows the state diagram for a finite state machine that implements the control for a *tic-tac-toe* player. Figure 8.4 shows the BDS file which specifies the logic part of this finite state machine. Note that in the BDS description, we specify default values for outputs. The synthesized logic will produce these default output values unless otherwise specified. The effect of default values can be seen in the WIN_STATE, where the WIN output is asserted but nothing is stated about the NextState outputs. The default assignment of "NextState = PresentState" indicates that by default the machine will stay in the WIN_STATE. WIN_STATE can only be exited by asserting the asynchronous reset to the state flip-flops.

Figure 8.5 shows the SDL wrapper file used to interface the BDS description to LAGER. Note that the structure-processor is Bds2stdcell. This is the script mentioned earlier which calls bdsyn and misII to synthesize the logic. The statement (depends-on "control.bds") in the parent-cell declaration places a dependency between this SDL file and the BDS file. If the BDS file changes then the structure_instance view will be remade whenever DMoct is invoked. The Bds2stdcell script uses the value of the bdsyn parameter in the structure_instance view as the name of BDS file to pass to the bdsyn program.

Figure 8.6 shows the SDL file used to connect the combinational logic block to the state flip-flops. In this particular case, we need three flip-flops to implement the five states in the finite state machine. The statement (bag FLATTEN_TO (CELLCLASS LEAF)) in the parent-cell declaration causes the hierarchy to be flattened before being passed to the Stdcell layout generator. This is necessary in this design because of the hierarchy created by having the logic block as a subcell along with the state flip-flops.

8.3. EXAMPLE DESIGN

Figure 8.7 shows the block diagram of the complete 3x3 *tic-tac-toe* player. The Move Generation block generates the computer move based on the current board status as indicated by inputs B0-B9. Each board location is implemented with two flip-flops and has three outputs which indicate if that location is player occupied, computer occupied, or empty. The Move Generation and Control

```
!Bds logic which defines control logic for tic-tac-toe player
MODEL control
! Outputs followed by inputs.
NextState<2:0>, LoadPMove, LoadCMove, Draw, Win
= PresentState<2:0>, Start, ,PlayerReady, IWin,
BoardFull,PositionFull;

ROUTINE main;
CONSTANT START_STATE= 0, MOVE_STATE= 2, CMOVE_STATE= 3;
CONSTANT WIN_STATE= 7,  DRAW_STATE= 1;
   ! Default output assignments.
   NextState = PresentState; Draw = 0;  Win = 0;
   SELECTONE (PresentState) FROM
      [START_STATE]:
        BEGIN IF (Start EQL 1) THEN NextState = PMOVE_STATE;  END;
      [PMOVE_STATE]:
        BEGIN
          IF (PositionFull EQL 0 AND PlayerReady EQL 1) THEN BEGIN
            ! Move is ready and valid.
            LoadPMove = 1;   NextState = CMOVE_STATE;
          END;
        END;
      [CMOVE_STATE]:
        BEGIN
          IF (PlayerReady EQL 0) THEN BEGIN
            IF (BoardFull EQL 1 ) THEN   NextState = DRAW_STATE
            ELSE BEGIN    LoadCMove = 1;
              IF (IWin EQL 1) THEN NextState = WIN_STATE
              ELSE    NextState = PMOVE_STATE;
            END;
          END;
        END;
      [WIN_STATE ]:  BEGIN Win  = 1; END;
      [DRAW_STATE ]: BEGIN Draw = 1; END;
   ENDSELECTONE;
ENDROUTINE; ENDMODEL;
```

Figure 8.4: BDS description of TicTacToe state machine

blocks were described through BDS files. The logic required for the board was straight-forward and regular so this logic was explicitly specified via an SDL netlist. The total standard cell count for the design was 225 standard cells and easily fit in the TinyChip padframe used by MOSIS for educational projects (2.2 mm x 2.3 mm).

```
(parent-cell control
  (SIVMASTER "control")
  (depends-on "control.bds")
)
(parameters (bdsyn "control.bds"))
(structure-processor Bds2stdcell)

(instance parent (
    ((terminal NextState (DIRECTION OUTPUT)) NextState (width 2))
    ((terminal LoadPMove (DIRECTION OUTPUT))LoadPMove)
    ((terminal LoadCMove (DIRECTION OUTPUT))LoadCMove)
    ((terminal Draw (DIRECTION OUTPUT)) Draw)
    ((terminal Win (DIRECTION OUTPUT)) Win)
    ((terminal PresentState (DIRECTION INPUT))PresentState(width 2))
    ((terminal Start (DIRECTION INPUT)) Start)
    ((terminal CompMoveFirst (DIRECTION INPUT)) CompMoveFirst)
    ((terminal PlayerReady (DIRECTION INPUT)) PlayerReady)
    ((terminal IWin (DIRECTION INPUT)) IWin)
    ((terminal IDraw (DIRECTION INPUT)) IDraw)
    ((terminal BoardFull (DIRECTION INPUT)) BoardFull)
    ((terminal PositionFull (DIRECTION INPUT)) PositionFull)
  )
)
```

Figure 8.5: SDL file for TicTacToe combinational logic

While this design is 100% standard cells, there is nothing to prevent the user from using standard cell blocks with blocks from the other LAGER libraries. In fact, most LAGER designs use a combination of different types of library blocks.

8.4. STANDARD CELL LIBRARY

Figure 8.8 lists all of the members of the standard cell library. Nominal and worst case timings for no load and 1 pF load are given in the documentation for each library member. The documentation also contains Spice decks for each library element.

Each standard cell is 66 lambda high, with cell width depending on the complexity of the function which the cell implements. Figure 8.9 shows a chip created using LAGER and the standard cell library. The chip illustrates the standard cell layout style of multiple fixed-height rows with the channels between the rows used for signal routing. In each standard cell, a Vdd bus runs along the top

```
(parent-cell controlfsm
   (SIVMASTER "controlfsm") (bag FLATTEN_TO (CELLCLASS LEAF))
)

(layout-generator Stdcell)
(subcells (control CONT)) ;combination logic block defined via bds
(dotimes  (i 3)
   (subcells (dfrf301 FF)) ;three flip-flops needed to hold state
)
(instance FF (
   (DATA1 NextState (net-index i))   ; NextState connected to
flip-flop input
   (Q PresentState (net-index i))        ;PresentState connected to
flip-flop output
   (CLK2 clock) (RST3 reset_b)
))
(instance CONT (
   (NextState  NextState (width 2))
   (LoadPMove LoadPMove) (LoadCMove LoadCMove)
   (Draw Draw) (Win Win)
   (PresentState  PresentState (width 2))
   (Start  Start) (PlayerReady  PlayerReady)   (IWin  IWin)
   (BoardFull BoardFull )
   (PositionFull  PositionFull)
))
(instance parent (
   ((terminal LoadPMove (DIRECTION OUTPUT))LoadPMove)
   ((terminal LoadCMove (DIRECTION OUTPUT))LoadCMove)
   ((terminal Draw (DIRECTION OUTPUT))Draw)
   ((terminal Win (DIRECTION OUTPUT)) Win)
   ((terminal clock (DIRECTION INPUT)) clock)
   ((terminal reset_b (DIRECTION INPUT)) reset_b)
   ((terminal Start (DIRECTION INPUT)) Start)
   ((terminal PlayerReady (DIRECTION INPUT)) PlayerReady)
   ((terminal IWin (DIRECTION INPUT)) IWin)
   ((terminal BoardFull (DIRECTION INPUT)) BoardFull)
   ((terminal PositionFull (DIRECTION INPUT)) PositionFull)
))
```

Figure 8.6: Top level SDL file for connecting combinational logic described using BDS of Figure 8.4 and the SDL file in Figure 8.5 with state registers.

of the cell and a GND bus runs along the bottom. These busses are joined whenever the cells are placed adjacent to each other. Vdd and GND terminals are then created at the end of each row, and will be joined together at the next level of hierarchy by Padroute or Flint.

Chapter 8   Standard Cell Design   95

8.5. LAYOUT GENERATION

The layout-generator program for standard cell designs is a UNIX shell script called Stdcell. Figure 8.10 illustrates the steps which Stdcell follows to produce a standard cell layout. The Stdcell script calls the wolfe program, which in turn calls the TimberWolfSC [1] and YACR [2] (Yet Another Channel Router) programs. TimberWolfSC performs the global place and route for the design, and YACR is called to perform the detailed channel routing for each channel. Wolfe converts the netlist from the OCT structure_instance view format to the input format required by TimberWolfSC and converts the output of TimberWolfSC and YACR to the OCT physical format. After wolfe is finished, the wolfepost program is called to add Vdd and GND terminals to the end of each row so that the block can be used a macrocell in a larger hierarchy. The last program called by Stdcell is oct2mag which is a filter program for producing Magic-format layout files from the OCT physical view.

Wolfe is the principal program called by the Stdcell script. In addition to performing input/output format conversion for the TimberwolfSC and YACR programs, Wolfe also scans each standard cell for implicit feedthrough channels (over-the-cell routing areas), and notifies TimberWolfSC that these are available for routing. Within a design, Wolfe looks for user-specified OCT properties or OCT bags which specify particular optimizations that TimberWolfSC can perform on the layout.

The OCT WOLFE-CLASS bag can be used to specify cells which should be placed on common rows. All cells that are within a WOLFE-CLASS bag, and that have a WOLFE-ROW property of the same value, will be placed in the same row with no other cells placed in that row. Cells with no WOLFE-CLASS bag can be placed on any row except one with cells having WOLFE-CLASS bags. In addition to the possibility of reducing routing area, this method is also useful if a standard cell family has cells which are not compatible (i.e., some cells with power/ground rails at non-standard heights) by allowing only compatible cells to be placed on a row.

An ACCESSIBILITY property can be placed on any leafcell terminal to inform wolfe how to connect to this terminal. The values of the ACCESSIBILITY property are TOP_ONLY, BOTTOM_ONLY, or BOTH. The HI-R property can be placed on a leafcell terminal to inform wolfe there is high resistance between the top and bottom implementations that terminal, meaning that it should

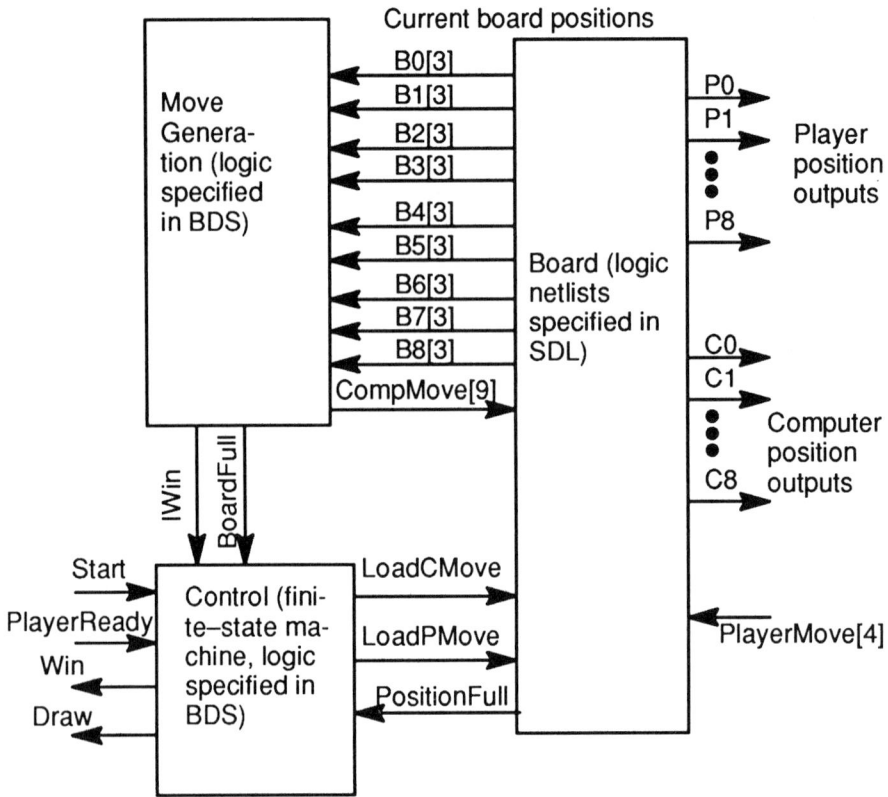

Figure 8.7: Complete Tic-Tac-Toe player schematic

not should not be used as a feedthrough for a net. The FEEDTHRU property can be attached to a pair of terminals to indicate that they are a feedthrough path for the cell. Wolfe will automatically find implicit feedthrough paths (vertical channels) but a cell may have a built-in jumper which is not vertical.

8.6. EXTENDING THE LIBRARY

If a user wishes to extend the current standard cell library or integrate a new standard cell library, then the user must be aware of certain rules which standard cells must follow in order to be compatible with the wolfe/TimberWolf-SC/YACR place-and-route package. The bounding box of a cell is the minimum sized box which surrounds all of the geometries in a cell (including the wells).

Chapter 8   Standard Cell Design

LIBRARY DESCRIPTION

Cell Name	Cell Description
aof2201, aof2301	2,2 AND/OR Mux; 2,3 AND/OR Mux
aof3201, aof4201	3,2 AND/OR Mux; 4,2 AND/OR Mux
aoif2201	2,2 AND/NOR Mux
blf00001, blf00101	2,1 AND/NOR Mux; 2,1 OR/NAND Mux
buff100	High Impedance Buffer
buff101	Non–Inverting Buffer
buff121	Tri–State Buffer
delf011	Delay Cell
dfnf311	D–Flip Flip with S, R, Q & QB
dfrf301	D–Flip Flop with Asynchronous R & Q
faf001	Full Adder
invf100	High Impedance Inverter
invf101 invf1013, invf104	Inverter, 3X,4X Inverting Buffers
invf121	Tri–State Buffer
invf201	Dual Inverter
labf111, labf211	NAND Latch, NOR Latch
larf310	Clock Latch
lrbf202	Logic Reference Cell
ldf001	RC Load Cell
muxf201	Data Select
nanf201, nanf211	2 Input NAND; 2 Input NAND/AND
nanf251	A OR B–not Decoder
nanf301, nanf311	3 Input NAND;3 Input NAND/AND
nanf401, nanf411	4 Input NAND;4 Input NAND/AND
norf201, norf211	2 Input NOR; 2 Input OR/NOR
norf251	A–not AND B Decoder
norf301, norf311	3 Input NOR; 3 Input OR/NOR
norf401, orf401	4 Input NOR;4 Input OR
oaif2201	2,2 OR/NAND Mux
pudf000, puuf000	Pull–Down, Pull–Up
swcf020	Transmission Gate
xnof201, xorf201	Exclusive NOR, Exclusive OR

Figure 8.8: Components of Standard Cell library

Figure 8.9: Complete Standard Cell design

The overlap box (also known as the tesselation box) is a rectangle placed on the OVERLAP layer to tell `wolfe` how to place the cell relative to its neighbors. If the OCT physical views of a cell library contain overlap boxes, then `wolfe` will place the cells such that their overlap boxes abut. Otherwise `wolfe` will abut the cells according to their bounding boxes. Using the overlap box allows cells to share common geometries such as well pieces. Figure 8.11 shows how cells might be placed using overlap boxes.

In the X direction, the overlap box must be designed such that any cell can be abutted to any other cell without causing design rule violations. Furthermore, the overlap box must be an integer multiple of the vertical layer pitch (currently

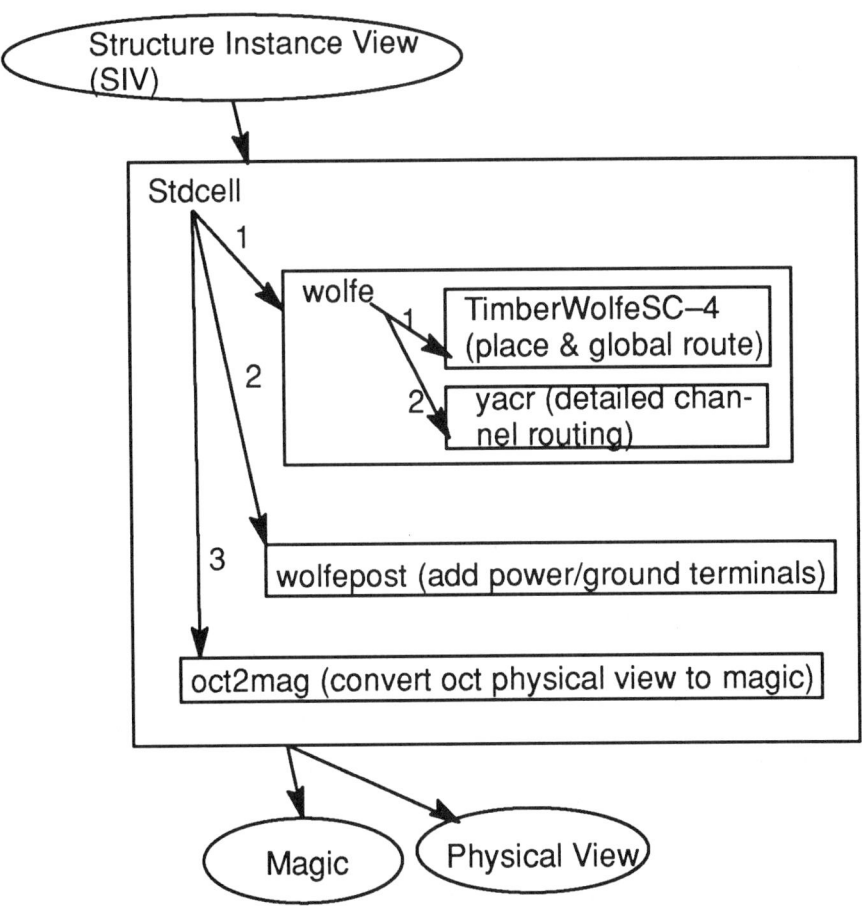

Figure 8.10: The Standard Cell layout generator sub-programs

eight lambda) in the X direction. If the vertical layer pitch (the minimum center-to-center separation of the layer used for vertical routing) is 8 lambda, then valid values for the width of an overlap box are 8*n lambda, where n is a positive integer. If this rule is not followed, then wolfe will generate designs which contain non-Manhattan channel routing. One should also observe the following:

1. Some policy regarding the extension of geometries beyond the overlap box is required to ensure that no inter-cell design rule violations will occur.

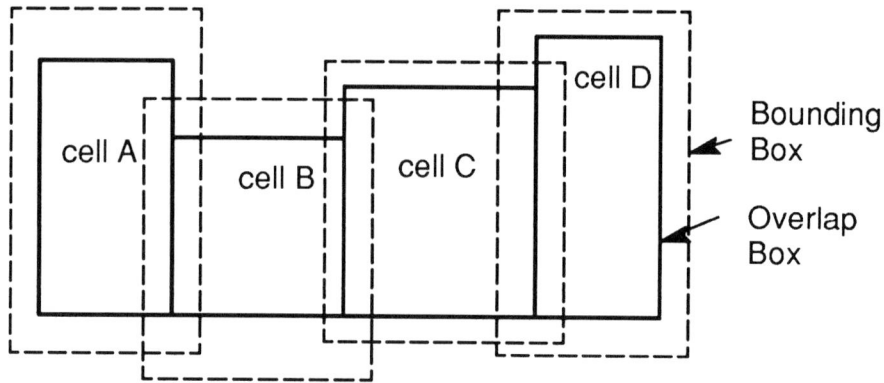

Figure 8.11: Cell placement

2. Important geometries (e.g. contacts) which are shared between cells must overlap exactly in order to meet design rule specifications.
3. Vertical metal1 which shares contacts between cells implies the use of metal 1 at the boundary, with the possible undesirable consequence of the feedthrough cell width being greater than the vertical layer pitch.

Wolfe places no restrictions on terminal placement in the Y direction. However, there is a policy for the terminal placement on the X axis: The terminals must be centered at (1/2 +n)* vertical-layer-pitch lambdas from either side of the cell's overlap box (n specifies the n'th possible vertical routing channel).

In Figure 8.12 the dashed lines indicate valid locations for terminal centers. Notice that the width of the overlap box for this example is an integer multiple of the vertical layer pitch. If this were not true, then the cell would not meet the requirements for terminal placement. Strictly speaking, the end terminals can be placed an integer multiple of the vertical pitch from the edge of the overlap box. However, this reduces the terminal density and is normally avoided.

8.7. SUMMARY

The Stdcell script controls the programs Wolfe, TimberWolfSC, YACR, Bds2stdcell, and misII, and along with the standard cell library implements a self-contained standard cell design framework. In typical LAGER designs, standard cell blocks are used to provide control for datapaths, glue logic for connecting TimLager blocks, or random logic for finite state machines. The

Chapter 8   Standard Cell Design   101

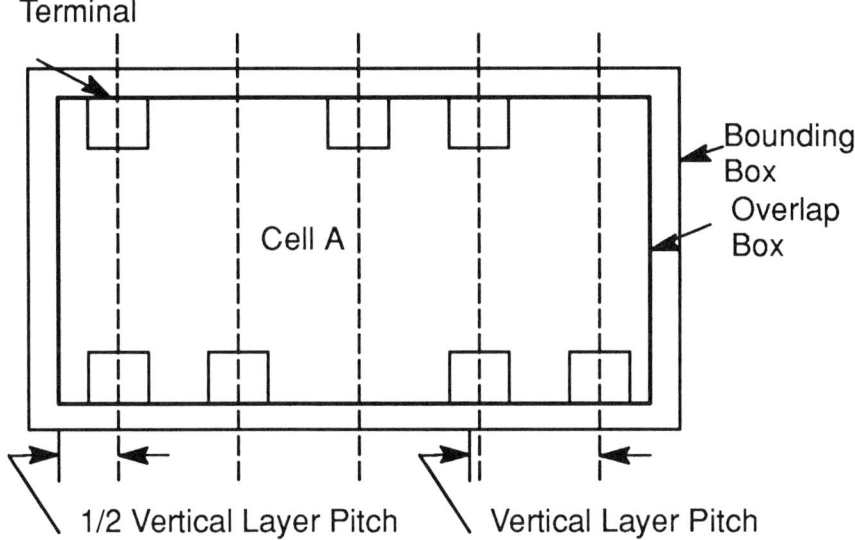

Figure 8.12: Terminal placement

standard cell design style provides a familiar metaphor for designers who are new to VLSI design. We have found most users new to LAGER create their first designs using standard cells exclusively. As they become more experienced with VLSI design and LAGER, they can begin to explore alternative methods of implementation, combining standard cell blocks with blocks created using dpp, TimLager and the other LAGER synthesis tools.

REFERENCES:

[Brayton87] R. Brayton, R.Rudell, A. Sangiovanni-Vincentelli, and A. Wang. MIS: A Multiple-Level Logic OPtimization System. IEEE Transactions on CAD, CAD-6(6):1062-1081, November 1987.

[Reed85] J. Reed, A. Sangiovanni-Vincentelli, and A. Santomauro. A New Symbolic Channel Router: YACR2. IEEE Transactions on CAD, CAD-4(3):208-219, July 1985.

[Sechen85] C. Sechen and A. Sangiovanni-Vincentelli. The TimberWolf Placement and Routing Package. IEEE Journal of Solid State Circuits, 20(2), April 1985.

[Segal90] Russel B. Segal. BDSYN Users' Manual Version 1.1. Octtools Distribution Tape.

9

Interactive Floorplanning

Seungjun Lee and Jan Rabaey

Floorplanning is the task of assigning relative positions on the plane to the components of the design so as to optimize the final chip layout with respect to a set of constraints defined on the interconnections and chip dimensions [Sangiovanni87]. The routing process on the other hand determines the absolute positions of the components and generates the interconnect wiring. Flint is an interactive floorplanning and routing tool for macrocell-based designs. Floorplans can be generated either interactively or automatically. The quality of a given floorplan can be evaluated instantly with the aid of a number of efficient placement and routing routines. The composing modules may be tiled macrocells, datapaths, standard cells blocks, or hierarchical subsystems, as generated by LAGER. The floorplans can be stored for reuse in later design iterations.

As input, Flint requires information about the interconnection between the modules, the physical dimensions of each module and the module terminal locations. This information is obtained from the structure_instance view (interconnections) and the physical views (locations). The resulting layout is stored as a physical view in the OCT database format.

9.1. OVERALL FUNCTIONALITY

`Flint` presents an interactive approach to the floorplanning task for macrocell-based designs. The floorplanning activity consists of the following consecutive steps: *placement, channel definition, global routing, absolute placement* and *detailed routing*. The first three steps can be considered as being part of the floorplanning process, and are executed as an interactive process. The floorplan realization (or routing) task consists of the absolute placement and the detailed routing, and is completely automated.

Besides the interactive process, `Flint` also supports automatic as well as batch oriented floorplan creation modes. In the batch mode, the floorplan topology is entered using a dedicated floorplan description language *fdl*. This allows smart module generators (such as `dpp`, the LAGER datapath generator) to generate a floorplan in *fdl* format and use `Flint` to generate the detailed layout. The floorplan description language is also used to store alternative floorplans, as generated during an interactive floorplanning session. The automatic floorplan creation approach is generally used to generate an initial solution. The user can then use, the initial solution as basis for interactive improvement. Automatic floorplanning can also be used when area minimization is not the highest priority. The algorithmic details of the automated procedures will be presented in the following sections.

The four steps in the floorplan creation process are placement, channel definition, global routing and power routing. Each of these processes appear as a separate mode in the `Flint` user interface.

The input database is loaded from the structure_instance view and the input netlist is converted into a cable list. A *cable* is defined as a group of nets that have the same source and destination (source and destination being the side of a module, see Figure 9.1). For example, RAM-BOTTOM:ALU-TOP represents a cable containing all nets connecting the bottom of the module RAM to the top side of module ALU. During the rest of the floorplanning, all nets in a cable are treated as a single object and will be routed along the same path. Although this approach may result in less than optimal routing, it serves user interactivity and floorplan generation efficiency in a dramatic way. In numerous examples, we have observed that the penalties inflicted by the routing restrictions are small, compared to the gains obtained from the user interaction in placement and global routing.

Chapter 9 Interactive Floorplanning 105

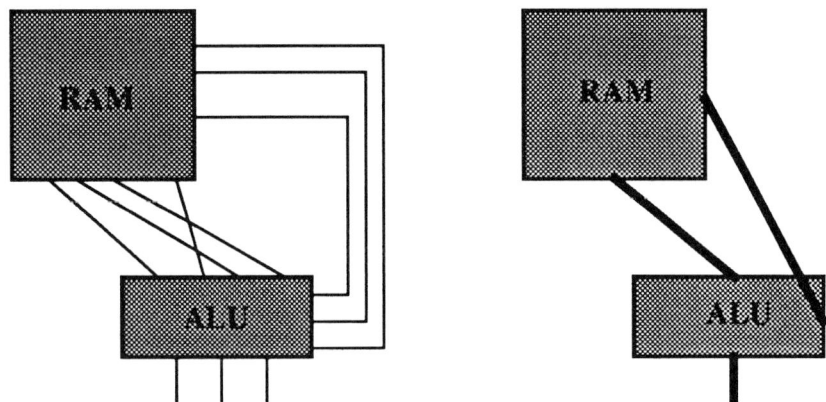

Figure 9.1: Nets versus cables

After the input phase, either a predefined floorplan can be loaded (in the library mode) or the interactive floorplanning mode can be entered. In the first step, the *relative placement* of the modules is defined. The initial placement, given by the automatic procedure, can be modified by moving, rotating, or mirroring each block. The final placement must have a *slicing structure* [Otten82] to guarantee routability by the channel router. The user's view of the placement process is shown in Figure 9.2. Once the placement is completed, channels are defined by dividing the area between the blocks into rectangular sections which are created or deleted by simple mouse operations.

The *global routing* step assigns to each a cable a routing path through adjacent channels. In the automatic mode, Flint routes the cables starting from the shortest. Cables which were already routed manually or in previous floorplanning steps will be preserved. At the end of the automated step, the user can interactively change the results through mouse operations (Figure 9.3).

Power nets and clock nets are treated individually and are not grouped into cables. Each of these nets can be routed one by one. Any other nets which need to be routed with special care may be defined as clock nets and routed during this *power routing* stage as well. Once again, a combined automatic/interactive approach is supported (Figure 9.4).

When a complete floorplan has been created or the pre-existing floorplan is loaded, the layout generation process can start. The generation step first checks

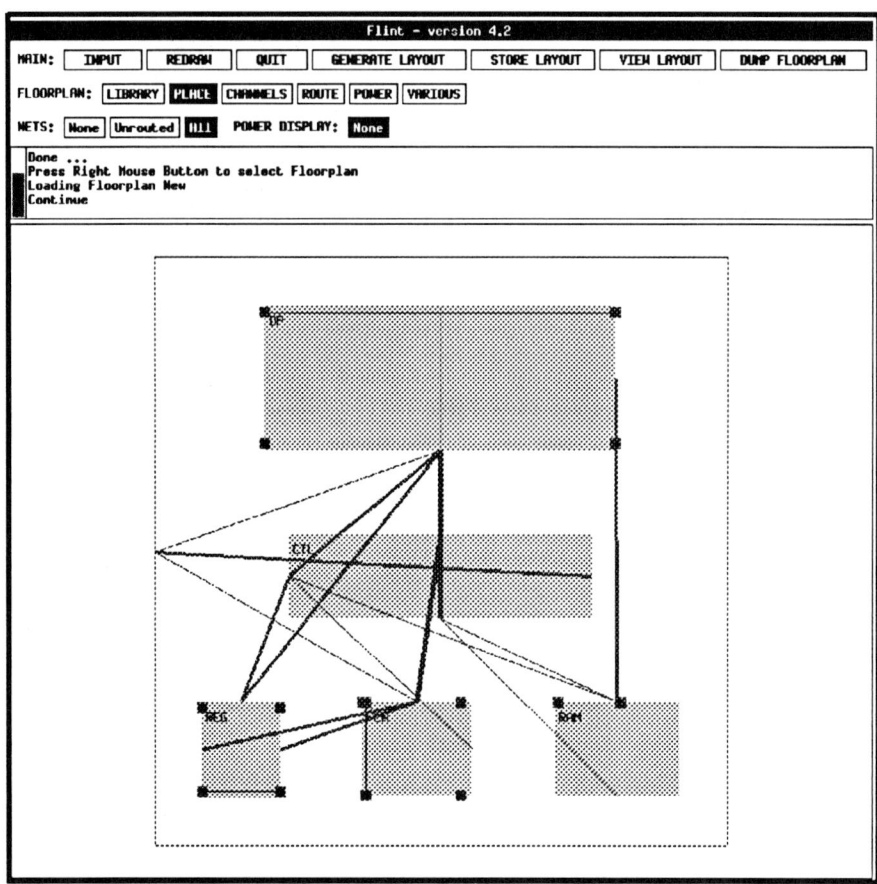

Figure 9.2: User interface during placement process

the consistency of the floorplan. If the floorplan is acceptable, the absolute placement and detailed routing processes will run automatically without user intervention. The final layout can be saved either in Magic files or in OCT format. The floorplan can also be stored in a file for later use. Figure 9.5 shows the floorplan after the detailed routing. The channels have been sized to their real dimensions and the total circuit area is displayed.

Flint also estimates the accumulated routing capacitance for each net. The capacitance values can be used to locate critical delay paths.

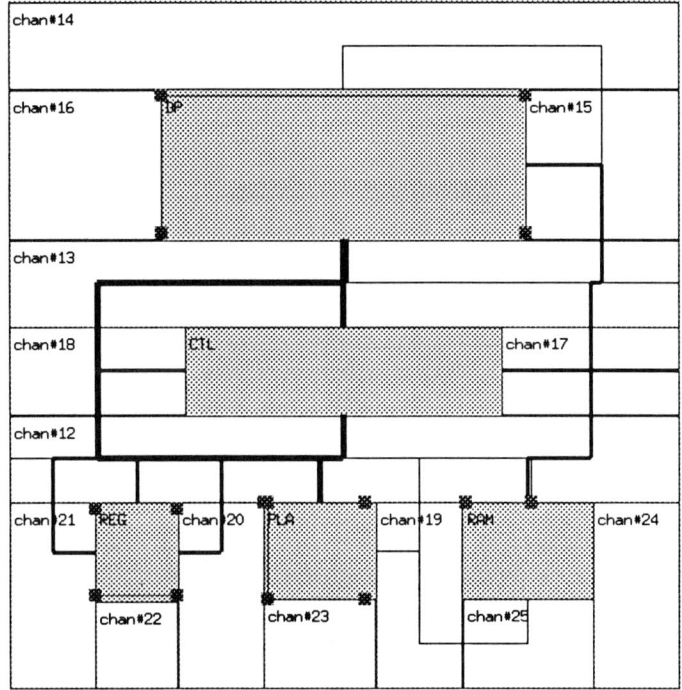

Figure 9.3: User interface during global routing phase

9.2. THE FLOORPLAN DESCRIPTION LANGUAGE

Before presenting in detail the algorithms and procedures used in Flint, a brief discussion of the floorplan description language *fdl* is appropriate. As mentioned earlier, *fdl* serves a dual purpose. Module generators and behavioral synthesis systems often have expert knowledge of the topological properties of the module under design. This knowledge can be translated into a floorplan topology which is far more efficient than what could be generated using automatic placement and routing programs. The *fdl* language allows those generators to formulate that structural knowledge. Examples of tools which use this approach are dpp and firgen. On the other hand, *fdl* is also used by Flint itself as a database format to store temporary floorplans, as generated by the designer during interactive floorplanning sessions.

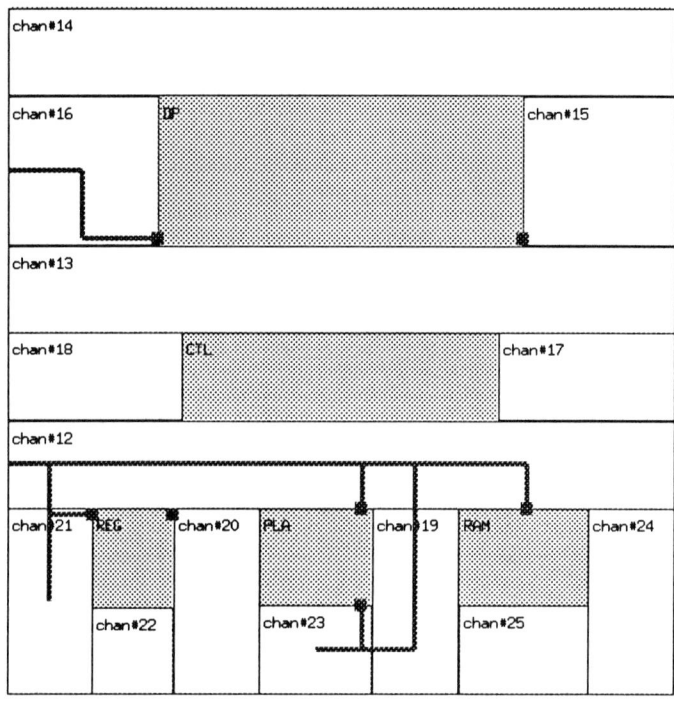

Figure 9.4: User interface during power routing

The four phases of the floorplan definition process (placement, channel definition, global routing and power routing) can also be related to the floorplan language. This is best demonstrated using a simple example, as given in Figure 9.6. The description contains for each macrocell (or module) its placement relative to the other cells. For a channel, extra information regarding the cables and signals entering its four sides is given. These data are sufficient to describe the global routing. Notice how the *cable* concept makes the description compact and easy to generate.

9.3. ALGORITHMS OF THE AUTOMATED PROCEDURES

9.3.1. Relative Placement

The placement heuristic uses a combination of *min-cut partitioning* and the *slicing* approach [Otten82, LaPotin86]. This approach has a number of advan-

Chapter 9 Interactive Floorplanning

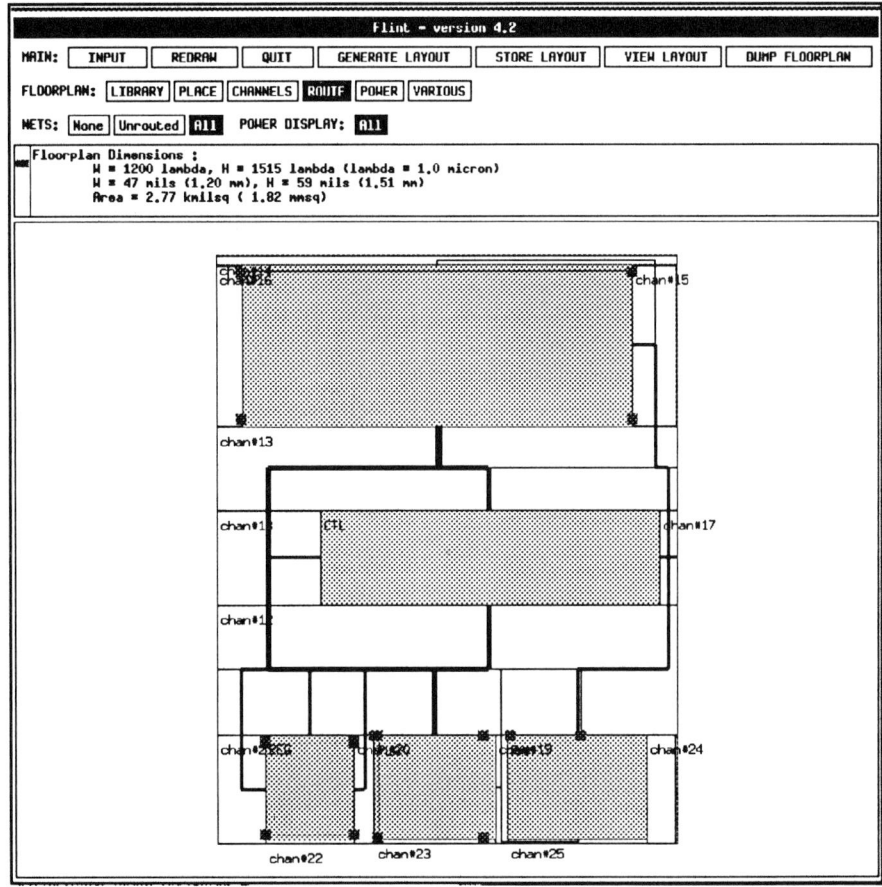

Figure 9.5: Floorplan after absolute placement and detailed routing

tages, particularly in terms of efficiency within an interactive floorplanning environment. First of all, it inherently produces a slicing structure, as required by Flint to perform channel routing. It also usually runs much faster than other heuristic methods, which suits the interactive style of Flint operation.

The placement algorithm is composed of three major parts, 1) a bipartitioning step, 2) physical positioning and orientation of each macrocell, and 3) a mirroring step.

```
//* Simple Floorplan */
module Root, ModA, ModB;
channel chan#1;
signal Vdd, GND;
Root (ModA, chan#1, ModB) {
        place (NULL, NULL, NULL, NULL);
}
ModA () {
        place (Root, Root, chan#1, Root);
        transform(bottom, left);
}
ModB () {
        place (chan#1, Root, Root, Root);
}
chan#1 () {
        place (ModA, Root, ModB, Root);
        route(ModA|right:ModB|left ModA|right:Root,

        ModA|right:Root, ModA|right:ModB|left, NULL);
        power(ModA, Root, ModB, NULL, Vdd);
        power(ModA, NULL, ModB, Root, GND);
        layer(metal2);
}
```

Figure 9.6: Simple floorplan description using *fdl*.

Bipartitioning Step

The first step of relative placement is to construct a binary decomposition tree where non-terminal nodes correspond to cuts (or slices), and leaf nodes are the macrocells (Figure 9.7). To construct the tree, an efficient heuristic/exhaustive min-cut bipartitioning scheme is applied recursively. Heuristics are used when the number of modules to partition is greater than a predefined constant. Otherwise, all possible subsets are enumerated to find an optimal partition.

The cost function for each bipartition is the weighted sum of the number of wires connecting two subsets and the absolute difference between the sum of the module areas contained in each subset. That is,

$$Cost = Ws \times S(A,B) + Wd \times D(A,B) \qquad (9.1)$$

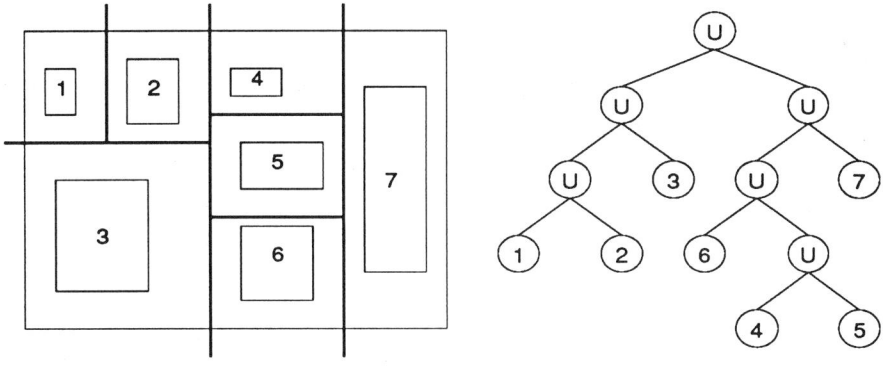

* U : directioin undecided

Figure 9.7: Construction of binary decision tree

where W_s and W_d are weight factors, S(A,B) equals the number of connections between subsets A and B and D(A,B) is the absolute difference in area between A and B.

The `Flint` placement scheme differs from other min-cut bipartitioning and slicing approaches in the following way: When a set of modules are divided into two subsets, it has to be decided if the slice should be horizontal or vertical and which subset goes on what side. In the *in-place partitioning* approach [LaPotin86], the latter problem is solved by slicing alternatingly in the horizontal and vertical directions. A tree-traversal operation [Stockmeyer83] is then performed to determine the dimension, orientation, and location of each macrocell within the slicing tree.

In `Flint`, the slice-line directions are not decided at the bipartitioning stage. They are determined at the next step along with the orientation and location for each macrocell, taking into consideration every possible configuration of the decomposition tree.

Location and Orientation of Macrocells

A modification of a standard placement algorithm [Stockmeyer83] is used for this task. In the process of finding the optimal orientation of each macrocell, the modified algorithm also determines the direction of the slice-line at each node of the decomposition tree.

There are two tree-traversal operations. First, a post-order traversal is used to create a list of possible dimensions for each node in the tree. Dimension is represented by a height-width pair. Each pair in the list is annotated with two pointers. After the list for the root of the tree has been constructed, and the total area associated with the desired aspect ratio of final layout has been minimized over all pairs in the list, the pointers are used for top-down tree-traversal to reconstruct the orientation that achieves the minimum area. In addition, each pair has a flag which specifies whether the dimension is obtained from vertical slicing or horizontal slicing. The direction of the slice-line for each node is determined according to the flags during the process of top-down traversal.

The algorithm begins by constructing a list of possible dimensions for each leaf node. To reserve a reasonable area for routing, the dimension of each macrocell is expanded proportional to the number of terminals at its bounding box. If the macrocell has expanded dimensions of a and b with $a > b$, the dimension list is set to $\{(a, b), (b, a)\}$ and the pointers are null. If $a = b$, there is just one pair in the list. The algorithm now works its way up the tree. In general, let u be a non-leaf node in the tree, with children v and v'. The two lists constructed for v and v' are combined to obtain a list for u by a list-merging procedure [Stockmeyer83].

The procedure is applied twice for both vertical and horizontal slicing, because the node u does not have a fixed direction. The two lists thus obtained are merged into one where each pair keeps the information on whether it resulted from vertical or horizontal slicing. A pair (a, b) is eliminated while merging if another pair (a',b') exists in the merged list, with $a > a'$ and $b > b'$.

At most, this algorithm keeps twice as many list elements for each node as previous methods, and it is easily seen that the number of pairs in the root node is $O(n^2)$ in a balanced binary tree. In the worst case, the number of pairs grows exponentially (Figure 9.8). Experiments show that the number of pairs in the root node grows almost linearly in a balanced tree, and that even in the worst case the growth rate is not exponential. This is caused by the fact that a considerable number of possible pairs are pruned away while traversing the tree.

The running time of the algorithm can be reduced using a simple heuristic. First, the maximum allowable dimension is computed with respect to the total area of the macrocells and the desired aspect ratio of the final layout. When two lists are combined, the pairs whose width or height dimensions are greater than the maximum dimension are dropped. This results in an important reduction in the size of the candidate list of the root node.

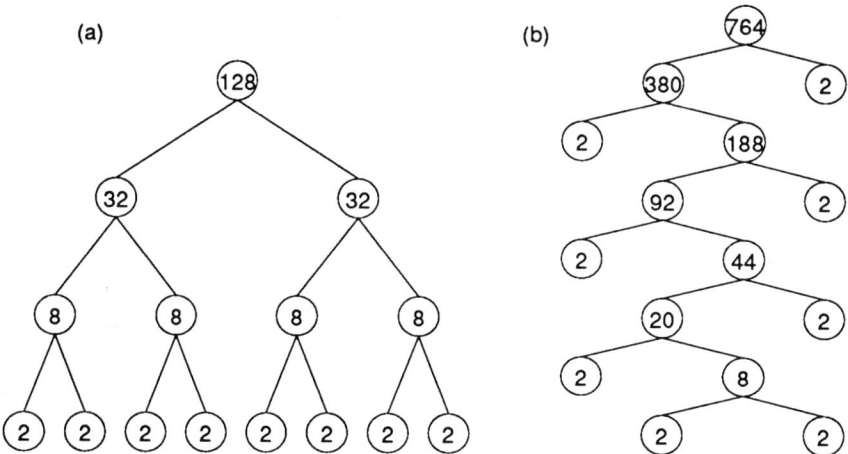

Figure 9.8: Growth of dimension lists for (a) binary and (b) unbalanced trees

When the decomposition tree is unbalanced, it may happen that no single pair can meet the given aspect ratio constraint. In that case we go back and modify the cost function, and rebuild the decomposition tree with respect to a new cost function.

Once the dimension-pair list of the root node of the tree is obtained, the optimal entry is selected with respect to area and aspect ratio. The direction of the slicing line is decided according to the flag in the entry, and the two pointers of the entry are used to traverse down the tree.

The relative positioning of the subblocks, with respect to the slicing line, is chosen considering the connections to modules contained in other partitions, as well as to the I/O pads. Whenever the position of a subblocks is fixed, the external connections to the other modules are updated. This idea is similar to in-place partitioning [LaPotin86].

Mirroring Step

Once the relative positions of the macrocells are fixed, the only freedom for each cell is the mirroring. This does not affect the area but can reduce the total wire length. To find the optimal orientation for every cell is another NP-complete problem. A simple heuristic is used to attack this problem.

For each cell the expected reduction of the total length of the connecting wires is calculated when mirroring with respect to both the x and the y axes. The solution with the maximal reduction is selected. This procedure is repeated until no further length reduction can be found.

9.3.2. Channel Definition

Routing Region Definition and Hierarchical Ordering

A channel is a rectangular area reserved for routing purposes. The task of the channel definition step is to divide the area between the macrocells into rectangular sections, such that standard channel routing techniques can be used to perform the detailed routing (Figure 9.9) This requirement sets some constraints on the way channels can be defined: a channel has a longitudinal direction along which the terminals have to have fixed positions and an orthogonal direction, along which the terminal positions can be chosen freely by the routing process.

- In   - Out

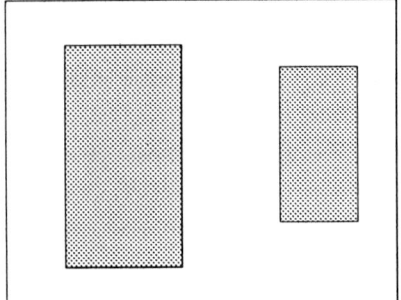 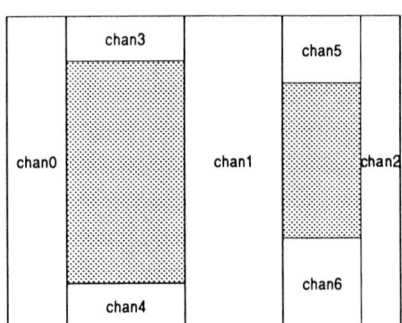

Figure 9.9: Channel Definition

After routing the channel, its dimensions can be expanded or contracted along the orthogonal direction. This must be done without affecting the previously routed channels. To avoid possible channel interferences [Dai85], we need a feasible routing order for the channels. With a placement having a slicing structure, this problem can be solved by routing each channel in the hierarchical order resulting from the slicing structures. Establishing the routing order is another task of this stage.

Scanning

To find feasible channels, scanning is performed along an initial direction which is selected to be the direction of the slicing line at the top level of the decomposition tree (Figure 9.10a). The initial scanning direction can be overridden by the user. Once channels are found, the areas between channels are defined as subblocks and the scanning is applied to those subblocks, recursively, until there remains no empty space not used as a channel area (Figure 9.10b).

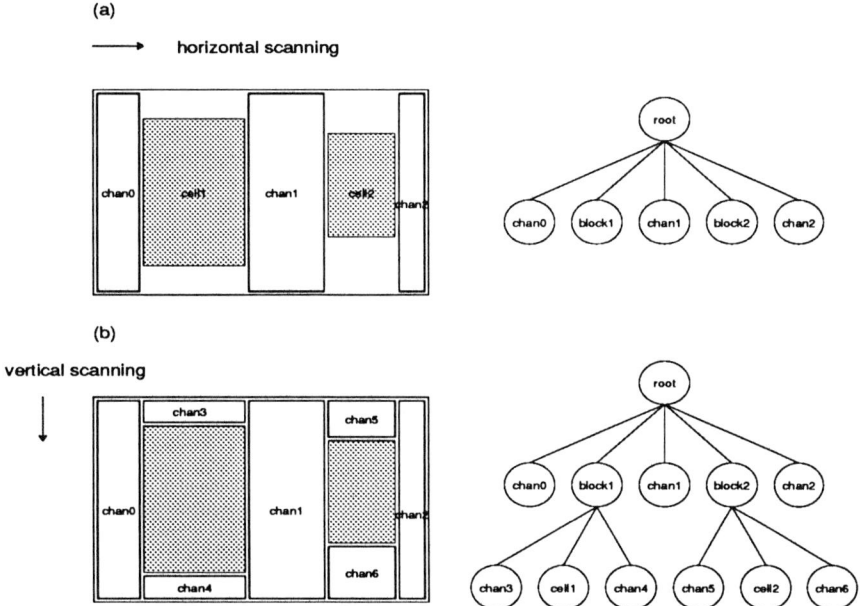

Figure 9.10: Scanning procedure and hierarchy building

This automatic channel definition procedure fails when there is a cycle in the placement. Two solutions to this problem can be formulated. The first approach uses more complicated routing channels (such as L-shaped or U-shaped channels). Routing such channels requires a switch-box router (or a specialized router for L and U shaped channels). Since this is not supported in `Flint`, the approach taken is simply to modify the existing placement. When a cycle is detected, `Flint` will issue a warning to do so. This will only happen if the placement is performed manually, since the automatic placement always results in a placement with a valid slicing structure.

Refinement

A final operation in the channel definition phase is the removal of redundant channels. When the placement is done manually, it is hard to align the macrocells accurately. The result is small, undesired channels. Therefore, when a channel is narrow enough to be considered redundant, it is removed automatically so that the cells can be aligned (Figure 9.11) The minimum allowable channel width is determined dynamically with respect to the overall sizes of the macrocells.

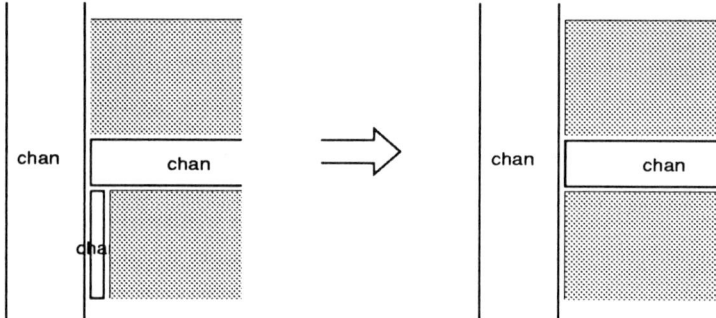

Figure 9.11: Macrocell alignment

9.3.3. Global Routing

The global routing phase defines the routing path for each cable. The global routing problem is basically a Steiner-tree problem, which is known to be NP-complete [Karp72]. However, the problem is reduced to a shortest path problem when the cables are dealt with one by one.

In this case, the goal is to find the shortest path for each cable while avoiding congestion of any specific channel. A graph is constructed where a node corresponds to an interface between two channels. An edge is present between two nodes if both interfaces have a single channel in common. Each edge is weighted with the manhattan distance between the connecting nodes (Figure 9.12a). Each node is annotated with the number of wires that have been routed through that node so far. The numbers are updated every time a cable is routed, so that channel congestion can be avoided when assigning a path for the next cable.

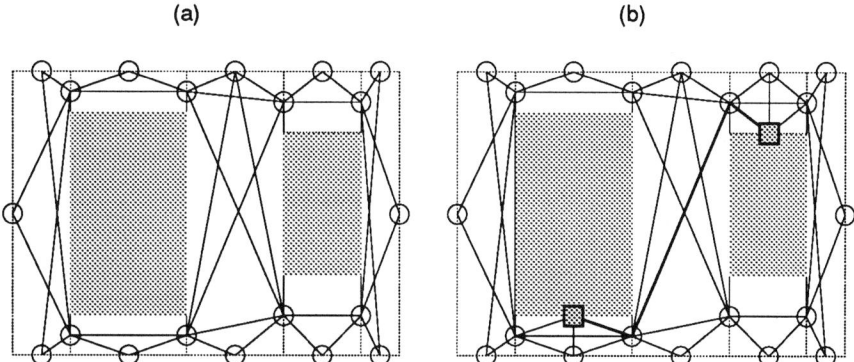

Figure 9.12: Routing graph representation

Algorithm

The procedure starts with the construction of the graph. Interfaces between channels and the outside world are also defined as nodes, but are treated differently: they can be used as terminal nodes but not as routing paths. The cables are then sorted according to the length divided by the number of wires in them. Thus, long cables and short cables with many wires have the highest priority. A short cable does not have many alternative paths, and a cable with many wires has to be routed efficiently or would create great area overhead.

Two sorted lists are generated simultaneously to handle multi-terminal nets. These nets are part of multiple cables (or cable clusters). A first list contains the so-called *essential cables*. Cables are *essential* when they have to be routed, i.e. there exists no alternative routing path. The second list compiles sets of cable clusters. Within a cluster, only a single covering set of cables has to be routed.

For each cable in the sorted list (in descending priority order), the shortest path through the routing graph is determined using Dijkstra's shortest path algorithm [Dijkstra59] in a modified formulation to accommodate the channel congestion constraint. The source and destination of a cable are defined as source node and destination nodes, respectively, and added to the graph (Figure 9.12b). The path that minimizes the weighted sum of the length of the arcs and the number of wires in the nodes along the path is selected. Once the optimal path for a cable is found, the node costs along the path are updated so that the path through those nodes becomes less favorable for the next cables.

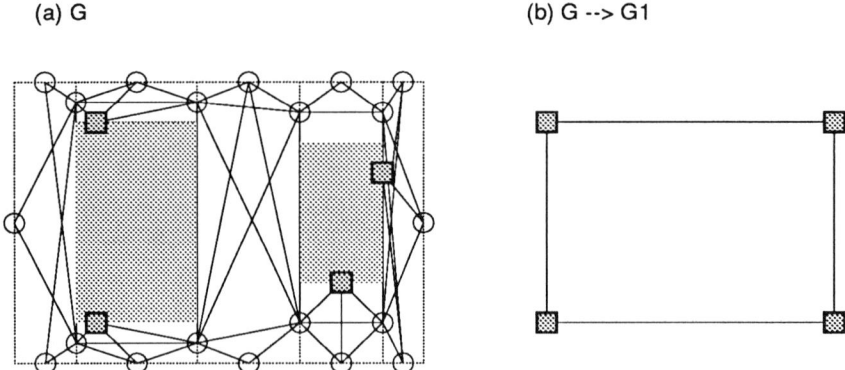

Figure 9.13: Power routing graphs

9.3.4. Power Routing

The power routing phase differs from the generic global routing in that power nets are treated on a net by net basis. Once again, this translates into a Steiner-tree problem, for which no polynomial time algorithms are known.

We have to differentiate between two classes of power nets: nets with and without external terminals. For instance, Vdd and GND nets always have external terminals, while clock nets and special nets might be completely internal to the current module. For the group of nets with external terminals the routing problem is a slightly more complicated, because all the terminals in such a net do not have to be connected into a single spanning tree, as long as they are connected to at least one external terminal (they can then be connected together at a higher level of hierarchy). In this case, the routing problem is more like finding the minimum spanning *forest* than finding a minimum spanning tree.

Another complication of the power routing problem is that a macrocell may have multiple equivalent (internally connected) terminals for the same net. In such cases, routing to only one of the terminals is sufficient, but the internal path should not generally be used as a feedthrough to reach other terminals in different cells[1].

---

[1] An internal connection may be used as a part of the overall routing network if it can sustain the current levels needed to drive the connecting macrocells.

Chapter 9　　　　Interactive Floorplanning

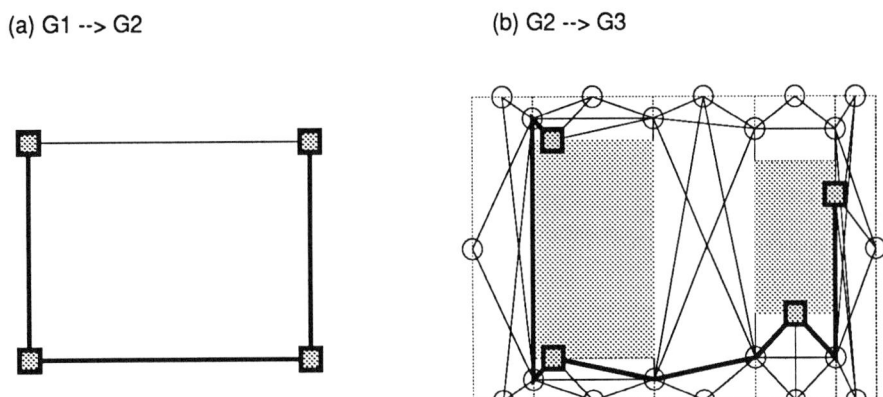

Figure 9.14: Construction of minimum spanning tree

Algorithm

The graph constructed in the global routing stage is used again for the power routing. Flint uses an approximation algorithm to solve the Steiner tree problem [Mehlhorn88].

Let $G = (V, E, d)$ be a connected, undirected distance graph, obtained by adding the set of terminals of a given net to the global routing graph. V is the set of nodes in G, E is the set of edges in G, and d is a distance function which maps E into the set of non-negative numbers (Figure 9.13a).

The algorithm consists of the following steps:

1. Construct the complete distance graph $G_1 = (V_1, E_1, d_1)$, where V_1 is the set of power terminals, and for every (v_i, v_j) in E_1, d_1 (v_i, v_j) is equal to the distance of a shortest path from v_i to v_j in G (Figure 9.13b).
2. Find a minimum spanning tree (MST) G_2 of G_1 (Figure 9.14a).
3. Construct a subgraph G_3 of G by replacing each edge in G_2 by its corresponding shortest path in G. (If there are several shortest paths, pick an arbitrary one) (Figure 9.14b).

Routing the internal nets follows Mehlhorn's procedure, but some modification is needed in step 2 to deal with the nets with external (also called *formal*) terminals. The MST algorithm described in [Berge65] has been modified to find the minimum spanning forests. The algorithm starts by adding to G_1 a set of nodes, each one of which is the formal terminal closest to at least one node in G_1. Formal terminals can be made either left and right, or top and bottom. Two dis-

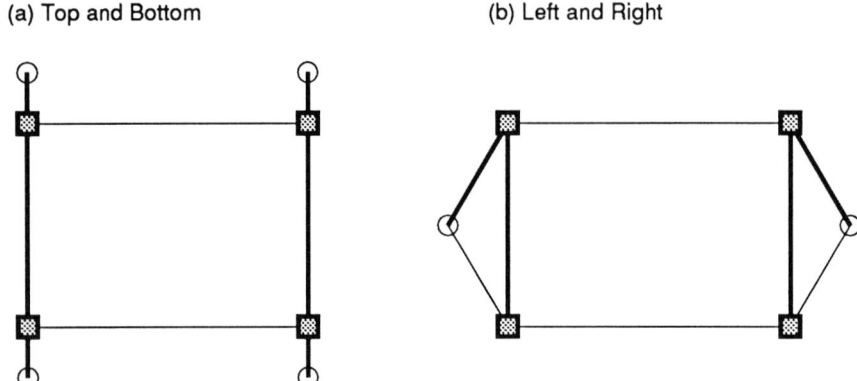

Figure 9.15: Graphs for a net with external (or formal) terminals

tance graphs are constructed for top/bottom and left/right, and the minimum spanning forests are found for each graph. Finally, the better forest is chosen (Figure 9.15).

To keep track of nodes being connected to a formal terminal node, every node carries a label. The nodes corresponding to formal terminals are initially marked as "ROOT" and the others are marked as "POWER". All the nodes of a subtree are marked as "ROOT" if the subtree has grown to be connected to any "ROOT" node during the tree growing step in the MST algorithm.

To avoid internal connections being used as a routing paths to other nodes, each node has another flag on it. Initially the flag of every node is set to "CANDIDATE" for connection. Whenever any two subtrees are merged into one by adding an edge, the flags of all the nodes in both subtrees are set to "NULL" except for the nodes corresponding to the ends of the edge. In the next tree-growing step, only "CANDIDATE" nodes are considered for connection, and that can prevent any node from being connected to a "ROOT" node through an internal connection.

9.3.5. Detailed Routing

The detailed routing phase uses a gridless channel router [Yoshimura82]. The algorithm has been modified extensively to eliminate fixed grid constraints, to handle cyclic constraints and to incorporate power, ground and clock routing.

Chapter 9   Interactive Floorplanning   121

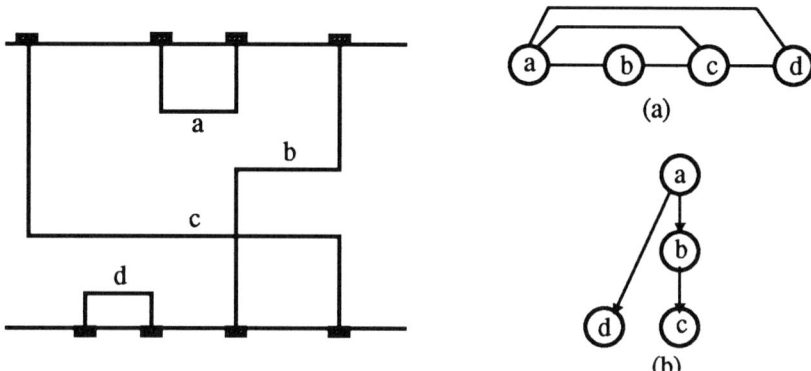

Figure 9.16: Horizontal (a) and vertical constraints (b) graphs for channel routing

The basic routing algorithm uses a so-called *left-right* approach. First, a set of constraint graphs are built. The *vertical constraint graph* expresses a set of relations between the vertical placement positions of the routing tracks for the different nets. The *horizontal constraint graph*, on the other hand, determines which nets can share the same track (i.e. which nets are non-overlapping in the length direction of the channel). Examples of vertical and horizontal constraints are shown in Figure 9.16. The left-right algorithm then assigns the nets to the routing tracks, starting from the left side of the channel. Heuristics are used to determine the allocation.

Gridless routers allow for denser routing, and can handle arbitrary locations of terminals on the macrocell boundaries. This is an essential requirement for a silicon assembly environment. The left-right algorithm is easily extended to fully gridless routing: the vertical constraints between opposing terminals are determined by checking the minimum distance between terminals, as dictated by the design rules, instead of just looking if they are on the same grid-line, as is done in grid-based routers. The design rules are stored in a user-provided technology file. The router has also been extended to handle three interconnect layers (polysilicon, metal1 and metal2). Wiring on the polysilicon layer is normally avoided, but is used for short internal connections in bit-sliced datapaths.

In order to be routable, the vertical constraint graph has to be acyclic. If not, no feasible placement (or ordering) of the routing tracks can be derived.

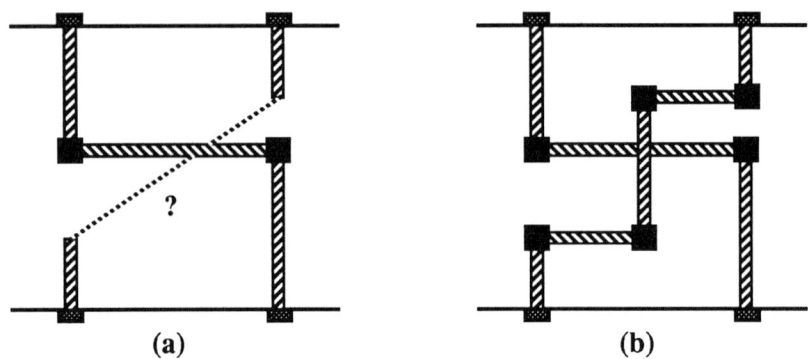

Figure 9.17: Routing graph cycles (a) and dog legs (b)

Cycles are unfortunately often present in typical routing problems. An example of a cycle is shown in Figure 9.17a. The most commonly used technique to eliminate such a cycle is to introduce a *dogleg* (Figure 9.17b). One of the nets causing the cycle is broken up and an intermediate jumper is introduced. This effectively creates an acyclic graph. The Flint channel router handles cycles in the following way: during the construction of the vertical constraint graph, the graph is checked for cycles every time a new terminal is added. If a terminal would cause a cycle, it is temporarily removed and stored in a waiting list. After the completion of the vertical graph construction process, all cycle generating terminals are revisited. For each of those terminals, a suitable dogleg position is determined by searching on both sides of the terminal. A dogleg position is feasible if it does not create any new cycles in the constraint graph. Out of all feasible positions, we select the one which minimizes the track length and the number of required tracks.

A main feature of the Flint router is its capability to simultaneously handle power, clock and signal nets. Power and clock nets are routed in the same way as normal signal nets. However some precautions are necessary: In order to minimize the channel area as well as the wire resistance, dog legs should be avoided as much as possible on power an clock nets. This is achieved by assigning a higher routing priority to those nets. Also, since power nets normally require wider wires, they should by preference be assigned to the same routing tracks.

The Flint global and local routing approach has some other features worth mentioning.

- The widths of the power wires are automatically scaled such that a constant current density is maintained. The current magnitude in each wire is determined by a traversal of the power network, starting from the connectors on the leaf modules up to the I/O power terminals. When no current information is available on the leaf modules, a simple "rule of thumb" approach is used, determining the width of the wire based on the number of wires connecting to it and their widths.
- `Flint` automatically extracts the parasitic capacitance of each wire during the routing process. This information is back-annotated into the database for usage by estimation and verification tools.

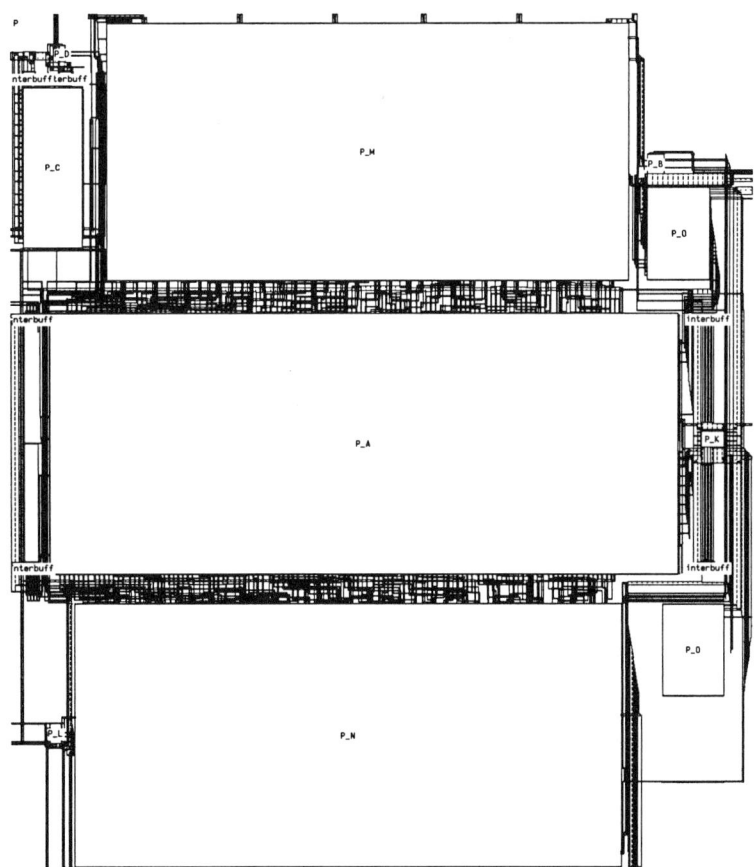

Figure 9.18: FIR Filter for pulsar signal recovery (top level view)

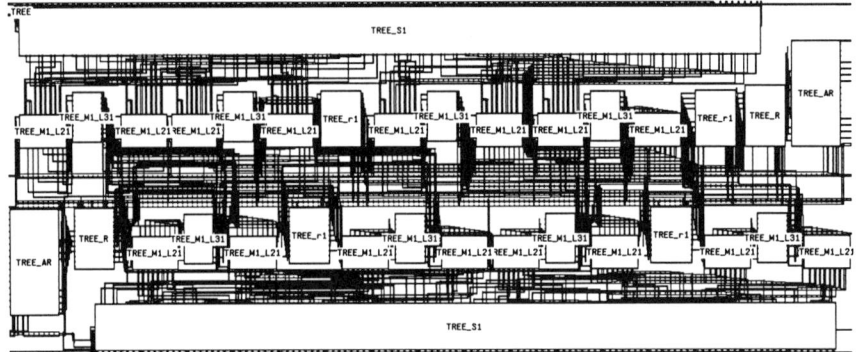

Figure 9.19: Interleaved accumulation trees of 1024 tap FIR filter.

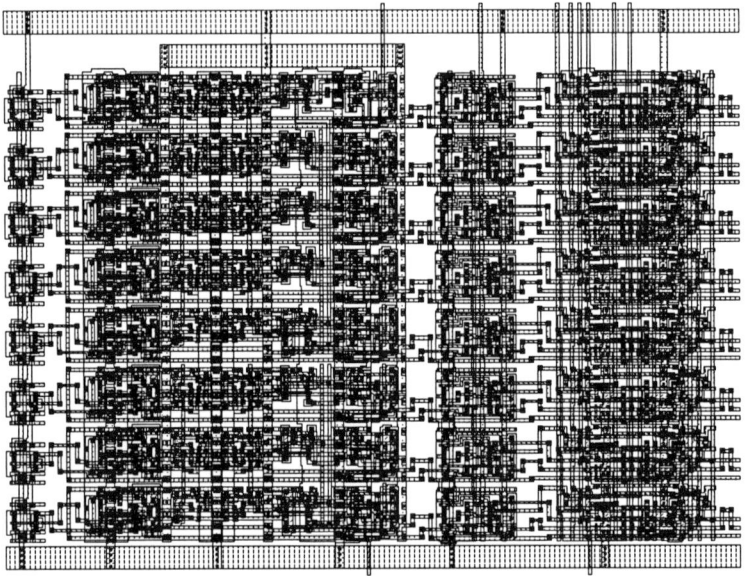

Figure 9.20: Datapath section of Viterbi search circuit

9.4. EXAMPLES

The floorplanning capabilities of Flint are demonstrated using a number of real examples. In the first example, the floorplan and layout of a 1024-tap FIR filter for pulsar signal recovery is examined. This complex circuit, which counts

140,000 transistors and is clocked at 32 MHz, was generated entirely using the LAGER tools. It contains a 2048 word shift-register, 64 complex multipliers and two data accumulation trees. The top layout view is shown in Figure 9.18. Especially hard to implement were the accumulation structures. Due to the tree-like structure, careful floorplanning was essential to avoid an explosion of wiring area. The interactive nature of Flint allowed interleaving the two trees in such a manner that the overall wire length was minimized (Figure 9.19).

A second example shows how Flint is used in a batch mode to route datapath structures. The datapath section under examination here is a part of a Viterbi search chip, used in a connected speech recognition system. In this case, a floorplan description in the *fdl* language was automatically generated by the datapath compiler dpp. Flint was then called to perform the absolute placement and detailed routing between the blocks. As can be seen from the layout in Figure 9.20, Flint exploits cell feed-throughs to reduce the routing area.

9.5. SUMMARY

Flint is an interactive floorplanner and router for macrocell-based designs. The floorplanner represents a close integration between interactive and automatic approaches. The automatic procedures can save a chip designer considerable effort by providing rapid initial solutions in every design phase The user can later modify the initial solutions so that fine tuning and spot design improvements are possible.

Flint has been used to generate a large number of complex designs, the largest ones containing up to 150,000 transistors. Designs with up to 5000 nets (at one single hierarchy level) have been routed in less than 5 minutes.

REFERENCES

[Berge65] C. Berge and A. Ghouilla-Houri, "Programming, Games and Transportation Networks", John Wiley & Sons, 1965.

[Dai85] W. Dai, T. Asano and E. Kuh, "Routing Region Definition and Ordering Scheme for Building Block Layout", *IEEE Trans. on CAD*, vol CAD-4, no 3, pp. 189-197, July 1985.

[Dijkstra59] E. Dijkstra, "A Note on Two Problems in Connection with Graphs", *Numer. Math.*, vol. 1, pp 267-271, 1959.

[Harrison86] D. Harrison et all, "Data Management and Graphics Editing in the Berkeley Design Environment", *Proc. IEEE ICCAD Conf*, Santa Clara, 1986.

[Karp72] R. Karp, "Reducibility among Combinatorial Problems", *Proc. Complexity of Computer Computations*, pp. 85-104, Plenum Press, New york, 1972.

[Lee89] S. Lee, "Automatic Floorplanning Techniques for Macrocell-Based Layouts", *UCB Master Report*, May 1989.

[Mehlhorn88] "A Faster Approximation Algorithm for the Steiner Problem in Graphs", *Information Processing Letters 27*, pp. 125-128, March 1988.

[Otten82] R. Otten, "Automatic Floorplan Design", *Proceedings 19th Design Automation Conference*, pp. 261-267, 1982.

[Ousterhout86] J. Ousterhout, "Magic Tutorial", UCB Computer Science Division. December 1986.

[Potin86] D. La Potin and S. Director, "Mason: A Global Floorplanning Approach for VLSI", *IEEE Trans. on CAD*, vol. CAD-5, no 4, pp. 477-489, Oct. 1986.

[Rabaey85] J. Rabaey, S. Pope and R. Brodersen, "An Integrated Automatic Layout Generation System for DSP Circuits", *IEEE Trans. Computer-Aided Design*, CAD-4, pp. 285-296, July 1985.

[Sangiovanni87] A. Sangvionanni-Vincentelli. P. Antognetti and G. De Micheli, *Design Systems for VLSI Circuits*, pp. 113-195, Martinus-Niijhoff Publishers, 1987.

[Stockmeyer83] L. Stockmeyer, "Optimal Orientation of Cells in Slicing Floorplan Designs", *Information and Control*, vol. 57, no2/3, pp 91-101, May/June 1983.

[Yoshimura82] T. Yosimura and E. Kuh, "Efficient Algorithms for Channel Routing", *IEEE Transactions on Computer Aided Design*, V. CAD-1, Jan. 1982, pp. 25-35.

10

Datapath Generation

Mani Srivastava

The processing power of application-specific VLSIs primarily comes from the use of dedicated datapaths with architectures tailored to the exact needs of the algorithm. This makes it imperative that a CAD environment provides the designer with the ability to quickly reconfigure a datapath and iterate on several designs while evaluating their area and performance.

The bit-slice datapath generator, dpp, provides this capability and is perhaps the most important utility for algorithm-specific ICs. General-purpose macrocell place-and-route tools, like Flint (Chapter 9) and Mosaico [Burns87], do not work well for bit-slice datapaths as they fail to exploit the regularity inherent in such structures. A special tool for datapaths was therefore found desirable. TimLager can be used to generate the individual blocks (adder, register,...) of a datapath, but cannot handle the routing between and through the blocks. Given the net-list of a bit-slice datapath, dpp does the placement, channel definition and global routing of the datapath and produces a floorplan file which Flint will later use to route the channels and generate the physical layout. dpp also back-annotates the structure_instance view of the datapath blocks with geometric constraints and feed-through specifications. This information is used by the block layout generator (TimLager) when generating the blocks. In LAGER terminology, dpp is a *structure-processor*.

10.1. THE BIT-SLICE DATAPATH MODEL

The datapath model adopted by dpp is characterized by a data width N and is composed of The blocks have two types of terminals: *data terminals* and *control terminals*. The data terminals represent N-bit wide busses. dpp assumes that the data signals flow horizontally and control signals vertically. Accordingly, terminals at the left and right edges of a block are assumed to be data terminals, and terminals on the top and bottom edges are assumed to represent control terminals.

The datapath itself is described as an interconnection of the blocks using data and control nets. The data nets connect data terminals of the blocks and optionally also formal data terminals of the datapath (terminals coming out at the left or right edges of the datapath). Similarly, the control nets connect control terminals of the blocks and optionally also formal control terminals of the datapath (terminals coming out at the top or bottom edges of the datapath).

An important point is that the data nets actually represent N-bit wide busses that connect to N-bit wide data terminals. The expansion into N-bit wide busses is done by dpp itself, instead of during the structure_master view to structure_instance view conversion phase of DMoct, and must not be done through the SDL syntax. This was done to enforce a bit-slice discipline whereby no inter-bit routing is allowed, because many heuristics used by dpp will not work otherwise. This idiosyncrasy of dpp is the source of the various differences in the way SDL files are written for datapaths as described later.

10.2. DESIGNING A DATAPATH

There are two distinct tasks that need to be done in order to design a datapath using dpp. The first is the design of the blocks used in the datapath which would be performed by a developer and the second is the design of the datapath by interconnecting the blocks which is done by users. Most users will not need to design their own blocks because most likely one block or a combination of blocks from the fairly extensive central library will suffice.

10.2.1. Designing a block for use in datapaths

A block for use in a datapath is a LAGER module generated by some layout generator, typically TimLager. The block has the property that it is parameter-

Chapter 10 Datapath Generation

ized by N, the number of bits. It has data terminals, which are N-bit wide, coming out on the left and right edges, and control and supply terminals on the top and the bottom. Due to limitations in `dpp`, the block cannot be hierarchical. In other words, the SDL file describing it may not have any subcells. In fact, the SDL file will just be a set of formal terminals with TERM_EDGE properties to indicate whether they are data or control terminals. Further, due to the idiosyncrasy in `dpp` mentioned earlier, the data terminals must not be specified as bus terminals.

Figure 10.1 shows the SDL file for *trist_inverter*, the tristate inverting buffer block shown in Figure 10.2, using `TimLager` as the layout generator:

```
(parent-cell trist_inverter (VERIFY_STOP) (verify-stop))
(parameters
    N
    (FEEDTHRUS 0)
    (BITHEIGHT 0)
    (LSBOFFSET 0)
)
(layout-generator TimLager)
(structure-processor dpp)
(terminal IN (TERM_EDGE LEFT) (TERM_EDGE RIGHT))
(terminal OUTINV (TERM_EDGE LEFT) (TERM_EDGE RIGHT))
(terminal CNTL (TERM_EDGE TOP) (TERM_EDGE BOTTOM))
(terminal CNTLINV (TERM_EDGE TOP) (TERM_EDGE BOTTOM))
(terminal GND (TERM_EDGE TOP) (TERM_EDGE BOTTOM))
(terminal Vdd (TERM_EDGE TOP) (TERM_EDGE BOTTOM))
(end-sdl)
```

Figure 10.1: The SDL file for trist_inverter.

A `dpp` block cannot use any arbitrary layout generator. This is because the layout generator is also called by `dpp` in a special estimation mode to obtain information about the physical characteristics of the block. There is a well defined protocol for this information exchange using OCT bags and properties. Also, the layout generator tool must be able to provide feedthroughs through the block, stretch it to a required height and add appropriate control logic at the top and the bottom, the information about which is passed through special parameters whose value is calculated by `dpp`. Only a layout generator tool with the above capabilities can be used, and at present `TimLager` is the only such tool in LAGER.

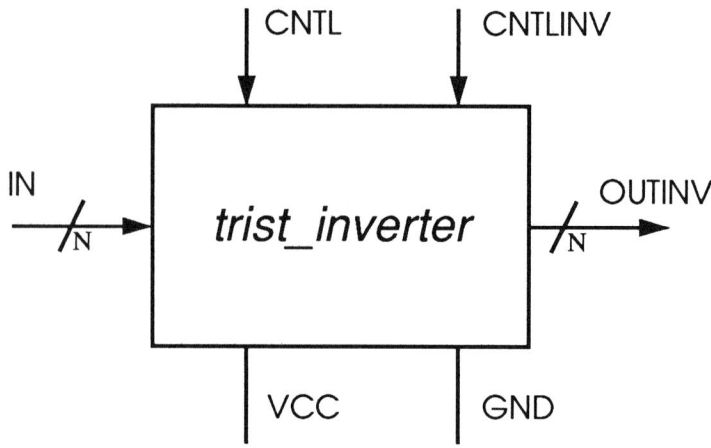

Figure 10.2: A tristate inverter block from the LAGER datapath block library

As described in the chapter on TimLager, for every block using TimLager as the layout generator, there is an associated file containing the TimLager tiling function. Special constructs were added to TimLager to support the stretching, feed-through and control logic features required by dpp and to pass information to dpp. The parameter N is used to decide the number of leafcells tiled in the vertical direction. Unfortunately, all the special constructs make a typical tiling function quite complicated to write manually. Therefore, a supporting tool called dppdotc has been provided to automate this task. It can handle most commonly encountered tiling schemes, although in cases like a carry-save adder or a carry-lookahead adder one still has to write the tiling function manually. Following is the tiling function generated by dppdotc for *trist_inverter*:

```
#include "TimLager.h"
#include "oct.h"

trist_inverter() {
    int i,nbits,lsboffset;
    nbits=Getparam("N");
    lsboffset=Getparam("LSBOFFSET");
    bit();
    Open_newcell(Read("name"));
    if (nbits<1) {
```

```
            Timerror("TimLager: FATAL ERROR:",
            " number of bits N < 1 in %s\n", "trist_inverter");
            exit(-1);
        } else if (nbits==1) {
            Addup("bit", BOTTOM|LEFT|RIGHT|TOP, LEFTINDEX, 0,
            RIGHTINDEX, 0, BOTTOMSTRETCH, lsboffset, END);
            if (!no_bot_stretch) {
                Addup(bot_stretch_cellname, BOTTOM, OFFSETX,
                bot_stretch_x_offset, OFFSETY,
                bot_stretch_y_offset, END);
            }
        } else {
            Addup("bit", BOTTOM|LEFT|RIGHT, LEFTINDEX, 0,
            RIGHTINDEX, 0, BOTTOMSTRETCH, lsboffset, END);
            if (!no_bot_stretch) {
                Addup(bot_stretch_cellname, BOTTOM, OFFSETX,
                bot_stretch_x_offset, OFFSETY,
                bot_stretch_y_offset, END);
            }
             for(i=1; i<=nbits-2; i++) {
                Addup("bit", LEFT|RIGHT, LEFTINDEX, i,
                RIGHTINDEX, i, END);
            }
            Addup("bit", TOP|LEFT|RIGHT, LEFTINDEX, nbits-1,
            RIGHTINDEX, nbits-1, END);
        }
        Close_newcell();
    }
    bit() {
        int nfeeds, height;
        nfeeds=Getparam("FEEDTHRUS");
        height=Getparam("BITHEIGHT");
        Open_newcell("bit");
        Addup("trist_inverter", LEFT|RIGHT|TOP|BOTTOM, HEIGHT,
        height, FEEDTHRU, nfeeds, -1, END);
        if (!no_feedthru)
            Addup(feed_cell_name, LEFT|RIGHT|TOP, END);
        if (!no_stretch)
            Addup(stretch_cellname, TOP, OFFSETX,
            stretch_offset, END);
        Close_newcell();
    }
```

There are some layout restrictions that need to be followed when designing leafcells for use in the blocks. As already mentioned, the data terminals need to come out at the left or right edges, whereas the control and supply terminals need to come out at the top or bottom edges. Due to restrictions imposed by `Flint`, the data terminals must come out on one of the following `Magic` layers: metal1, metal2, poly, via, polycontact. The control and supply terminals must come out on metal1, poly or polycontact. Neighboring terminals must be separated by a distance that meets the requirements imposed by `Flint`. In general, for MOSIS's SCMOS technology, terminals should be made 4λ wide and have a 4λ spacing. While designing the leafcells for the various bits of a block, one should try to minimize the variance in the heights in order to get better routing results. Finally, dpp tries to make use of horizontal feedthroughs already provided in the leafcell. This can be done in one of two ways. First, the leafcell designer can bring out a data terminal on both left and right edges. dpp will take advantage of this implicit feedthrough if the corresponding data net crosses the block. Second, the leafcell designer can provide explicit and uncommitted feedthroughs. In case of blocks using `TimLager`, terminals with names of the form FEEDi (i=1,2,...) are used to represent such explicit feedthroughs and are thus recognized by `TimLager`.

10.2.2. Designing a datapath

Just as with any other module designed in LAGER, a datapath is also described in the SDL syntax. As mentioned earlier, some idiosyncrasies in dpp result in restrictions and deviations from the norm in case of SDL files for datapaths. Figure 10.3 shows the block diagram of an example datapath which is the datapath *datapath* from the on-line LAGER tutorial *tut8*.

The SDL file of the example datapath is as follows:

```
; sdl-file to describe the data path for tutorial #6
    (parent-cell datapath)
    (parameters N)
; N is the number of bit-slices. The number of bit-slices has
; to be the same for the entire datapath.
    (structure-processor dpp)
; This declaration must appear for every datapath. It
; indicates to DMoct that the structural description must be
; processed by the program dpp before layout generation.
    (layout-generator Flint a)
; This declaration must always be present. The "a" flag
```

```
; specifies that Flint is to be run in automatic mode.
   (subcells
       (adder ADDER ((N N)))
       (mux2to1 Z_SAT ((N N)))
       (scanreg ACC ((N N)))
   )
; NOTE: any arbitrary subcells may be chosen, but for each
; cell the subcell parameter declaration "((N N))" must be
; given
; Data Nets
   (net ibus ((parent IBUS) (ADDER IN1)))
   (net out_feedback ((parent OBUS) (ACC OUT) (ADDER IN2)))
   (net adder_out (((ADDER OUT) (Z_SAT IN1)))
   (net z_sat_to_acc ((Z_SAT OUT) (ACC IN)))
   (net scanin ((parent SCANIN) (ACC SCANIN)))
   (net scanout ((parent SCANOUT) (ACC SCANOUT)))
   (net carryout ((parent CARRY) (ADDER COUT)))
   (net set_one ((parent ONEBUS) (Z_SAT IN2)))
; Control Nets
   (net cntl_GND ((parent CNTL_GND) (ADDER CIN)))
   (net cntl_Vdd ((parent CNTL_VDD) (ADDER CININV)))
   (net cntl_load ((parent LOAD) (ACC LOAD)))
   (net cntl_scan ((parent SCAN) (ACC SCAN)))
   (net cntl_keep ((parent KEEP) (ACC KEEP)))
   (net cntl_load_inv ((parent LOADINV) (ACC LOADINV)))
   (net cntl_scan_inv ((parent SCANINV) (ACC SCANINV)))
   (net cntl_keep_inv ((parent KEEPINV) (ACC KEEPINV)))
   (net cntl_sel1 ((parent SEL1) (Z_SAT SEL1)))
   (net cntl_sel2 ((parent SEL2) (Z_SAT SEL2)))
; Power Nets
   (net Vdd (NETTYPE SUPPLY) ((parent Vdd) (ADDER Vdd)
       (Z_SAT Vdd) (ACC Vdd)))
   (net GND (NETTYPE GROUND) ((parent GND) (ADDER GND)
       (Z_SAT GND) (ACC GND)))
; Clock Nets
   (net PHI1 (NETTYPE CLOCK) ((parent PHI1) (ACC PHI1)))
   (net PHI1INV (NETTYPE CLOCK) ((parent PHI1INV)
       (ACC PHI1INV)))
   (net PHI2 (NETTYPE CLOCK) ((parent PHI2) (ACC PHI2)))
   (net PHI2INV (NETTYPE CLOCK) ((parent PHI2INV)
       (ACC PHI2INV)))
(end-sdl)
```

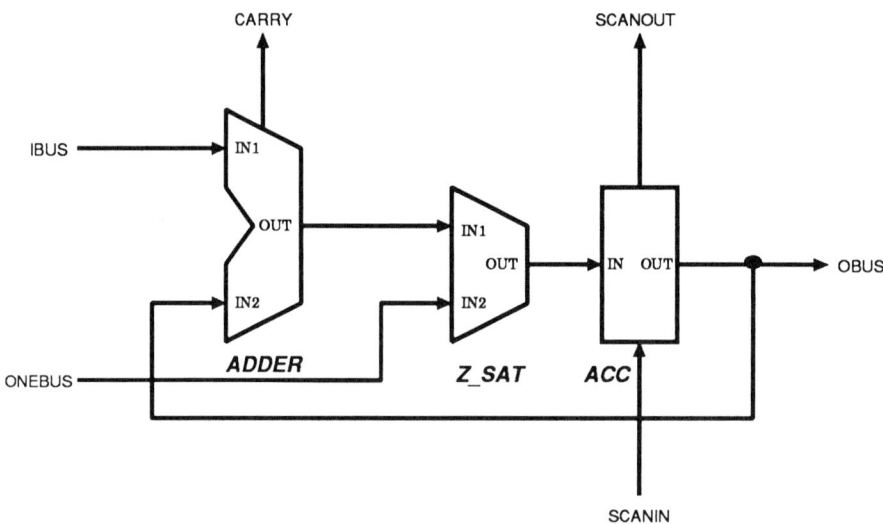

Figure 10.3: An example datapath

Every datapath SDL file specifies dpp as the structure-processor and Flint as the layout-generator. Flint is run in batch mode, using the option -a (automatic, as opposed to interactive mode). A variety of options can be specified for dpp to specify the amount of tool optimization versus user control, such as the ability to suppress the automatic placement. Every datapath must have a formal integer parameter named "N" whose value is used by dpp as the number of bits in the datapath. The value of N is also passed on to the subcells.

Although dpp requires that the datapath blocks be non-hierarchical and generated by a layout generator with some additional capabilities, one can use the flattening feature of DMoct to create libraries of datapaths that can be used in other datapaths as sub-datapaths, resulting in a hierarchical datapath specification. This was used to great advantage in the design described in Chapter 21.

10.3. IMPLEMENTATION DETAILS

As already mentioned, dpp is a *structure-processor* in LAGER terminology. This means that it takes a structure_instance view sub-tree as its input and converts it into another structure_instance view sub-tree (rooted at the same point) which is ready for a layout-generator. In case of dpp the structure_instance view sub-tree is a two level tree with the root being the datapath and the leaves being blocks from the library. dpp does the placement of the blocks, does global

Chapter 10 Datapath Generation

routing, creates the channel structure, allocates feedthroughs for over-the-block routing of data nets, and does bit-height equalization. The output is a floorplan file for `Flint` and a modified structure_instance view sub-tree where all the data nets and terminals are expanded, feedthrough terminals created and the leaf structure_instance views annotated with information for the block layout generator.

Bit-sliced datapaths are viewed by dpp to consist of blocks that are tiled in the vertical direction and placed linearly along the horizontal direction with the bottom edges of the blocks being co-linear. Routing channels separate blocks which are adjacent. Channels are also placed along the top edge of each block in order to equalize the heights of the blocks. Finally, global channels spanning the entire width of the datapath are placed at the top and the bottom of the datapath. Figure 10.4 illustrates the generic floorplan of datapaths created by dpp.

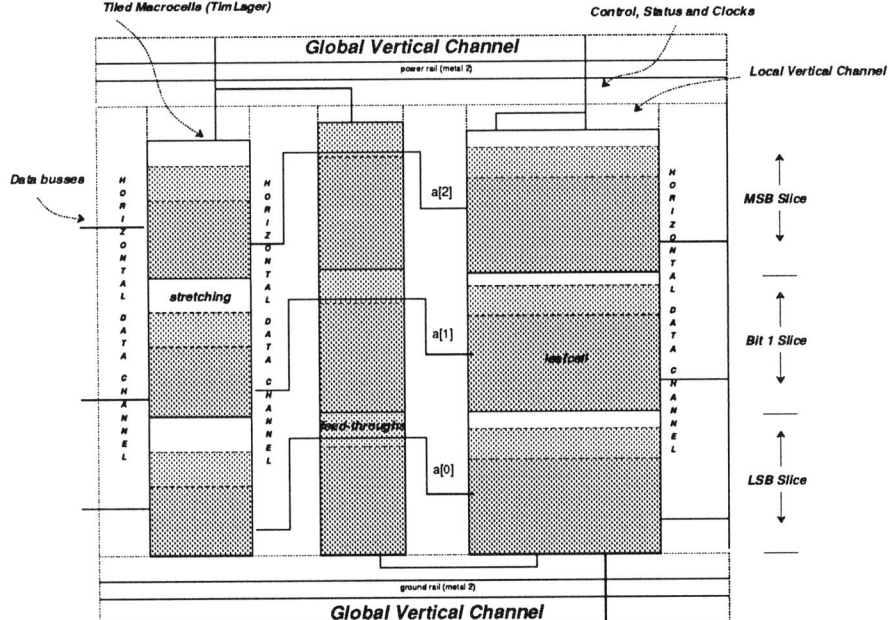

Figure 10.4: Bit-slice datapath by dpp. Each cell in the block consists of a *leafcell* (dark), an optional *feedthrough cell* (grey), and an optional *stretching cell* (white). Horizontal channels are used to route data signals between blocks. Global and local channels are used to route control, status, and clock signals. Each block is generated by `TimLager`. Routing of blocks is done by `Flint`.

The control, status, clock and supply nets run vertically. Within a macrocell they are routed implicitly by the abutment of leafcell terminals during the tiling process. The global routing of these nets between macrocells and to the outside is done by dpp using the global routing channels. The data busses flow in the horizontal direction and are routed explicitly using the routing channels between the adjacent macrocells. Data busses connecting non-adjacent macrocells are routed through the intervening blocks. This is done by back-annotation of the structure-instance views of the blocks with information for the block layout generator to generate enough feedthroughs for data busses going across the block. Feedthroughs already provided by the leafcell designer are used first before extra feedthroughs are generated.

The process of guiding the lower-level layout generation according to the requirements of the upper-level layout generator is a key feature of dpp. It makes the datapath blocks appear *porous* to Flint and saves the area wasted by the general-purpose macrocell place-and-route approach in routing the data busses around the opaque macrocells. As shown in Figure 10.5, this results in a 24% reduction in area from 4.2 mm^2 to 3.2 mm^2 in 2μ technology for the case of a 12-bit version of an example datapath from a chip for measuring position and velocity of a robot arm joint from the quadrature signals obtained from an optical incremental position encoder.

A problem with the approach outlined so far is that there may be a mismatch between the heights of adjacent blocks resulting in a *stair-case effect* or congestion in the interblock routing channels. Again, the process of back-annotation is used to direct the macrocell layout generator to artificially stretch the heights of each bit in all blocks to a uniform value. A further 32% reduction in area was obtained for the example datapath.

This feature of making the blocks porous and stretchable results in a total area reduction of 48% for a 12-bit version of the example datapath. The relative area penalty due to the stair-case effect and macrocell opacity increases with the number of bits. Consequently the relative area reduction obtained by making the macrocell porous and stretchable also increases as the number of bits increases and results in dramatic improvements in area. For example, a 24-bit version of the same datapath shows a 63% reduction in area. In general, using dpp results in an area that is a linear function of N as opposed to a quadratic function of N that results in the presence of stair-case effect and opaque (nonporous) blocks.

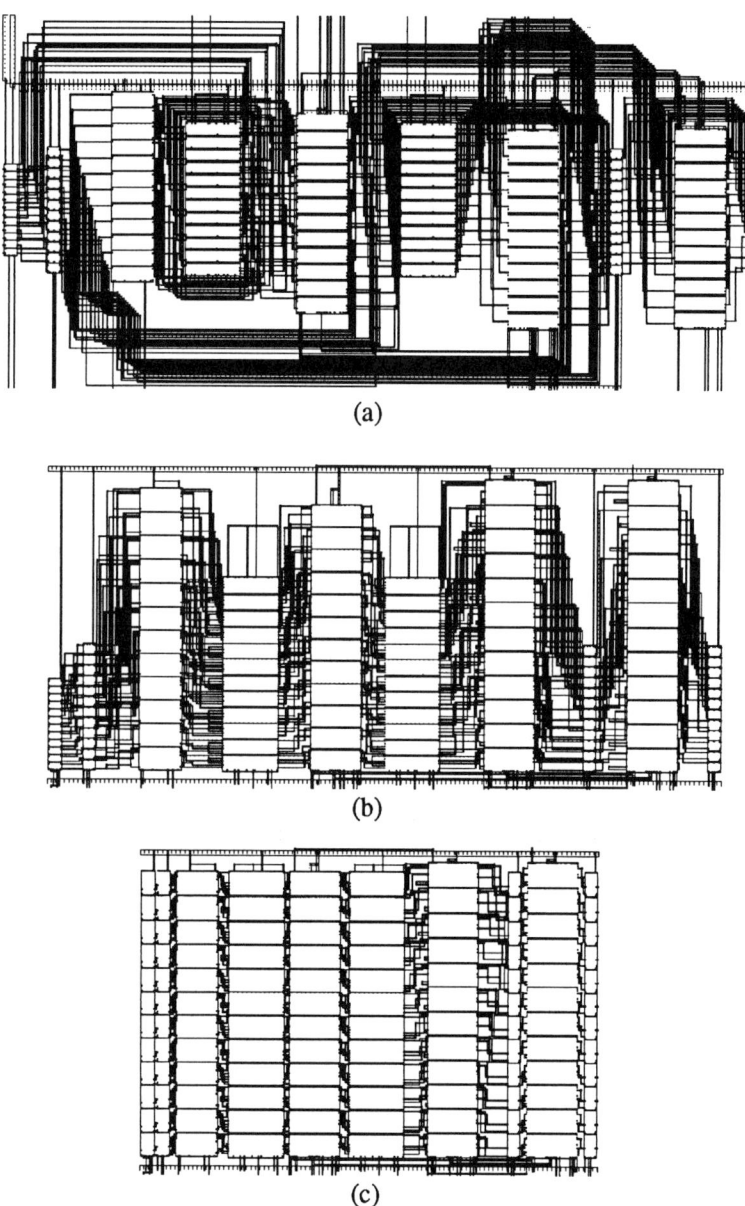

Figure 10.5: (a) Datapath layout with Flint; area = 4.2 mm^2. (b) Datapath layout with Flint after processing with dpp to add feedthroughs; area = 3.2 mm^2. (c) Datapath layout with Flint after processing with dpp to add feedthroughs and stretching of cells; area = 2.2 mm^2.

The other crucial step in dpp is block placement. The goal is to find a suitable ordering of the macrocells so as to minimize the area. With through-the-block routing of global busses and equalization of the block heights, the problem can be quite accurately modeled as minimization of the height of the tallest macrocell taking the extra feedthroughs required into account. dpp directly calls the macrocell layout generator in an estimation mode to get information about the physical characteristics of the macrocells. This information is then used by the placement procedure, which is based on the well known Kernighan and Lin's *Min-Cut* partitioning algorithm [Kernighan70]. It tries to minimize the height of the tallest macrocell taking the extra feedthroughs required into account, instead of just the number of nets crossing a partition.

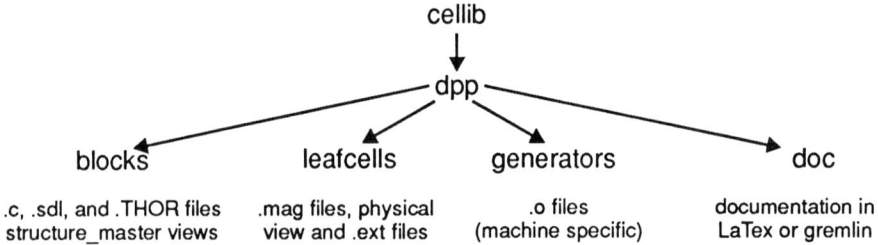

Figure 10.6: Organization of the dpp library in LAGER

10.4. DATAPATH LIBRARY ORGANIZATION

The datapath library consists of the primitive blocks used by the datapath designer and the leafcells used to generate those blocks. As mentioned earlier, all the current datapath blocks are actually TimLager modules following a very strict tiling discipline to cooperate with dpp. Every block is represented in the library by an SDL file describing its black-box appearance, a TimLager .c file describing the tiling structure, a .THOR file giving its functional simulation model, and one or more .mag files giving the layouts of the leafcells of the block. In addition, .o files, structure master views and physical views are automatically generated from the above files by the library management software. Figure 10.6 shows the directory organization of the main datapath library in LAGER. In keeping with the LAGER library management philosophy, the users can create their own datapath libraries.

adder	ripple carry adder with a dual carry chain for fast carry propagation
bufferbig	buffer with medium drive capability output is available in both positive and negative logic
bufferhuge	buffer with large drive capability output is available in both positive and negative logic
buffersmall	buffer with small drive capability output is available in both positive and negative logic
comparatorE	ripple type comparator for even number of bits
comparatorO	ripple type comparator for odd number of bits
compconst	compare with a hardwired constant
constant	implements a constant value in a datapath
csa	carry-select adder
inconstmux	output is same as the input or one of the four constants 0, 1, -1 and -2
inv2to1muxone	2-to-1 multiplexer with a negative logic output and a signal to force the output high
inv2to1muxzero	2-to-1 multiplexer with a negative logic output and a signal to force the output low
inverter	inverts the input
invpass	sets output to either the input or its complement
isoinvzero	sets output to either the input or its complement or zero (all bits 0)
isozero	sets output to either the input or zero
mux2to1	2-to-1 multiplexer with both positive and negative logic outputs
regconstant	hardwired constant value with output enable control
saturator	implements saturation in accumulators
scanmslatch	single-ported dynamic scan path register
scanmslatchmx	scanmslatch mirrored about a horizontal axis
scanreg	single-ported scan path register
scanregmx	scanreg mirrored about a horizontal axis
shift	parameterized shifter for logical up, logical down or arithmetic down shift by 1, 2, 4, 8 or 16. Several of these can be chained to form a log shifter.
signmag	converts a one's complement number into a sign magnitude number
trist_buffer	buffer with tristateable output
trist_inverter	inverter with tristateable output
xfer_gate	transfer gate
zero	conditionally pulls down a bus

Table 10.1 Datapath Block Library

Table 10.1 gives a listing of the various blocks currently installed in the main dpp library. One can add additional blocks to the library or create one's own library. The best way to do this is to follow the same organization as in the main library and copy the corresponding Makefiles for use with the GNU Make utility. The Makefiles are written such that they are independent of the names of the blocks and the leafcells, although appropriate UNIX environment variables need to be set. In general, for a block using TimLager, one needs to provide the .sdl, .c and .THOR files for the block, and the .mag files for all the leafcells used by the block. In addition, LaTex and gremlin files are used for documenting the block and the leafcells.

10.5. LIMITATIONS

dpp has had extensive use and during this process several limitations and problems have become known. The foremost problem is due to the confluence of two factors, a wide variation in the height of cells in the library and an unsophisticated placement algorithm, which produces layouts that have a substantial scope for improvement. A third factor contributing to this problem is that dpp does not have any knowledge of the locations of predefined feedthroughs in the leafcells, and as a result the data net routing is inefficient. This problem occurs because the only way for dpp to obtain physical information about the blocks is by running TimLager in the estimation mode and by also modifying the tiling functions to pass the desired information to dpp.

10.6. SUMMARY

By exploiting unique characteristics of datapaths, layout topologies can be found that are very efficient. dpp is a *structure-processor* which understands these optimizations and generates the required information and netlist modifications to generate datapath designs.

REFERENCES:

[Burns87] J. Burns, A. Casotto, M. Igusa, F. Marron, F. Romeo, A. Sangiovanni-Vincentelli, C. Sechen, H. Shin, G. Srinath, and H. Yaghutiel. *Mosaico: An Integrated Macro-cell Layout System*. VLSI Conference, IFIP, August 1987, pp. 133-149.

[Kernighan70] B. W. Kernighan, and S. Lin. *An efficient heuristic procedure for partitioning graphs*. Bell System Technical Journal, February 1970, vol. 49, pp. 291-308.

11

Pad Routing

Erik Lettang

The last stage of an integrated circuit design usually is to place a bonding-pad ring around the "core" portion of the chip and connect the nets from the ring to the core. The specialized nature of this routing problem requires algorithms optimized to the task, which are implemented in a program called Padroute. The pad ring is formed by one to four padgroups, each of which is a row of bonding pad leafcells from the pad libraries and assembled by TimLager. The bonding pads provide input buffering, input protection, and output drive capability for the chip. If desired, the pads can also implement boundary scan. Normally the pad ring is functionally transparent.

11.1. ROUTING STRATEGY

Figure 11.1 shows the general layout of a chip. The ring structure is employed to help prevent latch-up, which is most critical in the high-current CMOS devices present in the pad leafcells. N+ and P+ guard rings are placed around the devices in each pad. These rings are in turn connected to power and ground rails that appear on each side of the pad. The rails are connected by abutment to form the ring shown in Figure 11.1. Special power and ground pads feed current into the rails and to the core of the chip.

Padroute uses the *space* pads (cells with only power and ground rails) to equalize the size of mismatched padgroups or to build up an entire padgroup if it was not provided. Similarly, Padroute uses the *corner* pad pieces to complete the ring and connect the power and ground rails of the padgroups.

11.2. USER INPUT

The input to Padroute is the OCT structure_instance view of the top level of the chip hierarchy. This view must contain references to the structure_instance

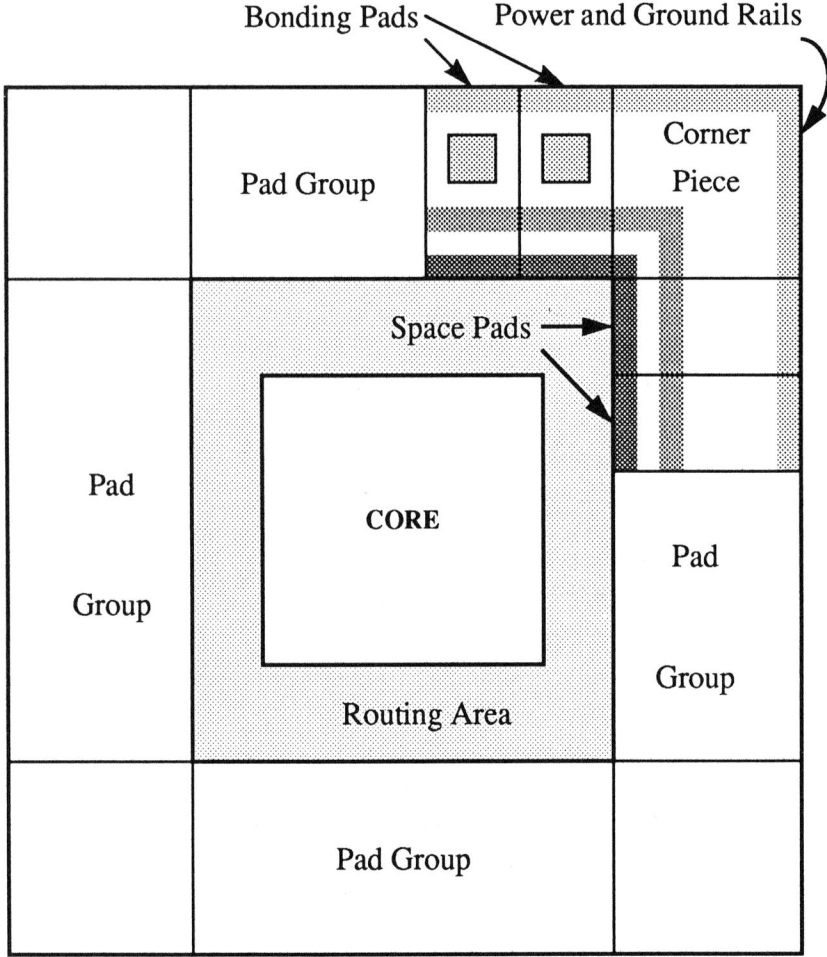

Figure 11.1: Padroute's View of an Integrated Circuit.

views of all of the cells in the next lower hierarchical level, consisting of the core and one to four padgroups.

Padroute assumes that the bonding pads have been assembled into padgroups by another layout generator (normally TimLager). Padroute also requires access to the OCT physical views of each chip subcell. The physical view contains the actual dimensions of all subcells and the locations of all terminals.

Padroute's command line must specify which the pad library was used to build the padgroups. With the library known, Padroute accesses the OCT physical views of the space pad and the corner piece. Padroute will use instances of the space pads, as necessary, when the chip core or the routing is too large to fit inside the ring. Space pads will also be used, if padgroups on opposite sides of the core are not the same length.

The proper filesystem path to the space pad and corner piece must be known so that both may be assembled into the final result. The paths are found using calls to special routines (Getpath) provided by LAGER. Padroute uses other LAGER utilities to determine the technology in which the chip is being designed. The technology specifies minimum wire widths and inter-wire spacings of the metal lines used to create the routing.

11.3. ROUTING ALGORITHM OVERVIEW

Padroute operates in two stages. The first stage is a simple placement of the core with the bonding pad ring arranged around it. The second stage connects together the signal pins that enter the channel, as specified by the structure_instance view. The second step is performed by Padroute's ring-shaped-channel router.

Padroute's router uses many of the algorithms described by Yoshimura and Kuh [Yoshimura,Kuh82], though tailored to function with a ring-shaped channel. The fundamental difference is that Padroute's channel is continuous and has no ends. The concept of tracks is as shown in Figure 11.2: The tracks form an unbroken circle around the core.

To operate in this topology, the vertical constraint graph of Yoshimura and Kuh is altered slightly (and given the name *radial constraint graph*), but both actually contain the same information. A major deviation from Yoshimura and Kuh is the introduction of a *circumferential constraint graph*. This graph shows which nets are not allowed to appear together in the same track. The radial con-

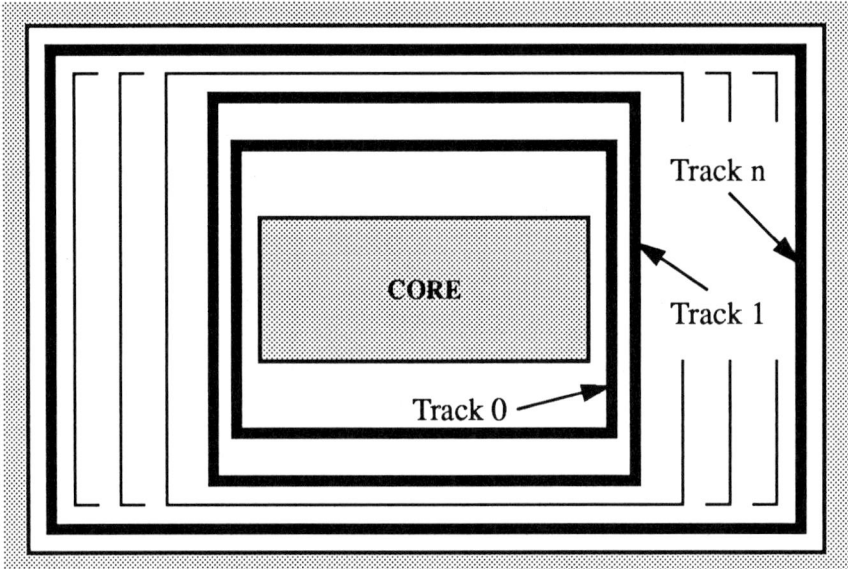
Figure 11.2: Circumferential tracks used in routing

straint graph is usually quite sparse, while the circumferential constraint graph is usually very dense.

The second major deviation from Yoshimura and Kuh is that Padroute's router is gridless, allowing terminals to appear at any point on the core or pad ring.

The radial constraint graph is used to detect cycles in the radial constraints. To remove them, Padroute adds doglegs to the nets involved. After all cycles have been removed, Padroute uses the circumferential constraint graph to combine multiple nets into single tracks. Then the physical geometries are generated and stored in an OCT physical view

11.4. PLACEMENT ALGORITHM

Padroute's placement algorithm is not very complicated. Every cell passed to Padroute has a parameter called "fplan" attached to it. The chip designer specifies fplan when creating the SDL files for the chip, the padgroups, and the core. As shown in Figure 11.3, the four padgroups are rotated and translated as necessary to form a ring. The core is then translated so that it is in the center of the ring.

Chapter 11   Pad Routing   145

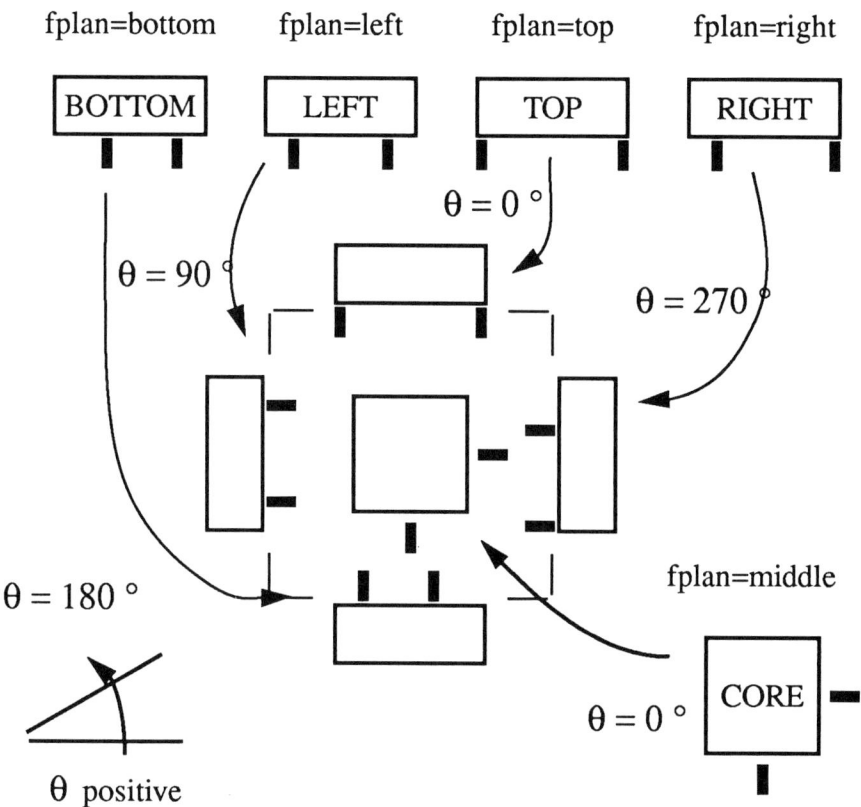

Figure 11.3: Rotation and Translation

If the core is too large to fit into the ring as provided, space pads are added so that the core will just fit inside the ring. Later, after the routing of the nets has been determined and if the padring is again found to be too small, more space pads are added so that the routing will fit inside the ring.

Space pads may be added after the routing is determined without any recalculation of the terminal positions by introducing the concept of fixed and sliding coordinates. The core is fixed in both the X and the Y direction. The padgroups are fixed in either the X or the Y direction. If the padring needs to be enlarged, the positions of the padgroups may be altered by sliding them away from the core (see Figure 11.3). Altering the positions of the padgroups in this way allows the channel width to be increased without requiring any routing to be redone.

11.4.1. Channel Cleanup

Once placement is complete, `Padroute` cleans up its net and terminal database. At this point `Padroute` performs the following operations:

1. Deletes invalid terminals. These are terminals not appearing on the bottom edge (inside) of a padgroup or one of a pair of terminals that are on the same net and appear in the same location.
2. Deletes invalid nets. An invalid net is one with only one valid terminal (the terminal is then deleted also).
3. Calculates the width of each net.
4. Finds the terminals that appear in the corner regions and finds the minimum distance from the core that the terminal's net may be routed.

`Padroute` considers all nets to be in one of three classes: signal, clock or power. The final routed width of a net is determined by the class it is in. Signal nets are always routed with the narrowest wire width. A clock net's width is determined by its widest terminal. A power net's width is the largest of the following values: the width necessary to handle the net's rated current (the sum of all the terminal currents), the width of the widest terminal on the net, or the default minimum wire width for a power net. The current through all terminals must be specified in the OCT views if the current summing feature is used. `Padroute` will check the terminals and sum their currents to calculate the width of each power net.

For clock and power nets, the radial wire that connects a terminal to the circumferential wire is never wider than the terminal it is coming from, but it may be narrower: The terminal on the pad ring for a power net is usually quite large. The width is usually not necessary for current requirements, and `Padroute` may make the radial wire narrower than what the terminal allows. In fact, the radial wire of a power net that connects to a pad is kept the same width as the circumferential wire for that net. Reducing the width of the radial wire reduces the number of nets that will have radial constraints with the power net. This makes the routing task easier later on.

In `Padroute`, terminals that appear in the corner region (the shaded areas in Figure 11.4(a)), and the nets that contain them, must are given special treatment. The reason they are special can be understood by studying Figure 11.4(b). This picture gives a view of a ring channel that has been unwrapped. The trapezoidal areas are part of the real channel. The triangular, darkly shaded, areas are not usable by the router since they do not actually exist in the real channel. Nets

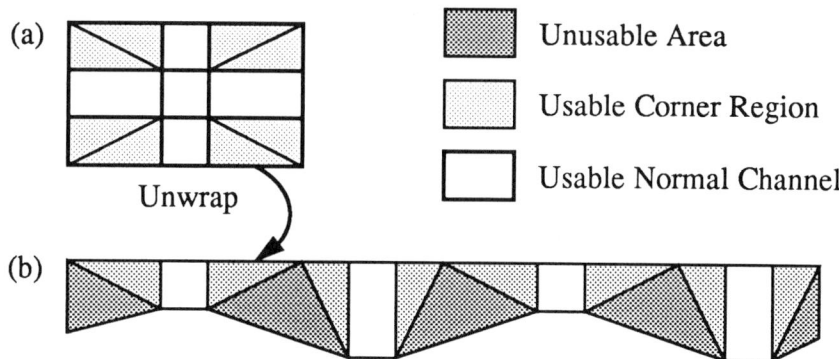

Figure 11.4: Corner Regions (a) and Unwrapping the Channel (b).

that just go around corners effectively cross the striped areas without actually occupying them, but nets that must have radial wires over the shaded areas (i.e. nets that have terminals in the corner regions) cannot enter them. This defines a minimum distance from the core at which a net with a corner terminal may be routed. Padroute uses separate values for the X and Y minimum radii. It is possible to define four different values for minimum radius depending on which part of the channel the net leaves for the corner, but this was not done in Padroute. The minimum radius is illustrated in Figure 11.5. A track which contains multiple corner nets will inherit the largest of the minimum radii. Tracks that do not contain corner nets have a minimum radius of zero.

11.5. ROUTING ALGORITHM

11.5.1. Organizing Terminals on a Net

Given a minimum radius, Padroute tries to route each net in the shortest distance. Padroute measures the distance between two terminals by projecting them to the outer edge of the channel and measuring the clockwise distance between them. Projection and clockwise distance is illustrated in Figure 11.6.

The ordering of the terminals on each net is determined as follows. A terminal in the net's terminal list is selected. The clockwise distance from that terminal to every other terminal on the net is calculated. The terminals are then put into the net's terminal list in the order of increasing clockwise distance. The next step is to measure the clockwise distance between each pair of adjacent terminals. The largest distance between adjacent terminals is found. The terminal on the clock-

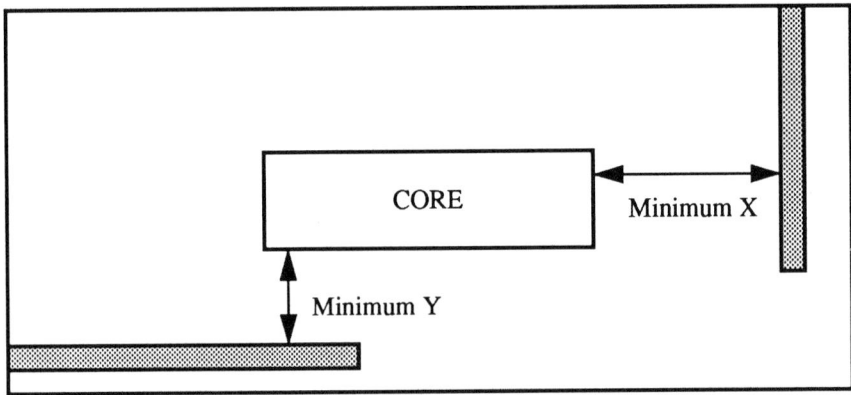

Figure 11.5: Minimum Radius of a Net.

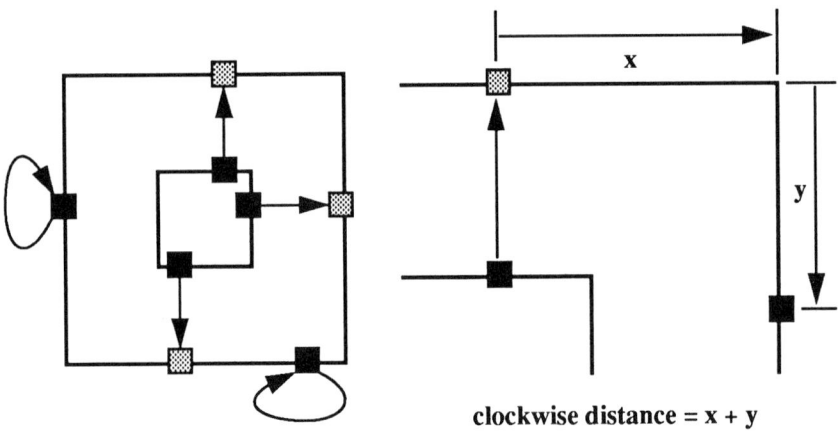

Figure 11.6: Mapping Terminals and Clockwise Distance.

wise end of this gap is made the first terminal in the net's terminal list. The terminal on the counterclockwise end of the gap is made the last terminal in the list. With the terminal list set up in this way, if the net's wire starts on the first terminal and runs clockwise to the last terminal, the total wire length is minimized, given a value for minimum radius. Figure 11.7 illustrates the terminal arrangement for two nets.

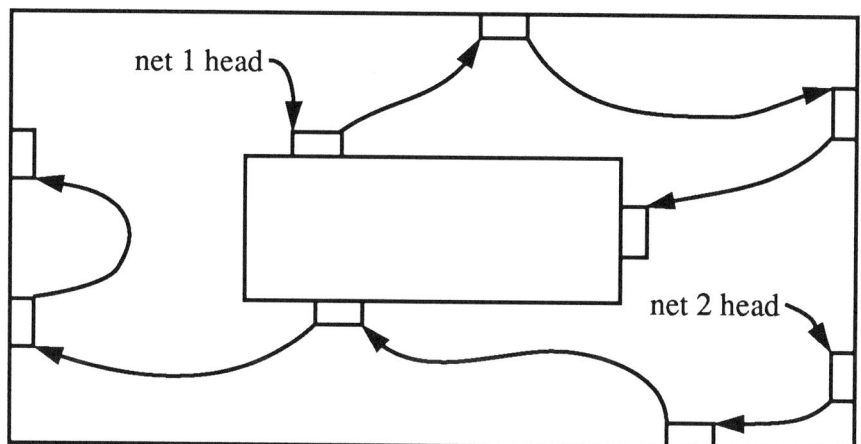

Figure 11.7: Arrangement of Terminals on Nets.

11.5.2. Radial Constraint Graph

The radial constraint graph is the equivalent of the vertical constraint graph in an ordinary channel router. In Padroute, the core is considered to be the top and the pad ring is the bottom of the channel. Even though the graph is formed for the entire channel, Padroute generates the graph by comparing the terminals in the top, bottom, left, and right sections independently. The terminals on the top edge of the core are compared with all of the terminals on the top padgroup. Similarly, the terminals on the bottom edge of the core are checked compared to the terminals on the bottom padgroup. The terminals in the left and right channels are checked in the same way. Figure 11.7 shows a sample channel and the radial constraint graph that is generated for it.

11.5.3. Circumferential Constraint Graph

The circumferential constraint graph shows which nets are not allowed to occupy the same routing track. A pair of nets are not allowed to coexist in the same track if any one of the following conditions is true:

1. Both nets are constrained with each other radially
2. Either net has a terminal in one of the corner regions
3. dist(net1.firstterm, net2.firstterm) + dist(net2.firstterm, net1.lastterm) < circumference of the inner edge of the pad frame

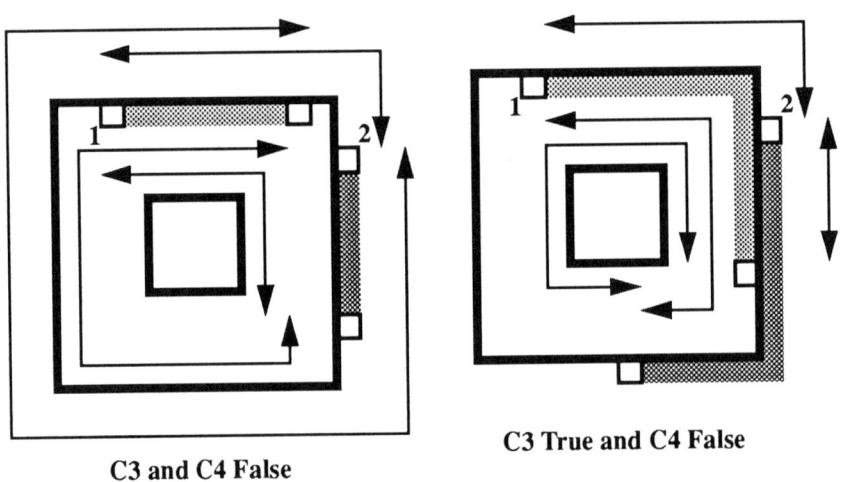

C3 and C4 False   C3 True and C4 False

Figure 11.8: Examples of conditions 3 and 4

4. dist(net1.firstterm, net2.lastterm) + dist(net2.lastterm, net1.lastterm) < circumference of the inner edge of the pad frame

Here, dist(terminal 1, terminal 2) means the *clockwise* distance from terminal 1 to terminal 2 after the positions of the terminals have been projected to the outer edge of the pad frame.

Examples of situations where rules C3 and C4 apply are given in Figure 11.8. The examples show the positions of the terminals at each end of the net. The distances are shown graphically, no values are given. The diagrams assume that all terminals have been projected to the outer edge of the channel. C3 is diagrammed on the outside of the circle. C4 is diagrammed on the inside of the circle. The arrows depict the distances calculated by the clockwise distance function. If any of the tests C1 through C4 are true, then, in the circumferential constraint graph, a bidirected edge is created between the nodes representing nets one and two.

In summary, the radial constraint graph contains information describing the relative stacking order of nets in the radial direction. The circumferential constraint graph contains information about which nets are not allowed to occupy the same routing track since they would occupy the same region in the channel.

11.5.4. Cycle Removal

The radial constraint graph may contain from zero to several cycles. Padroute uses a number of routines to detect and summarize all of the cycles present. Details of the cycle detection functions may be found in [Lettang89]. The end result is a concise list of the cycles, which is then passed to the cycle removal functions.

Padroute's cycle removal techniques are rudimentary. They do no more than add doglegs to nets involved in cycles. To break one cycle, Padroute picks one of the nets involved and puts in a dogleg. Adding the dogleg breaks the net into two separate nets, which may then be treated independently throughout the rest of the routing process. Every time a dogleg is added, Padroute re-creates the radial constraint graph and checks it for cycles. Padroute follows the procedure until no more cycles are found. The procedure just described has a possibility of not completing, so Padroute counts the attempts and will give up after a large number of iterations.

The procedure for placing a dogleg into a net is as follows. Every net involved in a cycle has at least two terminals that are constrained with other terminals on other nets in the cycle. Padroute selects, at random, a net involved in the cycle it is currently breaking. It locates the two terminals that cause the constraints in the cycle. Starting from the most counterclockwise terminal, Padroute searches for an opening to place the dogleg. A valid opening is one in which the radial jumper that forms the dogleg may be added to the channel *without adding an additional constraint* to the radial constraint graph. If such an opening cannot be located, Padroute discards the net and selects another one. Valid openings are generally available due to the nature of Padroute's task: there are many regions along the channel that are not populated by terminals at all. It is possible to produce a channel with a cycle that Padroute cannot break. However, such channels do not appear very often in practice.

Once a dogleg has been added to a net, Padroute views that net as two separate nets. A terminal is added to each new net where they were once connected together. The dogleg connects the two terminals, maintaining an electrical connection. The new terminals maintain pointers to each other to keep track of the end points of the dogleg.

11.5.5. Track Assignment

Figure 11.2 shows the routing tracks in which `Padroute` places the nets. The tracks are of variable width and run unbroken around the core of the chip. A feature of the channel is that a given track may be of varying distance from the core.

On the first pass of the track assignment phase, `Padroute` places one net per track. The track index acquires the net index. This preserves the indices of the constraint graphs. To help reduce the wire length of each net, `Padroute` initially assigns the longest nets to tracks closest to the core. The tracks are stored as a linked list, the first track in the list is closest to the core the last track is farthest from the core. At this phase, no attention is paid to either of the constraint graphs.

The second pass of track assignment is to merge multiple nets onto the same track. The algorithm starts at the first track and tries to pull nets from other tracks onto it. Each track farther down on the list is checked, in turn. Once the other tracks have been checked for inclusion, `Padroute` moves to the next on the list and fills it in the same way. This process continues until the track list runs out. Before two tracks can be merged, `Padroute` checks the circumferential constraint graph to make sure that the nets in the track will not overlap. `Padroute` also checks to make sure that merging the nets will not generate a cycle in the radial constraint graph. Ordering to satisfy the radial constraint graph is not verified until the next step.

The third pass of track assignment is to make sure that none of the tracks violate the radial constraint graph. This algorithm simple traverses the track list and makes sure that the successors of a track in the constraint graph also appear farther down in the track list. Any tracks for which this is not true are placed after their successor in the track list.

The final pass of track assignment is to set the physical location of the track in the channel. This step may rearrange the track order, but does not add to or violate any constraints in the radial constraint graph. This phase also pulls all the tracks as close to the core as possible.

As described in Section 11.4.1. (Channel Cleanup), each net has a minimum distance from the core. When a net was assigned to a track, that track inherited that net's minimum distance. When multiple nets were assigned to a track, that track's minimum distance became the largest minimum distance of all the nets in the track. A track may not be pulled closer to the core than its minimum radius. If the radius of each track was assigned paying attention only to the minimum radius of each track without rearranging the order of the tracks, the routing would contain unnecessary gaps between tracks. The final phase of track assignment places the tracks according to their minimum radii and moves the tracks that can move closer to the core.

To perform the final step, two interdependent algorithms are implemented. The first algorithm traverses the track list. It keeps two accumulation variables (one each for the X and Y axes). These variables maintain the value of the radius from the core that the current track should occupy. The accumulation variables are compared to the minimum radius of the track being processed. If the difference is large enough in both the X and Y directions (for a minimum width net), the second algorithm is used to select and insert into the gap tracks from farther down in the track list. When the second algorithm is finished filling the gap, the accumulation variables are updated to the minimum radius of the current track. The next track in the list is then processed.

The second algorithm operates as follows. It is given the size of the gap to fill and the current track from the first algorithm. It traverses the track list from the current track. A track which satisfies the following three conditions will be moved to fill in the gap:

1. The track fits physically into the gap.
2. Moving the track into this position will not violate the radial constraint graph.
3. The minimum radius of the new track will allow it to be placed into the gap.

The second algorithm completes when either the gap is filled or no tracks remain in the list. When the first algorithm runs out of tracks to be processed, all the tracks are fixed in position. All unknowns have been resolved and `Padroute` can report the routing.

11.5.6. Output and Cleanup

When nets have been assigned to tracks and tracks have been assigned positions in the channel, `Padroute` is essentially finished with routing. All that remains is to produce the rectangles of the two layers of metal and the vias that actually implement the routing. `Padroute` produces output in either OCT or `Magic` layout editor format.

Padroute traverses the net list and calls functions that produce either radial connections or circumferential connections between terminals. A radial connection will connect a terminal on a pad to the circumferential track or connect the two pieces of a net split up by a dogleg. The circumferential terminal connector works like the distance functions described above, but it does not report the distances. It does produce the proper rectangles in metal.

Padroute stores all the routing geometries internally before writing them out in either Magic format or in OCT. Padroute's final step is to write formal terminal information into the OCT physical view.

11.6. SUMMARY

Padroute uses specialized algorithms to perform the final step in generation of a complete chip. By modifying standard rectangular channel routing algorithms, it is possible to provide efficient solutions for the case of a circular channel surrounding a rectangular core.

REFERENCES:

[Lettang89] E. R. Lettang, "Padroute: A Tool for Routing the Bonding Pads of Integrated Circuits," Masters Thesis, Electrical Engineering Department, University of California, Berkeley, T7.49.1989.L436 ENGL

[Yoshimura,Kuh82] T. Yoshimura and E.S. Kuh, "Efficient algorithms for Channel Routing," IEEE Trans. on CAD of ICs and Systems, vol. CAD-1, n. 1, pp. 25-35, January 1982.

Part III

Verification and Testing

12

Design Verification

Wun-Tsin Jao and Rajeev Jain

Automatic layout generation of an integrated circuit, using a framework such as LAGER, should normally result in an error free design. However, in practice there are two possible sources of error. First, there is always a possibility that a layout error is introduced by a bug in one of the tools in the system, since users can integrate their own tools to the framework. Second, the designer may choose to manually alter the routing to achieve a better result than the automatic routers (a procedure which is strongly discouraged). In both cases it is desirable to check that the final layout actually implements the desired circuit as described by the SDL input description (see Chapter 3).

This chapter describes the approach used in LAGER for layout verification and gives some details of the program `DMverify` which can be used to verify designs generated using `DMoct`.

12.1. VERIFICATION METHODOLOGY

12.1.1. Logic Simulation

The silicon assembly is completely driven by the input netlist specified by the designer using the structure description language SDL. This description is stored in the OCT database and can be verified for functional correctness by simulating with THOR (see Chapter 13). The THOR simulation is based on models for each individual library cell used in the circuit. If these models are correct the THOR simulations can completely check the logic or functional behavior of the circuit description.

The correctness of the THOR models can be verified a priori for each cell and need not be repeated for each design. When a cell is created in the library, its layout is verified by extracting a transistor netlist (with Magic) and performing SPICE simulations. The THOR model for the cell can then be verified by checking that it conforms with the actual behavior of the cell layout as predicted by SPICE.

If a circuit description has been verified by THOR and the cell layouts have been verified by SPICE, then the only source of error in the layout generated with LAGER is in the interconnection of cells created by the automatic routing tools (and any subsequent manual modifications). The layout verification problem then reduces to checking that the interconnection of cells in the layout is identical to that described by the SDL circuit description. This layout connectivity check can be done with simulation or netlist verification techniques as discussed below.

12.1.2. Layout extraction and simulation

Layout connectivity can be verified by extracting a netlist of transistors from the layout and then performing circuit simulation with a switch-level simulator such as IRSIM (see Chapter 13). If the simulation results are not as expected one can conclude that a connectivity error occurred. However, the simulation gives no information as to the location of the error in the layout. Iterative simulation to detect the location of the error can be very time consuming.

An alternative for verifying the layout connectivity is to extract the connectivity of the library cells in the layout (as opposed to extracting connectivity of the transistors) and to compare this with the interconnect specified in the SDL input. With this approach, a chip designer needs to perform only logic-level simulation (with THOR) to verify the SDL input description and netlist comparison to check the layout. Circuit-level or switch-level simulation of the entire IC layout is unnecessary except for performance verification.

This approach works especially well for cell library based systems like LAGER, where the circuit design of library cells has been fully verified by cir-

Chapter 12   Design Verification   159

cuit-level simulation. The combination of functional simulation of the input circuit description along with netlist comparison of the resulting layout is the verification methodology presented in this chapter.

12.1.3. Routing Verification

The task of comparing a netlist extracted from a layout against its specification is known by various names such as network comparison, netlist comparison, interconnection check and connectivity audit. In this text, the term routing verification is adopted, because the purpose of this task is to verify that the routing done by layout generators in the silicon assembly phase is identical to the user specification.

12.2. NETLIST COMPARISON TECHNIQUES FOR VERIFICATION

To verify the routing of a design, netlist comparison must be performed between the actual layout and the original SDL netlist. Two approaches are generally employed for netlist comparison methods. One is to make use of labels placed on the layout corresponding to the node names in the reference circuit to compare the connectivity locally. The other is to apply graph isomorphism algorithms to compare the connectivity globally. The second approach is a more general one which does not require any pre-processing for assigning names to nodes in the circuitry. A lot of effort has been devoted to this second approach, as can be seen in systems like GEMINI [Ebz83], CCOMP [Tak82] [Tak88], YNCC [Shi86] and NECOM [Bar84].

However, due to the nature of the LAGER design environment, where all the terminals are named uniquely both in the user input schematic and in the actual layout, the first approach is adopted in this project. Graph isomorphism based approaches are essentially an overkill for the routing verification problem.

12.2.1. Layout Hierarchy

One major feature of the LAGER circuit design tool set is that entire chips are built up hierarchically from primitive cells. For those designs built up in a hierarchical cell structure, the simulation and verification tools should try to exploit the hierarchy. Unfortunately, all four of the verification systems mentioned above work on flattened layout only. This was advantageous for classical full custom handcrafted designs. Since LAGER designs are cell based and hierarchical, flattened netlist verification is both unnecessary and inappropriate, since the reference circuit (described by SDL) is a hierarchical cell netlist, rather than a transistor or gate netlist. To use conventional techniques mentioned above, the user input description has to be flattened. By contrast, the routing verification in LAGER is done hierarchically to exploit the approach that designs are assembled.

12.2.2. Labeling of Nodes and Nets

To compare the netlist extracted from the physical layout with the original netlist described in SDL, the hierarchical circuit must first be extracted, and the cells must be identified. During the identification of pins, gates, and modules, elements could be grouped as transistors/gates/modules. Some systems may also take care of pin permutation (Wombat [SpN83] [Spi83]) or functional isomorphism (YNCC [Shi86]). Most of these isomorphism algorithms are based on a signature calculation, where the signature of an element (a node, a net, or a device), given to identify elements uniquely, is usually a combination of information about the element itself and its neighbors. Some systems may allow elements to be labeled, where a label is the name of an element.

Since the LAGER policy ensures that all signal nodes are uniquely named in the original schematic netlist as well as in the physical layout, no extra effort need be devoted to assign unique labels or signatures to electrical nodes which appear in the original netlist.

However, a problem which does arise in LAGER is that nodes may be created in the layout that are not declared in the input netlist. These are nodes that are not created by routing but by implicit connections such as abutment and overlap of cells. The layout extractor assigns unique names to these nodes as well. Thus the extracted circuit contains two sets of unique node names a) those that correspond to nodes in the user specified netlist and b) those that correspond to nodes created by the layout generators without routing. For the comparison to work, the verification tool must distinguish between these two types of node names and only use those in set (a) for comparison. The nodes in set (b) usually do not need to be verified but they should be listed for visual checking by the user.

12.2.3. Comparison Scheme

Once the cells and nodes have been identified, the final step of netlist comparison is to compare the connectivity of the nodes from the extracted layout against the netlist specification of the original design. For those systems exercis-

Chapter 12　　　　　Design Verification　　　　　161

ing graph isomorphism algorithms (such as Gemini, CCOMP, YNCC, NECOM), the entire circuitry must be flattened and be represented by a graph at first. During the process of matching elements between two graphs (one from the extracted layout and the other from the design specification), some systems (such as Wombat, Gemini and NECOM) partition the graph repetitively until matching is found, while some others (like YNCC) group vertices of the graph repetitively until matching is found. These graph-theory operations constitute an important part of the system.

For the net comparison in LAGER, no grouping or partitioning is necessary, since nodes are well-labeled from the beginning and hierarchy is exploited. On the other hand, a hierarchical matching algorithm is needed to identify nets in the extracted layout and match them against nets in the user input.

12.3. ROUTING VERIFICATION

The task of routing verification is divided into three sequential steps (Figure 12.1):

1. Extract netlist information from the actual layout using conventional layout extraction methods.
2. Reconstruct a hierarchical netlist of library cells from the extracted data and record it back into the OCT database using an "extract" view (see below).
3. Compare the netlist in the extract view against that in the structure instance view.

These three steps are performed by three programs: `phy2ext`, `ext2oct` and `netdiff`, respectively. A master program, `DMverify`, integrates the flow control of the three steps in sequence.

In the first step, `phy2ext` reads the actual layout and uses `Magic` for extraction. Since the `Magic` layout does not contain all of the OCT database information, names of cell instances and the paths to leafcells are recovered from the OCT physical view.

In the second step, `ext2oct` constructs the hierarchical netlist from the extracted data and stores it back into the OCT database by creating an "extract" view. In some sense, the extract view is actually a subset of the structure_instance view and contains only cells, terminals, and nets. However, since names of nets are not recoverable from the layout, this extract view may not have the same net

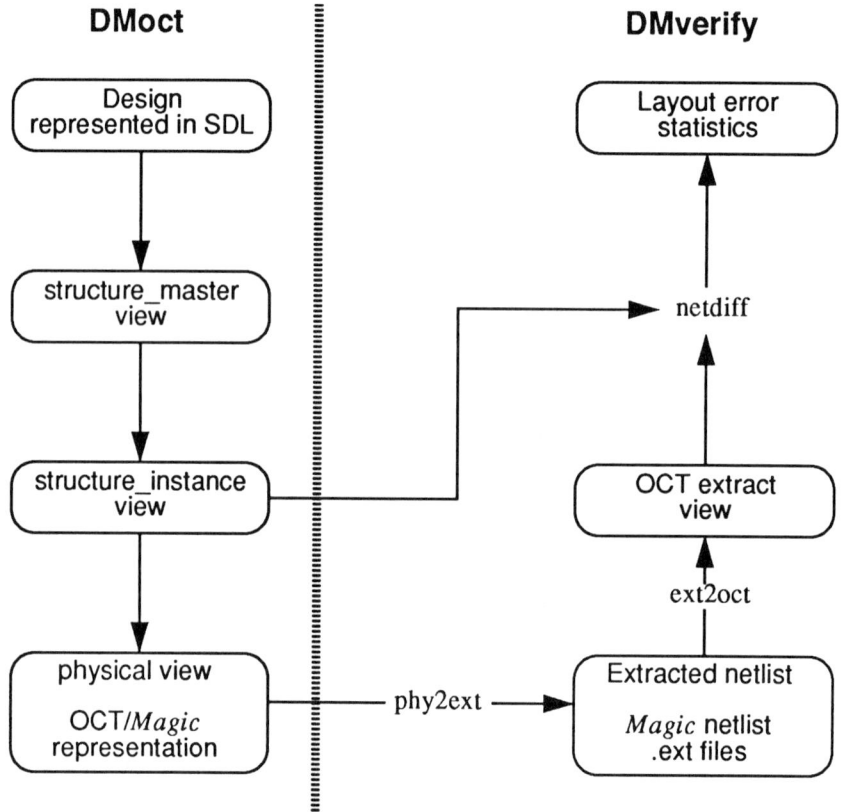

Figure 12.1: Routing verification

names as in the user input. Therefore the nets have to be compared by their composition.

In the last step, netdiff reads the connectivity described in both the structure instance view and in the extract view separately, and compares them. Being based on the concept of *nets*, this program matches nets hierarchically in the two views and checks if a net in the extract view has an identical counterpart in the structure instance view (and vice-versa).

12.3.1. Routing Information from the Extract Files

After extraction, a two-phase process is carried out to read routing information from `Magic` extract files. The first phase is a top-down traversal to record cell instances as well as electric nodes (terminals) in the internal data structure of `ext2oct`. This step is performed in a top-down manner since child cells are not known until consulting the extract file of the parent cell. The recording of nodes could be performed in a separate (top-down or bottom-up) phase if desired.

The second phase is a bottom-up traversal to record all the two-terminal connections described in the extract files. This is done in a separate stage since the connections can not be formed until all the cells and nodes are built up in the internal data structure. The second phase could be performed either top-down or bottom-up if only to build up these two-terminal connections. However, they are performed bottom-up to simultaneously convert the net representation, as will be discussed below.

12.3.2. Hierarchical Netlist Extraction

`Magic` extract data describes the routing by a sequence of connections between pairs of electric nodes (terminals). Such a connection is called a *link*. A multiple-terminal net is described implicitly by a collection of such links. Thus net information is not available explicitly and must be recovered.

A net consisting of n terminals is described by $n-1$ links in the extracted circuit. The links belonging to a net are created in a "chain" style, as shown in Figure 12.2(b). But, when creating the extract view for netlist comparison, it is preferred that these links are arranged in a "star" style as shown in Figure 12.2(a).

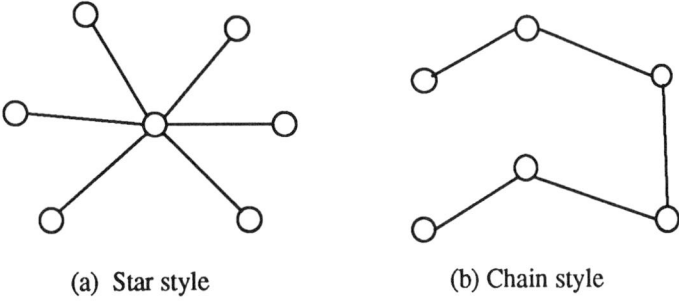

(a) Star style   (b) Chain style

Figure 12.2: Net representations (a) in OCT and (b) in `Magic` extract files.

This is more consistent with the OCT policy for representing nets. In the first phase of transformation, a general procedure constructs and removes links to convert the links of each net to a star style.

Taking advantage of the hierarchical nature of Magic extract files, the conversion can be performed in a bottom-up manner as the links are built up in the internal data structure. Whenever the bottom-up traversal stops at a certain cell, all the links specified by the Magic extract file of that cell are implemented in the internal data structure. With all of the links that have been constructed to that point, every set of chain style links associated with the currently traversed cell and all of its child cells is converted to an equivalent set of star style links. One of the typical examples is to convert a net as shown in Figure 12.3(a) to a net as shown in Figure 12.3(b).

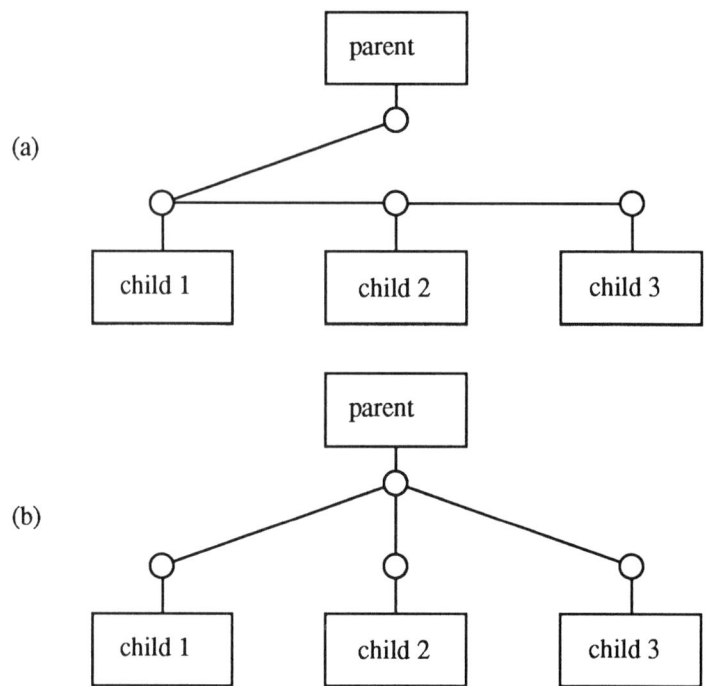

Figure 12.3: Hierarcical net representation (a) in an extract file, and (b) in an OCT extract view.

This kind of transformation should be performed in a bottom-up manner. Otherwise, a connection as shown in Figure 12.4(a) would be transformed as shown in Figure 12.4(b) in a separated top-down manner, which is certainly not what the structure instance view describes.

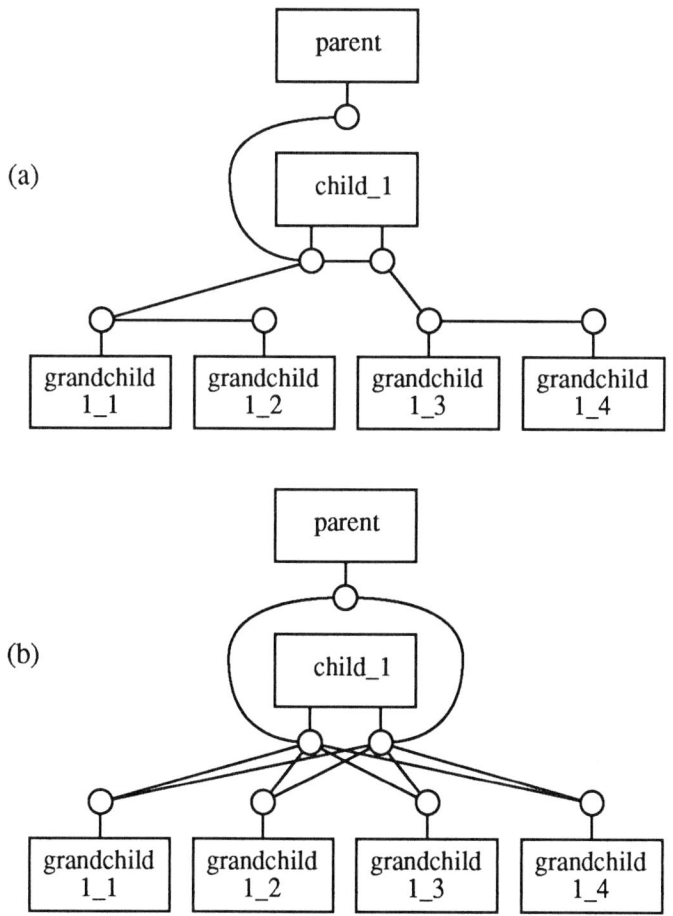

Figure 12.4: Undesirable top-down conversion of net in (a) to the representation in (b). Bottom-up traversal leads to the desired representation.

12.3.3. Isolating Implicit connections

Due to implicit connections created in the process of tiling macro cells, some of those links read from Magic extract files do not exist in the structure instance view since they are not part of the user input description (SDL). They are called *implicit links* and are not implemented in the extract view. The implicit connections always come from the abutment or overlapping of geometric rectangles in physical layout when blocks are tiled. Figure 12.5(a) shows an example of

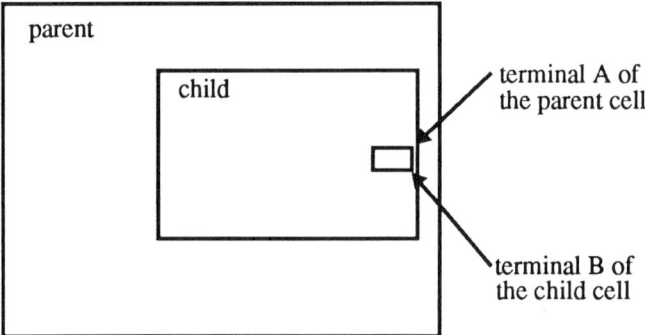

(a) An implicit connection of two terminals with overlapping geometry

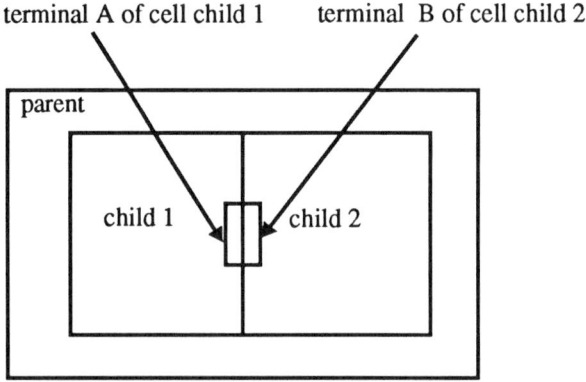

(b) An implicit connection of two terminals with abutting geometry

Figure 12.5: Two types of implicit connection

Chapter 12   Design Verification   167

implicit connections coming from geometric overlapping. Figure 12.5(b) shows a typical example of implicit connections coming from geometric abutment. Implicit links can not always be ignored since some of them may come from misrouting of layout generator programs. Therefore they are reported in a text file for users to examine manually.

12.3.4. Creating the Extract View in OCT (ext2oct)

Once all the links in the `Magic` extracted data have been processed, the hierarchical nets corresponding to cell interconnections have been identified and implicit connections isolated, the netlist information is stored in the extract view of the OCT database. All the cells, nodes, and connections are recorded in the contents facet of the extract view. Ideally all the cells, terminals and nets in the extract view would be identical to what can be observed in the structure instance view, except that names of nets could be different.

12.3.5. Netlist Comparison for Verification (netdiff)

To verify correctness of the routing, the connectivity in the extract view must be compared against the connectivity in the structure instance view. Names of nets are insignificant in the comparison since only the connectivity among terminals are of concern. There are four categories of mismatches that netdiff will identify:

1. A cell that appears in only one of the two views under comparison is reported as one mismatch.
2. A terminal that appears in only one of the two views under comparison is reported as one mismatch.
3. If there are nets in both views that have some terminals in common but not all. It is reported as one mismatch.
4. A net that appears in only one of the two facets under comparison, being unable to find a counterpart in another facet, is reported as one mismatch.

For example, as shown in Figure 12.6, say, a net N may contain terminals A, B, C and D in the first facet as shown in Figure 12.6(a), while N may contain only three terminals, A, B and C, in the second facet as shown in Figure 12.6(b). In this case, a mismatch is reported.

Netdiff first builds up two sets of internal data structures to record subcells, terminals and nets of the two facets, one set for each facet. With lists of subcells and lists of terminals pre-sorted alphabetically when the internal data structures are built up, mismatches of the first/second category are identified by performing a synchronous linear traversal on the two subcell-lists/terminal-lists to match pairs of subcells/terminals. Two subcells/terminals, one in each facet, are matched as a pair if they are named identically. This is based on a fundamental assumption that names of subcells and terminals are kept undistorted in the physical layout, which is what LAGER does. Any subcell un-matched is reported as a first-category mismatch. Any terminal un-matched is reported as a second-category mismatch.

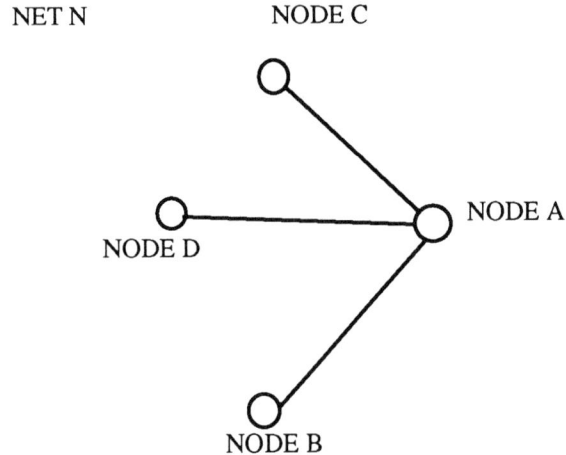

(a) A net N in one facet may contain four nodes, say , A,B,C and D.

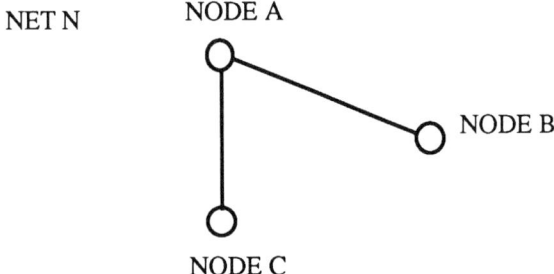

(b) It may contain only nodes A, B and C in another facet

Figure 12.6: Possible discrepency between two facets

To pair nets in the two facets, terminals attached to nets are examined. For each net to find a *partner* in the other facet, a linear traversal is performed on nets of the other facet. Two nets, one in each facet, having at least one terminal in common are defined as a matched pair and are marked to be exempted from further searching and matching. This matching scheme is based on a fundamental assumption that a terminal may belong to at most one net in a facet, which is a policy that LAGER holds. Any net un-matched is reported as a fourth-category mismatch. Any pair of matched nets not holding identical sets of terminals are reported as a third-category mismatch.

Since the design hierarchy is not completely recorded in the structure instance view, the contents facets of structure instance view are only available for user-defined cells. In other words, the structure instance view may not exist for library leafcells. Therefore, the connectivity comparison is performed only on user-defined cells. With the cell libraries being assumed to have been fully debugged, this does not cause any problem in the routing verification.

12.3.6. Net Comparison v.s. Facet Comparison

Although there is already a facet comparison tool called octdiff available with the OCT database, a the connectivity comparison tool netdiff was required since the loss of names of nets in the layout (and therefore the extract view) does not allow the comparison of nets by octdiff. This is because octdiff assumes two facets to be identical but facets in the extract view have only cells, terminals and nets. This may lead to several mismatches in the output of octdiff since any OCT object other than cells, terminals and nets would be reported as missing in the extract view. The major difference between netdiff and octdiff can be summarized as:

1. Netdiff checks only cells, terminals and nets. Any other OCT objects are ignored.
2. Netdiff checks only the connectivity among terminals, the names of nets are insignificant.

12.4. PERFORMANCE

All the programs are implemented on a 2 MIPS SUN 3/60 workstation where the whole environment of LAGER, OCT and Magic works. Listed above are the CPU time in minutes for processing designs of various sizes.

The overall complexity of DMverify (together with phy2ext, ext2oct and netdiff) is not clear without enough number of cases measured. However, the cases listed above give a rough idea that DMverify's overall complexity is worse than O(n), and better than O(n2), where n is the total number of links in the entire design.

no. of transistors	no. of levels	no. of cells	no. of nodes	no. of links	no. of nets	CPU time (min)	function of design
30705	3	2112	33335	21740	1274	1759	multiplier
11672	5	384	11787	6169	2816	320.6	filter
3584	4	452	6445	4070	1120	62.5	datapath module
1581	3	106	1590	678	172	8.27	random decoder
532	6	151	1223	911	213	3.51	processor
462	4	66	864	588	N.A.	1.52	latch
397	5	57	820	580	174	1.80	accumulator
322	4	43	629	419	121	1.53	multiplier

Table 12.1: Experimental Results

12.5. SUMMARY

The first stage in verification is to determine if the interconnection defined by the user is the same as that in the final layout. The netlist comparison programs described here are designed to efficiently perform this check.

REFERENCES:

[Bar84] E. Barke, " A Network Comparison Algorithm for Layout Verification of Integrated Circuits", IEEE Transactions on Computer Aided Design Vol CAD-3, No. 2, April 1984, pp 135-141.

[CIF80] C. Mead and L. Conway, "Introduction to VLSI systems", Addison-Wesley Publishing Co., 1980, pp 115-127.

[EbZ83] C. Ebeling and O. Zajicek, "Validating VLSI CIrcuit Layout by Wirelist Comparison", Proc. 1983 International Conference on Computer Aided Design, September 1983, pp 172-173.

[Lag88] Electronic Research Laboratory, University of California, Berkeley, "LagerIV Distribution 1.0 Silicon Assembly System Manual", June 1988.

[Lob84] C. Lob, "RUBICC : A Rule Based Expert System for VLSI Integrated Circuit Critique", UCB/ERL Memo M84/80, UC Berkeley Electronics Research Laboratory, September 28, 1984.

[LSN85] C. Lob, R. Spickelmier and A.R. Newton, " Circuit Verification using Rule based Expert Systems", Proc. 1985 International Symposium on CIrcuits and Systems, Kyoto, Japan, June 5-7, 1985, pp 881-884.

[Mag86] Electronic Research Laboratory, University of California, Berkeley, "1986 VLSI Tools : Still More Works by the Original Artists", Report No. UCB/CSD 86/272, December 1985.

[Oct88] Electronic Research Laboratory, Univerisity of California, Berkeley, "Oct Tools Distribution 2.1", March 1988.

[Pap88] A.C. Papaspyridis, " A Prolog Based Connectivity Verification Tool", Proc. 1986 International Conference on Computer Aided Design, pp 298-301.

[Shi86] Y. Shiran, "YNCC : A New Algorithm for Device Level Comparison between Two Functionally Isomorphic VLSI circuits", Proc. 1986 International Conference on Computer-Aided Design, pp 298-301.

[Spi83] R.L. Spicklemier, " Verification of Circuit Interconnectivity", UCB/ERL Memo 83/66, UC Berkeley Electronics Research Laboratory, October 21, 1983.

[SpN83] R.L. Spicklemier and A.R. Newton, " Connectivity Verification Using a Rule Based Approach". Proc. 1985 International Conference on Computer Aided Design, Santa Clara, USA, November 18-21, 1985, pp 544-550.

[Tak82] M. Takashima, T. Mitsuhashi, T. Chiba and K. Yoshida, "Programs for Verifying Circuit Connectivity of MOS/LSI Mask Artwork", Proc 19th Design Automation Conference, June 1982, pp 544-550.

[Tak88] M. Takashima, et. al., " A Circuit Comparison System with Rule Based Functional Isomorphism Checking", Proc. 25th Design Automation Conference, June 1988, pp 512-516.

13

Behavior and Switch Level Simulation

Lars Svensson, Lars E. Thon and Seungjun Lee

One of the most important characteristics of the LAGER set of tools is the tight coupling between the specification of the design and the simulation of that description. There are several levels of simulation which are employed. At the lowest level simulation is performed on a description which is obtained by extraction from the physical layout, using the simulator `IRSIM` [Salz89]. This is used to verify the functionality and timing of the actual completed layout. At a higher level, using models that are stored in the libraries along with the cell descriptions, is behavioral simulation using `Thor` [Thor88]. This chapter shows how `Thor` and `IRSIM` are integrated with LAGER.

13.1. THE THOR SIMULATOR

`Thor` is a compiled simulator, where functional descriptions of the hardware are converted into the C language, compiled, and linked with a number of libraries to form the executable simulator file. `Thor` offers fast simulation at several abstraction levels, as well as a convenient user interface.

The `Thor` simulator is capable of simulating a design at the behavior level, register-transfer level, or gate level. Mixed-mode simulation is also allowed; the modules of the design may be modeled at different levels of abstraction. The interconnection between the modules uses single signal lines, fixed width busses, or variable width busses. Each signal line carries one of four values: Low (0), High (1), Undefined (U), or Floating (Z). In addition, the value on a bus of signal lines may be represented as an integer. The integer value is the two's complement interpretation of the values on the individual signal lines. (The integer interpretation only works if the individual signal lines carry 0s or 1s).

The modules and the interconnection network are specified with two different description languages, CHDL and CSL. The main features of these languages are described below. After conversion into C and compilation with a standard C compiler, the hardware description is linked with an event-driven simulator kernel, a user interface, and libraries of commonly-used functions. The supplied function libraries include "standard" logic functions for simple gates, arithmetic operations on two's-complement numbers, module models for generation of test patterns (generators), module models for observations of the signals of some nets (monitors), and an extensive TTL model library as well.

The resulting simulator executable may be run in batch mode, with a command script that specifies the details of the simulation. The same executable also runs in interactive mode, letting the user set breakpoints, single-step a number of clock cycles, etc. An X-windows interface similar to that of IRSIM is also available. In each case, only signal values on nets which have a monitor model attached to them will be displayed. The input signals of the simulation may be generated algorithmically (as is the case for, e.g., clock signals) or read from input files (by virtue of a generator model that reads the file).

13.2. MODEL DESCRIPTION LANGUAGE

Functional models for Thor are written in a language called CHDL. This language is C with few extensions which is converted to true C by a pre-processorbefore compilation by the standard C compiler. Any C construct may be used in the model description, which has the advantages of being familiar to most users and allowing complex efficient descriptions.

A simple CHDL model of an eight-bit adder is shown in Figure 13.1. The MODEL(adder) line declares the model name as being "adder". The model name is used to refer to the model in the netlist specification (see below).

The initialization section of the CHDL model contains ordinary C code, enhanced with the special port constructs. It is executed once for each instance of

Chapter 13   Behavior and Switch Level Simulation

```
MODEL(adder) {
    /* list of the input connections */
    IN_LIST
        SIG(Cin); GRP(In1, 8); GRP(In2, 8);
    ENDLIST
    /* list of output connections */
    OUT_LIST
        SIG(Cout);
        GRP(Out1, 8);
        /* temporary variable, invisible from outside */
        TGRP(TOut, 9);
    ENDLIST
    /* No bidirectional ports (biputs) for this block */
    INIT
    /* No initialization needed; purely combinational */
    ENDINIT
    /* Model body starts here */
    /* Compute temporary result. Sub-bus addressing is
    an extension to C */
    TOut[8:0] = In1[7:0] + In2[7:0];
    /* Write values to outputs*/
    Out1[7:0] = TOut[7:0]; Cout = TOut[8];
    /* exit model */
    EXITMOD(0);
}
```

Figure 13.1: A simplified CHDL example of an adder

the model, before the actual simulation begins. Since any C statement is allowed, it is possible, for example, to read a file that determines the behavior of the model. (This is used in the LAGER RAM and PLA models). For purely combinatorial models such as the one shown in Figure 13.1, no initialization is necessary.

The model body implements the actual hardware behavior. The body is executed whenever the signals on the nets connected to the input and biput ports of the model change value. Typically, the body code reads the signal values from the input and biput ports, decides what to do (maybe using the state variables), and writes new values to the biput and output ports (and maybe to its state variables).

The macro EXITMOD() is used to exit the model; any argument but 0 indicates that an error has occurred.

13.3. NETLIST LANGUAGE

CSL is the netlist description language used to specify the interconnection of the instances of the CHDL models. The CSL syntax is very different from that of CHDL; it does not look like C at all, even though it is converted to C when the simulator is built, just like the CHDL models are. An example of a CSL file is given in Figure 13.2.

```
/* An instance of the adder of Figure 13.1. The nets connected
to the "i" and "o" ports of the adder are given in the same
order as in the CHDL model. The delay is zero simulator time
units for all outputs. */
(f=adder)(n=black_adder)(i=carry_in,in_a[0-7],in_b[0-7])
(o=carry_out,out[0-7])  (do=0,0,0,0,0,0,0,0,0);

/* A counter is used to generate data */
(g=COUNT)(n=in_data)(o=carry_in,in_a[0-7],in_b[0-7]);

/* Simulation data dumped on a file for later examination */
(m=Banalyzer)(n=data_dumper)(i=carry_out,out[0-7]);
```

Figure 13.2: A simple CSL (THOR netlist) example

Each instance of a CHDL module is declared in exactly one CSL statement. The statement specifies the element type (which refers to the model name in the CHDL file header); a model instance name; lists of nets to connect the input, output, and biput ports to; and the delays associated with the output and biput ports.

The second type of CSL statement is the subnetwork definition. This is essentially a macro definition; instances of the subnetwork look like CHDL model instance declarations, but are expanded into groups of model instances according to the macro definition. The ports of the subnetwork are declared as inputs, outputs, or biputs, just as the ports in the CHDL models.

These are the only types of CSL statements. Nets are not explicitly declared. There is no explicit interface to the outside world, since all input and output is performed with generators and monitors, which use the same syntax as other model instances.

13.4. BUILDING A SIMULATOR

Building the simulator is almost automatic, through use of the Unix utility Make. A Thor utility program, gensim, generates a makefile for the simulator and then calls make. Make will run mkmod on each of the CHDL files; this program generates true C from the model descriptions, and feeds them to the system C compiler to produce object files. In a similar fashion, the CSL files are converted into C and compiled. The objects are then linked with Thor libraries to produce the simulator. Gensim finally spawns the newly built simulator and exits.

Note that since make is used in the compilation process, a complete recompilation is not needed if only a few of the CHDL or CSL files have been changed. Only the updated files need to be recompiled. The link step must also be redone.

13.5. DATABASE GENERATION

In addition to the main objective, some extra requirements were set up for the MakeThorSim interface between Thor OCT representation:

- CHDL models for the standard library cells and module generators should be stored in the leafcell libraries.

- The interface must place very little restriction on the flexibility of LAGER.

- The user should be able to recognize the circuit, in order for interactive simulation to be useful. In practice, this requirement forces a one-to-one correspondence between LAGER net names and Thor net names. Also, the names of the Thor nets, as well as those of terminals and building-blocks, should be as close to the corresponding LAGER names as possible.

These requirements have strongly influenced the design of the interface. The main idea of the interface is simple and obvious: subcells of a design should be modeled in CHDL, and the interconnection network should be an automatically generated CSL version of the OCT netlist.

13.5.1. Parameterization

The flexible parameterization mechanism is an indispensable feature of the LAGER system. Actual parameter values may be strings, integers, floating-point values, or arrays of these. In contrast, parameterization of Thor models is more limited: the actual parameter values are in practice integers. Also, the terminal specification of a Thor model allows only one flexible-width port of every type, while the number of parameterized ports in a LAGER module is unlimited. Mapping several parameterized LAGER ports onto one flexible-width Thor port would in principle be possible, but it would violate the one-to-one requirement, making interactive simulation impractical.

In the MakeThorSim interface, LAGER-style parameterization is provided by the introduction of a CHDL model template. The template is a CHDL model, where the LAGER parameter names are used at any place where the parameter value would be used. The actual parameter values are substituted for the parameter names before compilation of the CHDL model.

To make it easier to identify the formal parameters in the CHDL templates, an exclamation mark, !, is prepended and appended to the parameter name. Thus: to access the value of the parameter BUS_WIDTH, !BUS_WIDTH! is inserted in the template; at parameter substitution, this construct is replaced by (say) 8. This (integer) parameter would typically be used in the dimension specifier of a GRP port, in limit expressions for loops in the code, and to specify dimensions for local array variables.

13.5.2. Storage and instantiation

The CHDL template itself is stored as a string-valued OCT property, named THOR_TEMPLATE, in the structure_master view. There is no way to specify Thor information in the SDL representation of a cell; the property is inserted by a utility program that works directly with the OCT database.

Since DMoct copies properties from the structure_master view to the structure_instance view, a copy of the THOR_TEMPLATE property is present in the structure_instance view. The structure_instance view also contains the actual values of all the parameters. An instantiation procedure performs the parameter substitutions on the THOR_TEMPLATE property and writes the results to a CHDL file. It also stores the instantiated model in the structure_instance view as another string property, THOR_MODEL. For a cell without parameters, the instantiation is of course unnecessary; all instances of the cell are identical, and identical CHDL models may be used for them all. For such a cell, the CHDL model is stored directly in a THOR_MODEL property.

13.5.3. CSL generation

Generation of a CSL version of the OCT netlist is conceptually easy. Some practical quirks complicate the procedure somewhat.

In OCT, a design is represented in a hierarchical manner. CSL offers a subnetwork construct, which also makes hierarchical descriptions possible. It seems natural to map one hierarchy onto the other; however, this would force the user to specify the signal direction of each terminal of each module, that is, whether the terminal is an input, an output, or a biput. This information is usually not available in the OCT representation.

Therefore, another approach is used here: the hierarchical OCT representation is flattened into a one-level netlist before the conversion to CSL. The signal directions of the terminals of the CHDL models are specified in the model itself: input ports are given in the IN_LIST, output ports in the OUT_LIST, and so on. This means that the designer need not care about the signal direction at all, if the design only includes leafcells with CHDL models already in the system library.

The flattening is performed by a standard OCT library routine. The flattening starts from the top-level SIV and stops when an SIV with a THOR_TEMPLATE or a THOR_MODEL is found.

Once a flat netlist representation has been constructed, the CSL file generation may start. For each cell instance of the flattened netlist, the THOR_MODEL is extracted; if necessary, it is constructed from the THOR_TEMPLATE by replacing the formal parameters with their actual values.

When a cell description is parameterized, separate CHDL files are written for the instances. On the other hand, if the cell has no parameters, one copy of the code is sufficient for all the instances. This makes a big difference in the size of the simulator executable for a standard-cell design, which might contain hundreds of instances of the same non-parameterized cell.

For each port of the CHDL model, the corresponding terminal is then extracted from the OCT representation of the design, and a CSL statement is written to the output file. When all the cells have been processed in this manner, `MakeThorSim` exits; the CSL file should now be complete except for the generators and monitors.

13.5.4. Restrictions

The `MakeThorSim` interface generates a collection of CHDL and CSL files that represents the structure of the OCT design. Some pieces of information are not handled by the interface, or at least not handled well. These restrictions are addressed by the `MakeThor` interface, which is the subject of the next section.

First, generators and monitors have to be inserted in the CSL file by hand. Actually, the program is capable of inserting a monitor that provides graphical display of the waveforms of all the external connections of a module. This still leaves the generators.

Furthermore, in `Thor`, output delays are specified in the CSL file. This makes it possible to adjust the module delays according to wire length and capacitance of the interconnecting nets. For a gate in the standard cell library, a numerical value of the propagation delay could be stored as an OCT property. However, for a complex module like a PLA, the delay depends heavily on the parameters of the module. Therefore, the delay figure would need to be computed anew for each parameter set, as part of the instantiation procedure. Because of this complication the solution used with `MakeThorSim` was to usea unit delay of zero everywhere, and such a delay specification is inserted in the CSL file. This means that a well behaved network always evaluates in one `Thor` time unit.

13.6. IMPROVED INTERFACE

The `MakeThor` interface is an improved version of `MakeThorSim`. It provides a direct interface between `Thor` and the structure_instance view in LAGER. Just like `MakeThorSim`, it extracts interconnection data from an structure_instance view and fetches copies of the `Thor` models for the subcell instances of the structure_instance view In contrast to `MakeThorSim`, `MakeThor` builds the `Thor` interconnection structure directly from the structure_instance view i nstead of generating a CSL description. It then compiles the models along with the connectivity information and links them with `Thor` simulation kernel and library files to build the executable simulator. The two approaches are illustrated in Figure 13.3.

13.6.1. Generators and monitors

Compared to `MakeThorSim`, `MakeThor` has a slightly different view of the simulation process. The CSL that `MakeThorSim` generates from the SIV

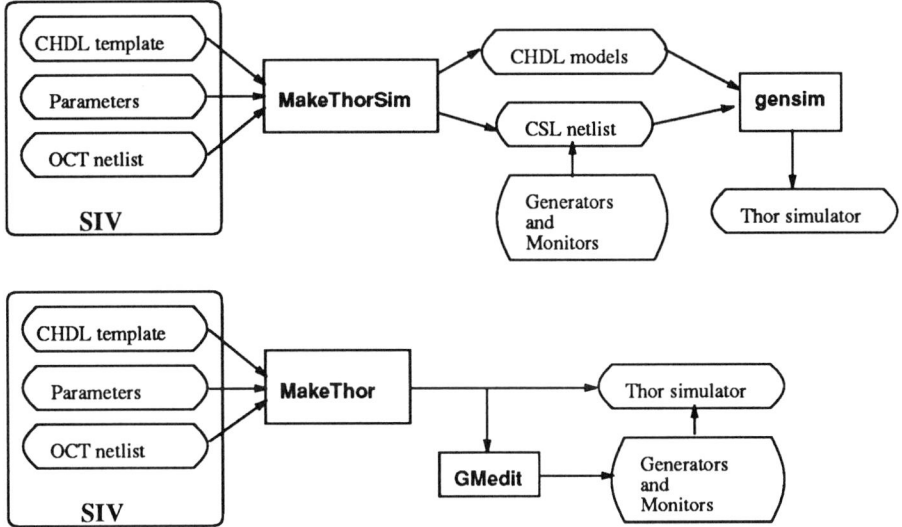

Figure 13.3: The approaches of MakeThorSim and MakeThor

does not contain any generators and monitors. They have to be inserted into the CSL file by hand. Then, gensim builds the simulator from the hand-patched CSL file. Whenever generators and monitors are added or deleted, gensim has to go through the full procedure all over again to generate the simulator corresponding to the modified CSL file. This procedure may take some time, since it requires recompiling and relinking.

In contrast, in the MakeThor interface, generators and monitors are added at runtime. The executable simulator which MakeThor generates contains no generators and monitors. It starts its execution by reading information about generators and monitors from a specified input file. Adding or deleting generators and monitors is done by editing this input file. To ensure consistency, a special-purpose editor (GMedit) is used for this task. GMedit makes recompiling and relinking unnecessary when the only changes are to the generators and monitors. The front end of the Thor kernel has been modified to accommodate this feature.

13.6.2. Delay handling

The simulator generated by MakeThor can take an option that specifies the internal delay specification. The internal delays of all the modules get the constant value provided by the user. If no value is given, the `MakeThorSim` behavior is used, that is, a delay of zero is assumed everywhere.

If `Thor` had allowed delay to be specified in the CHDL model, the INIT section of the model would be the ideal place to include it. A powerful programming language, C, would then be available, as well as all the parameter values for the module, so even complicated functions of the parameters could be used as delay values. Parameter-dependent delays would be computed once, before the actual simulation, thus insuring efficiency.

13.7. SWITCH LEVEL SIMULATION

`IRSIM`[Salz89] is an event driven switch level simulator that provides first order timing information for NMOS and PMOS transistor circuits. Two transistor models are supported by `IRSIM`: A switch model where each transistor is modeled as a voltage controlled switch and a linear model where each transistor is modeled by a voltage controlled switch in series with a piecewise linear resistor, and where each node in the circuit has a capacitance attached to it.

13.7.1. Input data

`IRSIM` requires two input files:

1. *The parameter file* contains electrical parameters for the circuit technology, mainly properties of the transistors, such as threshold voltages, gate area capacitances and dynamic channel resistances (for the piecewise linear model). There are also area capacitances for various layers, but this information is not used by `IRSIM` at present. `IRSIM` provides a calibration facility where a spice model (e.g. as provided by MOSIS) can be used to tune the parameter file so that there is close correspondence between `IRSIM` and spice delay values. The difference between spice and `IRSIM` is often less than 10% when using tuned models on a moderately sized circuit (say, a full adder with 50 transistors). Sample parameter files (untuned) for MOSIS SCMOS technologies are provided. An example is shown in Table 13.1.

2. *The circuit file* is a flat circuit description in the `Magic` sim file format. This file is typically generated directly from the circuit layout by using the `Magic` extract facility and then running the `ext2sim` program to flatten the hierarchical .ext description into a flat .sim file.

Additional command files and/or interactive entry can be used to specify input patterns, timing and simulation commands. The simulation results are displayed through a graphical interface called the *analyzer*.

13.7.2. Using IRSIM

IRSIM is the basic tool to check a layout for connectivity, functionality and timing. Any missing connections or shorts in the layout will be revealed if properly exercised by your test patterns. Functionality can likewise be established

Name	value	comment	
capga	.000816	gate capacitance pF/sq-micron	
lambda	1.0	microns/lambda	
lowthresh	0.4	logic low threshold (normalized)	
highthresh	0.6	logic high threshold (normalized)	
Channel resistances	width(λ)	length(λ)	resistance (Ω)
n-chan dynamic-high	10.0	2.0	3847.0
n-chan dynamic-low	10.0	2.0	1856.0
n-chan static	10.0	2.0	2472.0
p-chan dynamic-high	20.0	2.0	2020.0
p-chan dynamic-low	20.0	2.0	3969.0
p-chan static	20.0	2.0	2011.0

Table 13.1: The primary IRSIM model parameters

by applying appropriate input sequences. The timing accuracy is also quite reasonable as long as the wiring resistance in the circuit can be ignored. IRSIM does not model resistance except as part of the transistor model, so it will not give accurate results if, for example, a layout contains long polysilicon lines.

13.7.3. IRSIM example: The PUMA chip

The PUMA chip [Chapter 19] is a relatively complex chip containing about 44,000 active transistors and about 16,500 nodes in the flattened circuit. The pur-

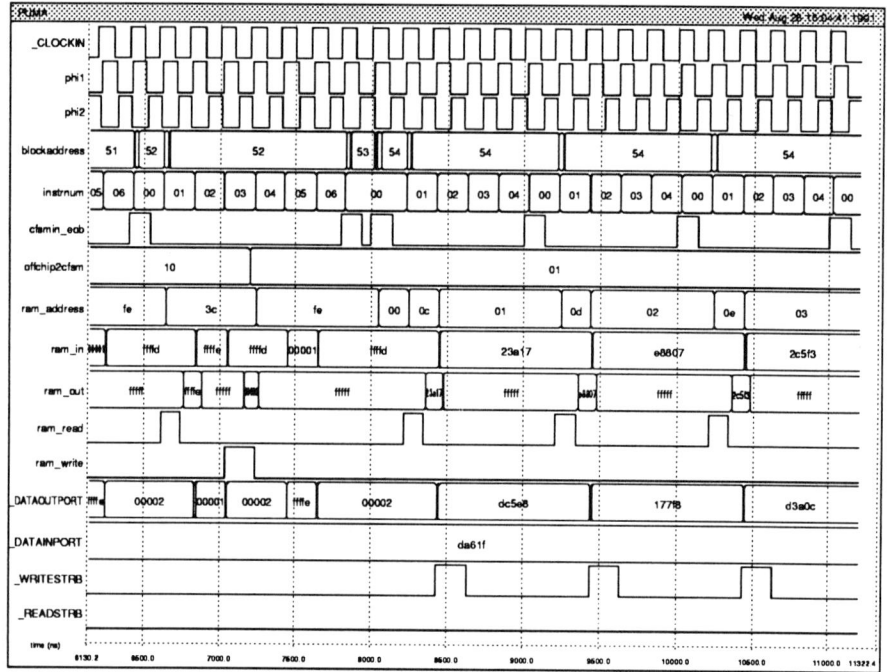

Figure 13.4: IRSIM analyzer display of PUMA chip simulation

pose of our simulation of PUMA with IRSIM was to go through a complete execution of a 700 line C program that had been compiled into silicon, using the techniques described in Chapter 17. This required simulation of more than 18,000 clock cycles of the extracted layout.

The input to the simulation consists of resetting the processor, starting the program execution, and waiting for the chip to signal that it is ready to read its inputs. We then feed the inputs to the processor and then simulate until it signals that it has finished. The chip is then given the input sequence which outputs the results. The outputs are then examined and compared to Thor simulation results.

Figure 13.4 shows parts of the analyzer display just as the PUMA chip is starting to output the first few numbers on the DATAOUTPORT terminal. The signal WRITESTRB (next to last in the display) goes high for one cycle to signal that a new value is available on DATAOUTPORT.

13.8. SUMMARY

The objective of the `Thor` interfaces was to automate the `Thor` simulation of LAGER designs. The high level simulations made possible with `Thor` are the primary means of debugging the input specification. On the other hand `IRSIM` is used to verify chip connectivity, functionality and timing and has been found useful for quite large designs, in spite of the low level of the simulation.

REFERENCES:

[Salz89] Arturo. Salz and Mark. Horowitz, "IRSIM: An incremental MOS switch-level simulator," Proceedings of the 26th ACM/IEEE Design Automation Conference, June 1989, pp. 173-178.

[Salz90] Arturo Salz, Stanford University. *Irsim manual*, 1990.

[Thor88] CAD Group, Stanford University. *Thor tutorial*, 1988.

14

Chip and Board Testing

Kevin T. Kornegay

Upon receipt of the chip from a fabrication foundry, testing is required to exercise the chip to determine whether it implements its intended functions. If an incorrect response is observed, a second objective of testing is to diagnose why the chip behaved incorrectly. Furthermore, in order to meet the tight design constraints imposed on today's chip designers, such as reduced chip to market time and reduced cost, testing must be considered very early in the design process.

The use of design techniques specifically employed to guarantee that a chip is testable is commonly known as design for testability (DFT). In this chapter, we will describe how to integrate two DFT techniques, namely scan path [Funatsu75] and boundary scan [IEEE90], into LAGER. Examples of how these techniques are used for chip-level and board-level testing are also presented. We also describe the `oct2tgs` tool which translates an OCT structure_instance view into a netlist format required by the test generation system (TGS) [USC88]. Finally, the architecture for the Test Controller Board, which is used for control of and access to the scan path and boundary scan devices, will be described. Currently, since LAGER is primarily focused on fast prototyping of circuits, these testing techniques are used only to provide us with diagnostic information about the location of a failure. High speed testing for high volume fabrication has not been addressed.

14.1. CHIP-LEVEL TESTING

Most chip-level structured DFT techniques are built upon the concept that if the values in all of the latches can be controlled to any specific value, and if they can be observed with a straightforward operation, then test generation can be reduced to that of doing test generation for the combinational circuits between the controlled latches.

14.1.1. Methodology

The scan path methodology is probably the most widely used technique for testing those parts of the circuit that are constructed of clocked flip-flops interconnected by combinational logic[Funatsu75] As illustrated in Figure 14.1, it is based on converting the circuit's flip-flops into a serial scan path chain denoted by the thick black line threading the circuit flip-flops. When the circuit is put in the test mode, it is possible to shift an arbitrary test pattern into this serial register. By returning the circuit to normal mode for one clock period, the contents of the scan register and primary input signals act as inputs to the combinational circuitry and new values are stored in the scan path register. If the circuit is then placed into test mode, it is possible to shift out the contents of the scan register for comparison with the correct response.

14.1.2. Scan Path Register Cell

By using test points, one can easily enhance the observability and controllability of a circuit. The scan path register effectively provides such test points. A

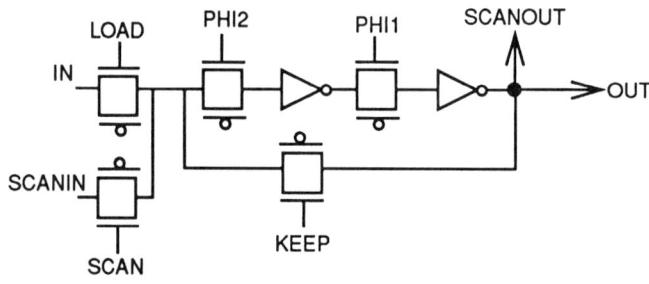

Figure 14.1: Scan path register cell.

Chapter 14   Chip and Board Testing   189

scan path register is a serial cascade of scan path register cells whose inputs and outputs are connected to the internal logic of a chip. A schematic of our scan path register cell is shown in Figure 14.1. Operation of the cell requires two non-overlapping clocks, namely *Phi1* and *Phi2*. During normal operation, the *LOAD* signal is asserted and the logic value at the input (*IN*) reaches the output (*OUT*) after one clock cycle. When the signal *SCAN* is asserted during test, the logic value at *SCANIN* arrives at *SCANOUT* one clock cycle later. The *KEEP* signal is used to refresh the value stored in the cell.

14.2. BOARD-LEVEL TESTING

To better address problems of board-level testing, several standards have been developed. The primary goal of such a standard is to ensure that chips contain common circuitry that will make test software development and the testing of boards containing these chips significantly more effective and less costly. One such standard is called the JTAG Testability Bus Standard [IEEE90], also known as Boundary Scan, 1149.1 or simply JTAG.

14.2.1. JTAG Boundary Scan Architecture

The JTAG standard consists of a dedicated serial test bus which resides on a printed circuit board, a protocol which controls the I/O pins that connect the chip to the test bus, and control logic that resides on chip to interface the test bus to the DFT circuitry residing on the chip. The primary reasons for boundary scan are to allow efficient testing of board interconnect, and to facilitate isolation and testing of chips either via the test bus or with additional circuitry.

With boundary scan, chip-level testing can be supported at the board-level by simply connecting boundary scan register cells to the chip's scanin and scanout pins as shown in Figure 14.2. There are two major components associated with this standard, namely boundary scan registers and the finite state machine test access port (TAP) controller. Scan path registers are accessible from the test data input (TDI). The normal terminals of the application logic are connected through boundary scan cells to the chip's I/O pads. The rest of the test circuitry consists of a 1-bit bypass register and the instruction register.

The TAP bus consists of four lines, namely the test clock (TCK), the test mode select (TMS), the test data in (TDI) and the test data out (TDO) lines. Boundary scan cells can be interconnected forming a single scan-chain. Various tests such as scan path, interconnect test, or snapshot observation, can be carried out with this architecture. To implement these tests, three test modes exist, namely internal test, external test, and sample test. While in the internal test mode, multiple internal scan path operations can be activated to test a chip. During external test mode, printed circuit board interconnect tests can be carried out. In the sample test mode, internal nodes of the chip can be sampled during normal system operation.

14.2.2. Boundary Scan Register Cell

The boundary scan register tests circuitry external to the chip package (primarily the board interconnect). It also permits the signals flowing through the chip's I/O pins to be sampled and examined without impacting the operation to the system logic. The boundary scan register is single shift register based path containing cells connected to all system inputs and outputs of the circuit and the system logic.

A block diagram of an output boundary scan register cell is shown in Figure 14.3. During normal operation, the *MODE* signal is disabled and the cell

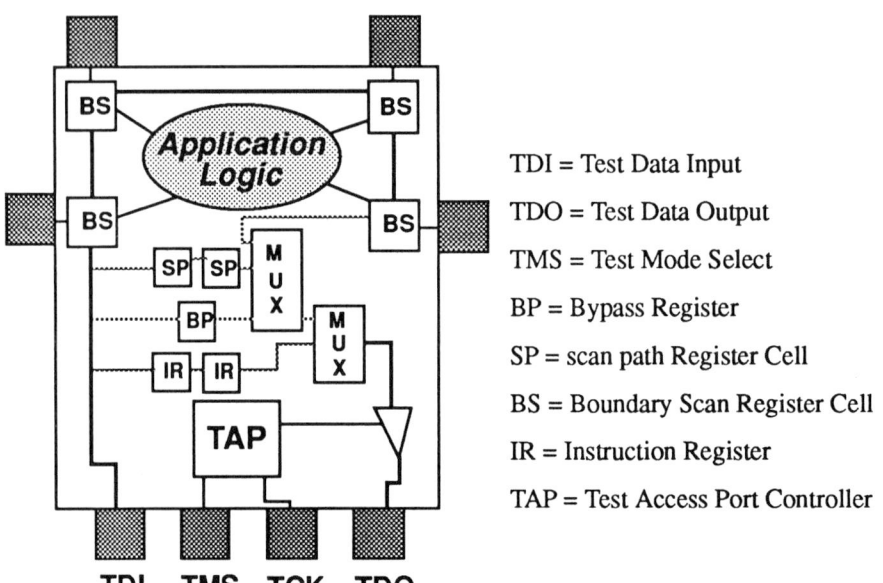

Figure 14.2: JTAG Boundary Scan Architecture.

becomes transparent and data is passed directly to the output pin. When a test session is invoked, the *ShiftDR* and *ClkdrDR* signals are asserted until all the test data is loaded into the boundary scan chain after which, the *UpdateDR* signal is asserted and the test data is applied to the output pins. Meanwhile, the *Mode* signal is asserted during the entire test session. Test results are then captured at the input pin of an adjacent chip and shifted out for comparison with that of a good circuit.

14.3. IMPLEMENTING SCAN PATH AND BOUNDARY SCAN

An example of a chip containing both scan path and boundary scan, called TESTCHIP1, is illustrated in Figure 14.4. TESTCHIP1 is partitioned into four boundary scan register blocks, one TAP controller block, and one data path and scan path block. It was generated from a hierarchical SDL file comprised of two lower-level sdl files which are called JTAG_MACRO and DATA_PATH. JTAG_MACRO contains the minimum amount of circuitry necessary to implement the JTAG Boundary Scan standard. DATA_PATH contains the data path and scan path circuitry. The widths of both the scan path and boundary scan registers are determined by the user specified design parameters.

For designs which are limited by the core area, we have designed dedicated chip pad library whose primitive cells contain boundary scan register cells. A list and brief description of these pads is described in Table 14.1. An implementation which utilizes these pads is illustrated in the TESTCHIP2 example shown in Figure 14.5

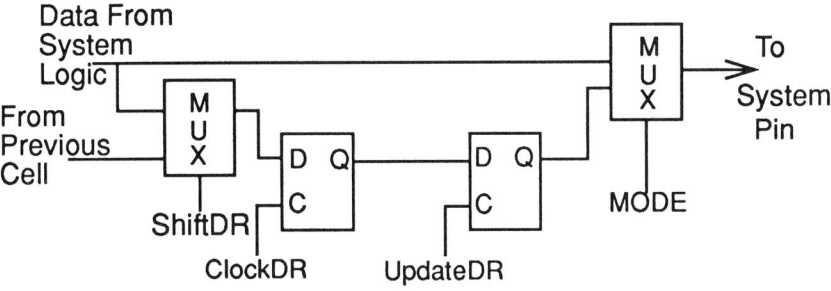

Figure 14.3: Block diagram of output Boundary Scan register cell.

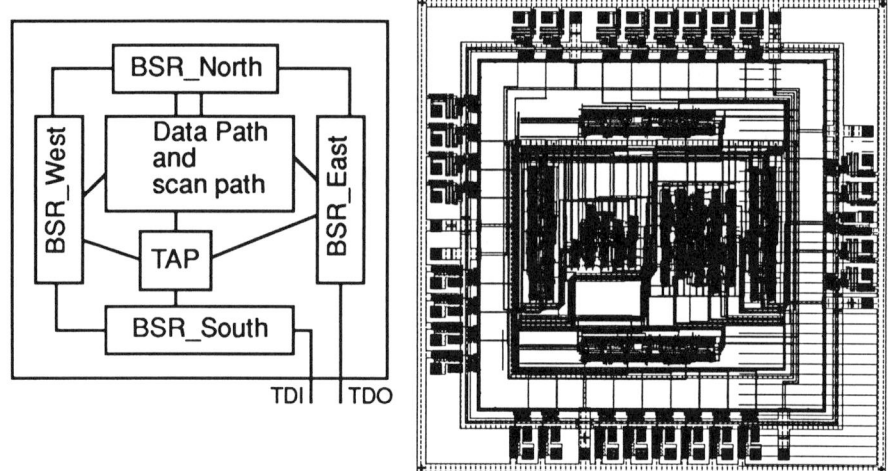

Figure 14.4: Block diagram and plot of TESTCHIP1.

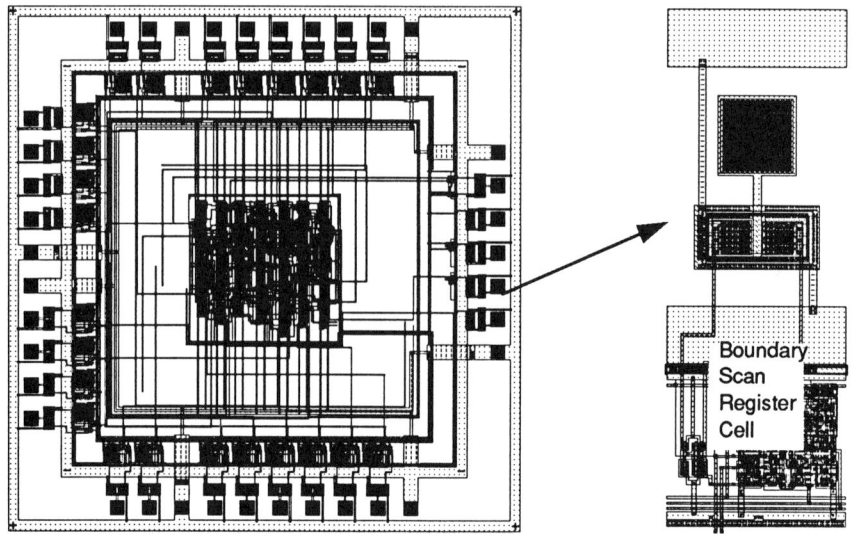

Figure 14.5: Plot of TESTCHIP2 and Boundary Scan I/O pad.

Chapter 14 Chip and Board Testing

Padname	Function
in	Boundary Scan input pad.
out	Boundary Scan output pad.
io	Boundary Scan bi-directional pad.
tdi	Test data input pad.
tdo	Test data output pad.
miscellaneous	Supply and ground pads.
nobs_in/out	Non-Boundary Scan I/O pads.

Table 14.1 Boundary Scan Pad Library Cells

14.4. TEST CONTROLLER BOARD

When testing printed circuit boards, it is often useful for purposes of testing to be able to isolate one chip from the other. Hence, provisions must be made in order to access and control the scan path and boundary scan hardware implemented on these chips. This can be accomplished using the Test Controller Board, which implements the JTAG Boundary Scan bus master protocol.

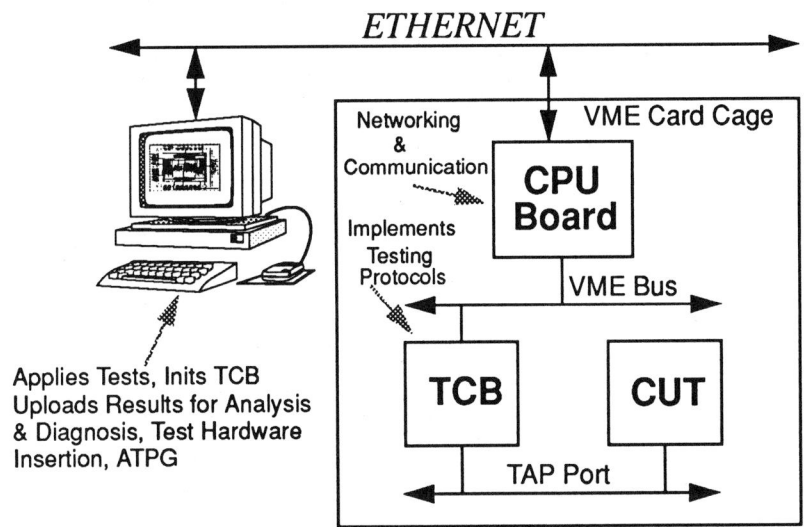

Figure 14.6: Block diagram of system hardware.

A block diagram illustrating how the Test Controller Board is used in our system hardware development environment is shown in Figure 14.6, where the Test Controller Board, CPU board, and the chip under test (CUT) all reside in a card cage.

14.4.1. Test Controller Board Architecture

The TCB provides us with a means to exercise the test hardware implemented on the CUT. Assuming every chip on the printed circuit board conforms to the JTAG Boundary Scan standard, we can test the internal logic or test the interconnect ion between the chips on the printed circuit board.

A block diagram of the TCB is shown in Figure 14.7. It contains VME interface logic, which handles communication between the HOST and the Test Controller Board, control, status, and data registers, all of which are used to initialize and configure the Test Controller Board for a test, a test controller, which implements various test algorithms, test data memory, which stores the test stimulus and response, and the test interface, which connects it to the DFT structures implemented on the CUT.

14.4.2. Test Controller Board Software

The software, called SCANTEST, controls an entire testing sequence by providing a user on a host system with the means to apply test data to a circuit, to

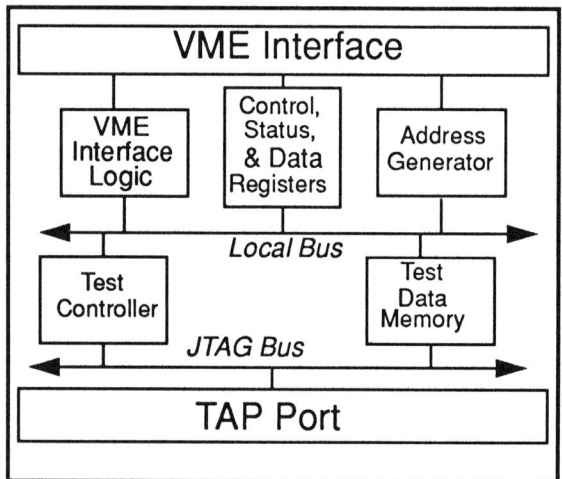

Figure 14.7: Test Controller Board (TCB) architecture.

execute a particular test, and to capture the response for analysis. This software essentially controls the Test Controller Board and is capable of supporting three types of test, namely scan path, boundary scan, and data acquisition. SCANTEST was written in the C-programming language and it is composed of three primary functions. First, the board is initialized, and its status is checked, after which test vectors are downloaded. Next the board is operated, and finally, the data is uploaded back to the host.

14.4.3. Testing using the Test Controller Board

During a test, the scan path or boundary scan registers are exercised and the following sequence of events occur:

1. the TCB scans in the test data.
2. the CUT runs for one clock cycle.
3. the TCB then scans out the test results.

Once the results are obtained, the functionality of the CUT can then be verified. For example, the functionality of the Viterbi processor [see Chapter 21] was verified using the scan path test method described in Section 14.1. A block diagram of one its data paths with its scan path register is shown in Figure 14.1. By issuing the scan path instruction, a scan path test sequence was executed, and the functionality of the logic surrounded by registers REG[1] through REG[5] was verified. The software prompts for the type of Boundary Scan test to be performed, and any of the public instructions outlined in [IEEE90] can be executed during a Boundary Scan test.

14.5. TEST PATTERN GENERATION

The test pattern generation process provides a test vector (consisting of 0, 1 or dont-care values) for any given detectable fault. In a combinational circuit, a specific stuck-fault can be tested with a single input vector. Most automatic test pattern generation (ATPG) programs work extremely well for combinational circuits, and are usually some derivative of the D-algorithm or PODEM [Goel81]. Testing a fault in a sequential circuit, in general, requires a sequence of vectors. Test generation for sequential circuits is much more complex than that for combinational circuits. In fact, none of the known sequential circuit test generators provide satisfactory performance. This motivates the use of the scan path method. By using such DFT techniques, test generation for sequential circuits can then be treated as test generation for combinational circuits.

Test generation is performed in two steps. First, an interface tool, called oct2tgs, translates the design into an input format which is understood by the ATPG program. Second, the test patterns are generated using TGS [USC88]. TGS generates a set of test patterns for combinational logic circuits described at the gate level using the PODEM [Goel81] test generation algorithm. The gate types supported by the system are AND, OR, NAND, NOR, INV (inverter) and BUFF (buffer). Fault collapsing, whose main objective is to classify the set of all possible faults into one set, and fault simulation, which attempts to identify all faults that can be detected by a given input vector, are also supported by TGS.

14.6. SUMMARY

Most techniques deal with either the resynthesis of an existing design or the addition of extra hardware to the design. The approaches described in this chapter require modifications to the design and affect such factors as area, I/O pins and performance. Hence, a critical balance must exist between the amount of DFT used and the gain achieved when employing these techniques. Furthermore, these techniques provide us with adequate diagnostic information and were found to be extremely suitable for a rapid-prototyping environment.

Without DFT and ATPG, tests have to be generated manually, with DFT and ATPG, they can be generated automatically. The implementation of the Boundary Scan and scan path cells have also been described. These cells help provide support at both the chip and board levels. Finally, the Test Controller Board which is used to access the DFT devices implemented on the chips was described.

RERERENCES:

[Funatsu75] S.N. Funatsu, N. Swkatsuki, T. Arima, "Test Generation Systems in Japan", 12th Design Automation Symposium, pp. 112-114, June 1975.

[Goel81] P. Goel, "An Implicit Enumeration ALgorithm to Generate Tests for Combinational Logic Circuits", IEEE Trans. on Computer, pp. 215-222, 1981.

[IEEE90] IEEE Std 1149.1-1990, *"IEEE Standard Test Access Port and Boundary Scan Architecture"*, February 15, 1990.

[USC88] USC Test Group, "Test Generation System (TGS) User's Manual", Department of Electrical Engineering-Systems, University of Southern California, June 27, 1988.

Part IV

Behavioral Synthesis

15

DSP Specification Using the Silage Language

Paul Hilfinger and Jan Rabaey

Digital signal processing (DSP) algorithms are most easily described using a block diagram or schematic representation of the data flow in the algorithm. This can be captured in a graphical format[Lee89, Goering90], which is convenient if the number of blocks is relatively small and the interconnection is not complex. For more complex systems it is often more convenient to use a text representation. However, the commonly used procedural languages such as C, impose an ordering on the operations, and the variables actually represent memory locations, rather than true arithmetic values from which meaningful arithmetic equations can be written. An applicative (functional or single assignment) language accurately represents the computation which is to be performed, rather than *how* it is to be calculated. Silage is such a language with the addition of constructs (e.g. delay) that are commonly found in DSP applications.

15.1. BASIC CONCEPTS

A popular form for high-level descriptions of signal-processing algorithms is a graphical signal-flow representation in which nodes represent instances of functions (such as addition, multiplication, and delays) and arcs represent the path followed by data for, and results from, these functions. The semantics implied by such a representation are what are known as *data-flow* semantics, in which the emphasis is on the paths followed by the inputs and intermediate results of a computation, rather than the sequence of imperative operations performed on memory locations, the *control-flow* semantics. One potential advantage of the data-flow representation is that no explicit statements are made about the order or concurrency of the operations specified, and simple graphical properties of the data-flow graph correspond to opportunities for parallel computation.

A natural textual representation for a data-flow semantics is an *applicative language*: a language whose fundamental operation is function application, and that has no variables or assignment operator. A program consists of a set of definitions of functions and named values. Each function definition is a set of equations relating the values of the inputs, outputs, and intermediate values of the function as if they were static, timeless quantities. Hence, in a typical applicative language, one would *not* write

```
S := 0;
for i := 1 to 10 do S := S + A[i];
```

but rather

```
PS[0] = 0;
(i: 1 .. 10):: PS[i] = PS[i-1] + A[i];
S = PS[10];
```

This looks very much the same, of course, but all of the statements above are equations, rather than assignments; a statement such as

```
i = i+1;
```

is meaningless. Furthermore, imagine that each of the PS[i] is the result of a separate functional unit (an adder in this case). Then the second program describes a pipeline with i indexing the stages of the pipe, which is a bit difficult to extract from the first description. Of course, i does not have to index pipeline stages; it can also be thought of as indexing the steps of an iteration, as in the first program.

Expressions in Silage generally denote infinite streams of values. For example, in an expression such as A+B, the operands A and B denote infinite streams of numbers, and the result is the infinite stream of sums of corresponding

Chapter 15 DSP Specification Using the Silage Language

elements in the two sequences. This is basically the same interpretation one places on signal-flow graphs, cast in a textual form. More specifically, if we use subscripts to denote the individual items in a sequence, then by definition

$$(A+B)_i = A_i + B_i \tag{15.1}$$

for all i. The dummy variable i may be thought of as ranging over the integers. It is implicit; with few exceptions, one does not normally refer to individual members of a sequence. Unless explicitly defined, values at negative indices have arbitrary values.

Inputs to a program are certain designated streams having no explicit definition. In general, a Silage program is driven by the receipt of inputs. Physically, a set of values from a group of input streams typically arrives together. In keeping with standard usage, we call each such set a *frame*.

Many operations, like addition, are *elementwise*—each element of the stream they produce is purely a function of the corresponding elements of their operand streams. This is not sufficient to define systems that have state, and there are other operators that perform more complex manipulations of streams. The most common of these is the delay operator, denoted by '@', and defined

$$(E@n)_j = E_{(j-n)} \tag{15.2}$$

for an expression E, and non-negative integer expressions n and j.

The introduction of the delay operator makes it easy to write expressions that are rather difficult to implement. For example, consider the following pair of definitions.

```
i = i@1 + 1;
A = A@1 + A@(i div 2);
```

Here, the operator **div** performs integer division. Ignoring for the moment the problem of getting initial values for i and A, it should be clear that computing the values of A will require ever-increasing amounts of storage. Therefore, although one can make perfect semantic sense out of definitions of this sort, Silage implementations will place restrictions on the use of delay and related operators that allow most cases that are useful in practice without complicating implementations too much.

These restrictions are most easily motivated by considering implementations. In many applications, the indices of the elements of a stream of data are related in some simple fashion to the times at which those data arrive or are computed—they are *synchronous*. When two streams, A and B, are synchronized, with each A_i arriving at the same time as B_i, the computation implied by A+B, for example, is simply a matter of repetitively receiving values from the two streams and sending off their sums to some consumer. When the streams are not synchronized, the implementation must in general maintain a dynamically-changing amount of additional storage in the form of queue or buffer of values. To avoid the costs of providing this general facility at execution time for application areas where it is seldom needed, an implementation may place restrictions that avoid it.

15.2. LANGUAGE FEATURES

This section informally describes the major features of Silage. Because it is not intended as a reference manual, sections of the language that duplicate familiar features of other common programming language (arithmetic expression syntax, for example), as well as obscure points that are not important to a general understanding of the language, have been skipped in the following discussion.

15.2.1. Definitions of Values.

A Silage program consists of an unordered sequence of definitions of values (or, as discussed in the preceding section, streams of values) and of functions that act on streams of values to produce one or more streams of results. Names and expressions that yield values are statically typed, as in most modern programming languages. A value may be an integer with a specified width (*width* refers to the number of bits in the representation, including sign), a fixed-point number with specified width and scaling, a floating-point number with specified significand and exponent widths, a boolean (true/false) value, or a one-dimensional array of values having identical type. A multidimensional array type is simply an array type whose element type is itself an array type.

Definitions of values are equations having one of the following forms.

```
A = E;
A[i_0]...[i_{m-1}] = E;
(A_0, ..., A_{m-1}) = E;
```

where A is a name of the quantity being defined and E is a defining expression. The first form defines A to be the single stream of values produced by E. In the second form, A denotes an array of at least m dimensions; the equation defines the value of one single element of A. The third form simultaneously defines *m* quantities; it is intended to facilitate the use of functions that return multiple results.

These constructs should all be thought of as definitions or equations—static assertions of relationships between values, rather than assignments to variables. In particular, the order in which definitions are written does not generally matter. Likewise, anything that would constitute a multiple or conflicting definition is illegal. Without further restriction, a compiler would find this latter restriction extremely difficult or impossible to check in the case of arrays, as illustrated by the following equations.

```
A[i] = 1;
A[i * j div k] = 0;
```

Without a general-purpose theorem prover for the integers, a compiler would have no way of knowing whether there is any conflict in these two definitions. Hence, Silage imposes the additional restriction that the indices of array elements being defined in an equation must be manifest (compile-time computable) expressions.

Since Silage is single assignment, an expression such as:

```
M = 3*M + X;
```

is invalid. Recurrences in which the definition of any element of the stream X depends on previous values of that stream, such as commonly found in digital filter descriptions, are allowed and compactly written as follows:

```
X = C * X@1 + Y;
```

In general, the right-hand side of a definition is allowed to refer to the stream being defined as long as this reference is delayed.

15.2.2. Expressions with Elementwise Operators.

As the examples in previous sections suggest, Silage provides the usual arithmetic operators, which operate element-by-element on streams of numeric values. The full set includes the usual relational operators (such as '<'), arithmetic shift operators ('<<' and '>>' as in C), and the bit-wise logical operators—and ('&'), or ('|'), exclusive or ('^') and not ('!')—on numeric and boolean values.

The definitions of the arithmetic operators are non-deterministic in their overflow behavior and rounding. For cases where these must be specified more precisely, the language provides functions add, sub, mult, divide, and round, taking the expected operands plus an extra argument that designates the specific behavior desired—such as saturating or wrapped arithmetic on overflow, or truncation or rounding for real results.

The only real complication in the semantics of arithmetic involves the fixed-point operations. A fixed-point quantity represents a rational number with a fixed *scale*, or number of binary digits to the right of the binary point. As is only too familiar to those dealing with such values, the intricacies arise from the need to define the widths and scales of the results of all possible arithmetic operations, as functions of the scales and widths of the operands. Silage provides default rules for such results, so that, for example, the result of multiplying two fixed-point numbers of identical width w and scales d_0 and d_1 has width w and scale $d_0 + d_1 - w + 1$. These defaults are not always adequate, and so Silage provides explicit coercions (or *casts*) that indicate the desired type of a result. For example, two values x and y might range over the interval (-2,2), both representable by the type fix<16,14> (the fixed-point type with a width of 16 bits with a 14-bit binary fraction). The programmer might know that their product is always bounded by the same interval, and accordingly write their product as

```
fix<16,14> (x*y)
```

As illustrated, coercions are denoted by prefixing the value to be coerced with the target type.

Manipulation of arrays is always a problem in applicative languages, because the familiar operations on arrays from languages such as FORTRAN or C change an element at a time. Since a change in any one element of an array changes the value of the array as a whole, this means that in an applicative framework, a sequence of N changes corresponds to N distinct values of the array, and naively, to N copies of the array. While careful analysis by the compiler can diminish this problem, it is nevertheless very helpful—both for programmers and compilers—to have operations that succinctly describe operations affecting or referencing an entire array at once. The *array constructor* in Silage defines all values of an array either according to some enumeration—as in

```
{ 1, 3, 5, 7, 9, 11 }
```

which lists the elements of a vector, beginning with element 0—or by a rule, as in

Chapter 15   DSP Specification Using the Silage Language

```
{ i : 0 .. 5:: 2*i + 1 }
```

or, equivalently,

```
{ i : 5:: 2*i + 1 }
```

which map the integer indices in some range (here, 0 to 5) to values at those index positions. The dummy variable i, which is local to the constructor, is called a *control variable*.

In addition to the constructor, there are several *reduction operators* that operate on entire arrays. The standard function calls sum(A), max(A), min(A), and innerprod(A,B) return, respectively, the sum, maximum, and minimum of the elements of A, and the inner product of the elements of identically-indexed arrays A and B. Together, these operations are quite useful; for example

```
sum({ i : 1 .. 5:: A@i})
```

is the sum of the previous 5 values of A.

Corresponding to the conditional statements found in conventional programming languages, Silage has a conditional expression, like those found in the Algol dialects, Lisp, and C. The expression

```
if C₀ -> E₀ || C₁ -> E₁ || ... || Eₙ₋₁ fi
```

yields the value of E_i for the first i such that the boolean condition (or *guard*) C_i is true. That is, given one value from each of the streams C_j and E_j, the conditional expression uses the values of the C_j to select the value of one of the E_j, throwing all others away. For convenience, conditional expressions can yield tuples of values, which may in turn be used to define tuples of values. For example, in the following definition, a quantity P10 is defined that holds each tenth value of P for ten frames.

```
(k, P10) = if k@1 == 9 -> (0, P) || (k@1+1, P10@1) fi;
```

15.2.3. Expressions with Stream Operators.

The operators described so far have all been elementwise—they can be described as extending some operation on individual values to streams of values in the obvious fashion. The more interesting operators cannot be so described. They explicitly manipulate streams of values.

The most common of these is the previously-described delay operator, denoted by '@'. Element i of the stream E@n is defined to be element i-n of stream E, for positive n. Since E is usually undefined at negative frame indices, the first few elements of E@n would ordinarily be undefined. Where this is undesirable, the language permits the explicit definition of values at particular positions in a stream. For example, in the previous example, defining quantities k and P10, it is desirable to insure that the initial value of k@1 is well-defined. The declaration

```
k@@1 = 9;
```

defines the '-1'th value of k, and insures the proper values for the k and P10.

The implementation of E@n presents a problem if the right operand is not restricted. If n is allowed to be unbounded, then any value in the entire stream of values E may be called for at any time, requiring that the storage required for the computation increase linearly with time. To prevent this, the second operand is required to be *manifestly bounded*—its upper bound must be a fixed number determinable at translation time. This condition is obviously met when n is a constant. For other situations, the language provides the expression bnd(n, L, U), which simply yields n as long as L ≤ n ≤ U, and is otherwise erroneous. Thus, when L and U are constants, the legal values of this expression have known bounds, and it can be used to specify a delay.

The most elaborate and specialized of the stream-manipulation operators are those for converting the frequency of a stream of values. The most general is the switch function, which "re-packages" the values in a stream of vectors into another stream of vectors having a different length. Specifically, for a stream of vectors A, the stream B defined by

```
B = switch(A, N)
```

is computed by first converting A into the single stream of values

$$A[L]_0, A[L+1]_0, \ldots, A[U]_0, A[L]_1, \ldots$$

where L = lwb(A) (the lower index bound of A) and U = upb(A), and then in effect renaming the stream

$$B[0]_0, \ldots, B[N-1]_0, B[0]_1, \ldots$$

Chapter 15   DSP Specification Using the Silage Language

Letting $M = U - L + 1$, this may also be expressed

$$\alpha_{i+km} = A[\text{lwb}(A) + i]_k$$
$$B[i]_k = \alpha_{i+kN}$$

The sampling rate of each of the individual elements `B[i]` is therefore M/N times that of each `A[i]`.

For expressive clarity and compiler convenience, two common special cases of "switch" have their own designation. The expression `interpolate(A)` yields the stream X defined by

```
X = Q[0];
Q = switch(A,1);
```

That is, it "flattens" A into a single stream of values; if A is of size N, this stream is N times faster than any of the elements A[i]. Going in the other direction, the expression `decimate(Y,M,p)` yields the stream Z defined by

```
Z = R[p];
R = switch(YVEC,M);
YVEC[0] = Y;
```

In other words, this decimates the stream Y by a factor of M, selecting sample p out of each M samples. It follows that for A with bounds 0 and N-1,

```
A = { i: 0 .. N-1:: decimate(interpolate(A), N, i) }
```

Here, for example, is another definition of P10 from page 205.

```
P10 = interpolate({ j: 10:: decimate(P, 10, 0) });
```

15.2.4. Iteration and Blocks.

When defining a set of interrelated arrays, it is convenient to be able to have some construct to give rules for their construction, as for individual array constructors, without having to enumerate their contents element-by-element. Silage provides iterative constructs for this purpose. These allow entire definitions or groups of definitions to be subject to a control variable. For example, the definition

```
V = { i : 0 .. 5:: 2*i + 1 };
```

may also be written using either of the iterated definitions

```
(i: 0 .. 5):: V[i] = 2*i + 1;
(i: 5):: V[i] = 2*i + 1;
```

either of which is equivalent to the sequence of definitions

```
V[0] = 2*0 + 1; V[1] = 2*1 + 1; ...; V[5] = 2*5 + 1;
```

The repeated clause may also be a block—a group of definitions bracketed by **begin** and **end**. For example,

```
N = 10;
delta = (X - X@1)/N;
S[0] = 0.0;
(i: 1 .. N)::
    begin
        p[i] = X + delta * i;
        S[i] = S[i-1] + C[i]*p[i];
    end;
```

Frequently, iterators accumulate values, as in the case of this last example, where the S[i] are partial sums, of which the last, S[N], is probably the only one of interest. This is so common that the delay ('@') operator is preempted inside iteration blocks. The iteration above is equivalent to the following.

```
N = 10;
delta = (X - X@1)/N;
(i: 1 .. N)::
    begin
        S@@1 = 0.0;
        p[i] = X + delta * i;
        S = S@1 + C[i]*p[i];
    end;
```

In addition, if the quantity p is not needed except to define S (and is included, therefore, to break up the definition for clarity), it can be made local to each iteration of the loop as follows.

```
(i: 1 .. N)::
    begin
        local p;
```

```
            S@@1 = 0.0;
            p = X + delta * i;
            S = S@1 + C[i]*p;
    end;
```

The bounds in all the preceding iterations are required to be static. This is not a semantically necessary restriction, but is a reasonable restriction for the sorts of algorithms and target architectures at which Silage is aimed. Sometimes, however, iterations without predetermined bounds are necessary. For these cases, the language provides an *indefinite iterator*. A representative example will convey the semantics. The following defines m to be the first index in the array A at such that A[m+1] is less than or equal to A[m], if the fictitious element A[N] is taken to be $-\infty$.

```
    do
            i@@1 = -1;
            i = i@1 + 1;
            m = exit i == upb(A) -> 0;
                 ||     A[i] >= A[i+1] -> i;
                 tixe;
    od;
```

Again, the delay operator here refers to prior loop iterations. The guards of the **exit** clause enumerate the conditions under which the loop terminates, with a syntax resembling that of conditional expressions. The guarded clauses in this case are alternate definitions of m, one of which is selected at the end of the iteration.

15.2.5. Function Definitions

As do most high-level languages, Silage provides a way of abstracting a computation and giving it a name. The most general form of a function definition is as follows.

```
    func f(p_0: T_0; ...; p_{n-1}: T_{n-1})
           r_0: RT_0; ...; r_{m-1}: RT_{m-1} =
    begin
            definitions
    end;
```

where f, the p_i and the r_i are unique identifiers and the T_i and RT_i are type designators. The p_i are the formal parameters (inputs) of the function and the r_i, which must be defined in the definitions in the body of the function, are the outputs. In the case of a single output, the r_i may be elided, yielding the more familiar form

```
func f(...): RT = ...
```

In this case, there is a single output of type *RT* named return, which must be defined in the body.

A function is called using the usual prefix form:

```
f(A_0, ..., A_{m-1})
```

and may be used as a value in an expression in the case of a single return value, or to define a tuple of values in the case of multiple return values. The semantics of this call are that it is replaced with a copy of the body in which the formal parameters are first defined to have the values of the corresponding actuals, and local names are changed as necessary to prevent conflicts. That is, it is by definition implemented as a form of macro-expansion, but with (in effect) automatic name changes to prevent the name conflicts that macro expansion tends to cause. The actual parameters of the function may not be defined using its outputs.

The types of the actual parameters must match those of the corresponding formals. The formal parameters, however, are allowed to have *generic types*, which match any of a class of actual types. Thus, the type int<w> denotes the specific type of w-bit integers. The generic type int matches any integer type. Likewise, the generic type fix matches any fixed-point type (fix<w,s>), and float matches any floating-point type. Furthermore, the type num matches any numeric type at all, and any matches any of the non-array types. The meaning of any particular call on a function with generic types is therefore determined in part by the actual parameters (as a result, certain type errors become apparent only at the point in the text that a function is called, rather than at its definition). Finally, to allow parameter specifications that relate the types of two parameters, the type expression typeof(E) denotes the type of the expression E.

Normally, macro-expanded functions do not allow recursion. However, recursively-defined structures are sometimes useful in signal processing applications, and Silage provides a simple form of *primitive recursion* for describing them. Specifically, a function is allowed to be recursive if it has an integer argument that is always called with a manifest (compile-time determined) value, that is always smaller in inner calls of the function, and that is never negative. For example, to define the inner product of two 0-based vectors computed by a tree--structured collection of additions, we may write the following.

```
func innerprod2(A: num[]; B: typeof(A)): typeof(A) =
begin
    func F(Q: typeof(A); R: typeof(A); L: int; N: int):
        typeof(A) =
    begin
        M = N div 2;
        return =
            if N == 0 -> 0
            || N == 1 -> Q[L] * R[L]
            || N > 1 -> F(Q, R, L, M) + F(Q, R, L+M, N-M)
            fi;
    end;
    return = F(A, B, 0, upb(A));
end;
```

The arithmetic operators are, in effect, pre-defined functions with a special syntax. It is notationally convenient to be able to use them to denote addition, subtraction, and so forth, even when the programmer needs to be more specific about the behavior of these operations by using the previously-described arithmetic functions (add, sub, etc.). Therefore, Silage provides a way of locally redefining the arithmetic operators for this purpose. For example, the definition

```
oper +(x,y) = add(x,y,saturating);
```

is essentially a macro definition that causes each occurrence of $E_0 + E_1$ in the scope of the definition to be replaced by

```
add(E₀, E₁, saturating)
```

(which specifies that overflows saturate).

15.3. EXAMPLES

To demonstrate the expressive power of the language, let us look at a number of simple examples, drawn from the signal processing literature.

15.3.1. The FIR filter

The FIR (Finite Impulse Response) filter, whose block-diagram is given in Figure 15.1, is one of the most common signal processing building blocks. The Silage description of an FIR filter is rather straightforward, as is demonstrated with the example of an 11th order $x/(\sin x)$ correction filter, required in front of a D/A converter.

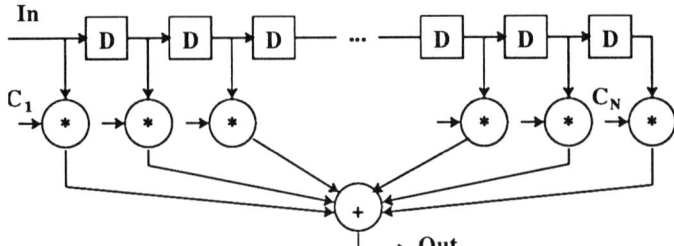

Figure 15.1: FIR signal-flow graph

```
#define word fix<8,0>
#define N 11
Coef = {  -0.001953125, 0.003906250, -0.007812500,
          0.01953125, -0.06640625, 0.75,
          -0.06640625, 0.01953125, -0.007812500,
          0.003906250, -0.001953125};
func FIR ( In : word ) Out : word =
begin
      acc[0] = 0;
      (i : 1 .. N ) ::
           acc[i] = acc[i-1] + word (coef[i-1] * In@(i-1));
      Out = acc[N];
end;
```

The cast operation after the multiplication explicitly denotes the introduction of truncation noise. This description can be simplified drastically using the reduction operator **sum**.

```
func FIR (In : word ) Out : word =
begin
      Out = sum({i:0..N-1:: word (coef[i] * In@i)});
end;
```

Chapter 15   DSP Specification Using the Silage Language   213

The frequency domain responses of the x/(sinx) filter, as obtained by the Silage simulator, are shown in Figure 15.2. The full line show the ideal response,

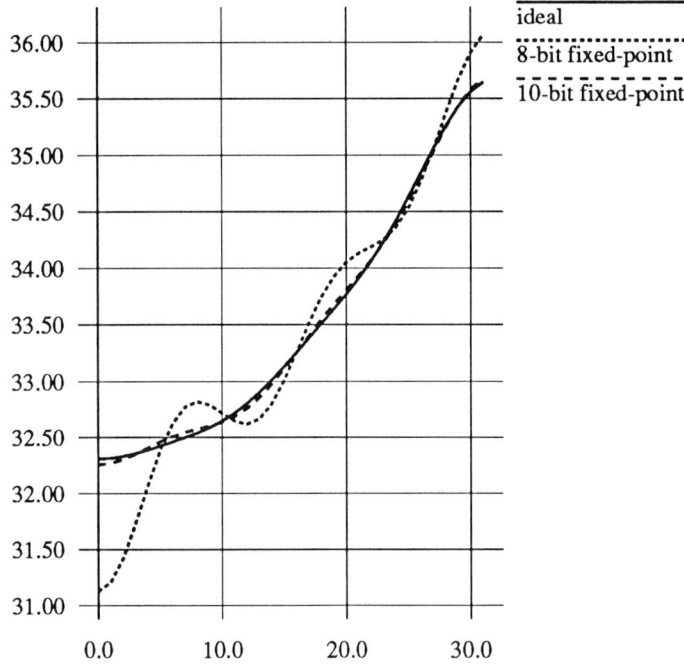

Figure 15.2: Frequency-domain response of 11th-order filter, showing ideal response and 8- and 10-bit fixed-point approximations. The x-axis gives relative frequency (64=sampling frequency), and the y-axis is in db.

as obtained from a floating point simulation, while the dotted lines are the results of the fixed point simulations with respectively 8- and 10-bit precision. This clearly demonstrates the importance of accurate data typing in signal processing applications.

The basic FIR filter is easily extended to an adaptive version:

```
func ADAPT_FIR(In, d : word) e : word =
begin
    e = d - sum( { i: 0..N-1 :: word (In@i * h[i]@1) } );
    h = { i : 0..N-1 :: h[i]@1 + K * e * In@i };
end;
```

15.3.2. CORDIC

The cordic rotation is a well-known arithmetic operation and is often used, for instance, to compute trigonometric functions in hardware accelerators. It has also been extensively used in signal processing applications such as FM-demodulation. A Cartesian to polar coordinate conversion is described using the following Silage code.

```
#define xyWord fix<22, 18>
#define phaseWord fix<22, 20>
#define N 20
#define Angle90 phaseWord (0.5)
#define Corrector xyWord(0.607253)

func CORDIC (Xin, Yin: xyWord; CorAngles: phaseWord[N])
        Amplitude: xyWord, Phase: phaseWord =
begin
    /* Rotate into the right-hand half of polar circle */
    (X[0], Y[0], Phi[0]) =
        if Yin >= 0 -> (Yin, -Xin, Angle90)
        ||           (-Yin, Xin, -Angle90)
        fi;

    /* Perform cordic iteration */
    (i : 1 .. N - 1) ::
    begin
    (X[i], Y[i], Phi[i]) =
        if Y[i - 1] >= 0 ->
                    (X[i - 1] + (Y[i - 1] >> (i - 1)),
                     Y[i - 1] - (X[i - 1] >> (i - 1)),
                     Phi[i - 1] + CorAngles[i - 1])
        ||          (X[i - 1] - (Y[i - 1] >> (i - 1)),
                     Y[i - 1] + (X[i - 1] >> (i - 1)),
                     Phi[i - 1] - CorAngles[i - 1])
        fi;
    end;

    Amplitude = X[N - 1] * Corrector;
    Phase = Phi[N - 1];
end;
```

Chapter 15   DSP Specification Using the Silage Language   215

Figure 15.3 plots the consecutive values of X, Y and Phi for the case of Xin and Yin equal to 1. Fast convergence is obtained towards the final values (Phase = $\pi/4$). The final X value has to be multiplied with a constant corrector value to give the exact amplitude.

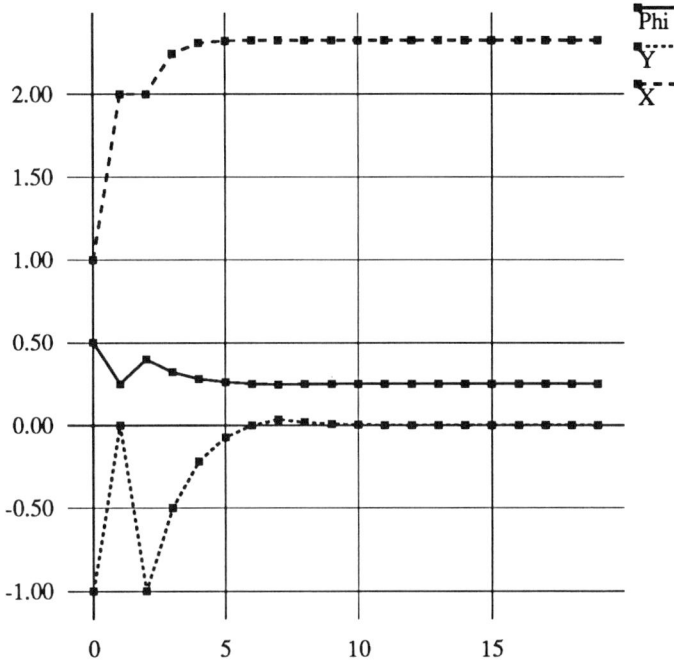

Figure 15.3: Time-domain simulation of cartesian to polar conversion, for Xin=Yin=1. The x-axis shows the number of iterations. The graph plots Phi (with 1.0 corresponding to π) and the amplitudes of X and Y.

15.3.3. PCM-FDM Transmultiplexer

The last example demonstrates the usage of the stream-operators such as switch and decimate. A PCM-FDM transmultiplexer takes as input a set of telephone channels (16 in this particular case, each with a bandwidth of 4KHz) and multiplexes them in the frequency domain, resulting in a single signal with a bandwidth of 64 Khz. The is realized by using a cascade of filter banks, each of them operating at different sampling rate, as diagramed in part (a) of Figure 15.4.

The sampling rate is doubled after each stage. Each of the boxes in the diagram is a lattice wave digital filter of the type shown in part (b) of Figure 15.4.

216 Behavioral Synthesis Part IV

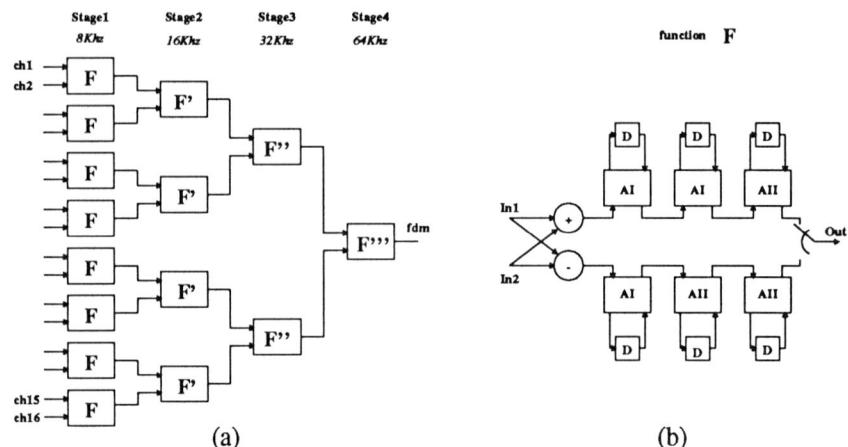

Figure 15.4: PCM-FDM transmultiplexer overall view (a) and basic filter structure (b).

The Silage description of this complete system is very compact. For the sake of brevity, only one of the filter blocks is described here.

```
#define wdata fix<30,8>
#define wcoef fix<16,14>

func TRANSMULTIPLEXER(ch : wdata[16]) fdm: wdata =
begin
        (i : 0 .. 7) ::      /* first stage */
          (Out1[i], Out1[8 + i]) = stage1(ch[2*i], ch[2*i+1]);
        In2 = switch(Out1, 8);

        (i : 0 .. 3) ::      /* second stage */
          (Out2[i], Out2[4 + i]) =
             stage2(In2[2*i],In2[2*i+1]);
        In3 = switch(Out2, 4);

        (i : 0 .. 1) ::      /* third stage */
             (Out3[i], Out3[2 + i]) =
                stage3(In3[2*i], In3[2*i+1]);
        In4 = switch(Out3, 2);
```

Chapter 15 DSP Specification Using the Silage Language

```
        /* fourth stage */
        (Out4[0], Out4[1]) = stage4(In4[0], In4[1]);
        fdm = interpolate(Out4);
end;

func stage1(in1, in2 : wdata) out1, out2 : wdata =
begin

        /* Butterfly Network */
        but1= in1 + in2;
        but2= in1 - in2;

        /* adaptor calls*/
        (n1, state1) = adapt1 (but1, state1@1, 0.070579);
        (n2, state2) = adapt1 (n1, state2@1, 0.459054);
        (n3, state3) = adapt2 (n2, state3@1, 0.1925556);
        (n4, state4) = adapt1 (but2, state4@1, 0.2474406);
        (n5, state5) = adapt2 (n4, state5@1, 0.3485903);
        (n6, state6) = adapt2 (n5, state6@1, 0.0626826);

        /* delay and scaling at the end of the second */
        /* lattice branch F'' */
        out2 = wdata ( wcoef(0.5) * n6);
        out1 = wdata ( wcoef(0.5) * n3);

end;
...
```

Figure 15.5 displays the frequency domain response for two different channels, as obtained from the Silage simulator, showing an excellent channel separation of over 60 db.

15.4. SILAGE-BASED TOOLS AND ENVIRONMENTS

Silage has been used as the front end language for a number of DSP simulation and synthesis environments. It was originally developed to serve as a high level description language for early version of the LAGER silicon compilation environment. This early version targeted the automatic synthesis of dedicated multi-processor architectures and evolved in a later phase into C-to-Silicon compiler, which is described in detail in Chapter 17. A simulator and language

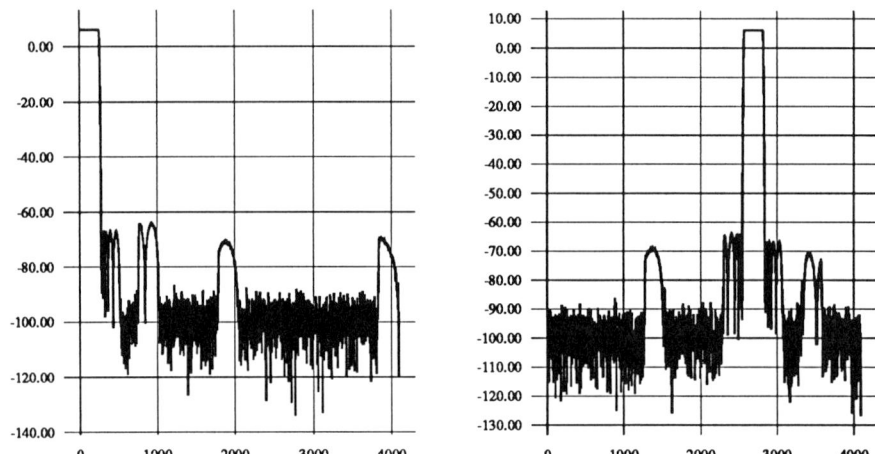

Figure 15.5: Frequency-domain response of PCM-FDM channels ch[1] and ch[16]. Frequencies on the x-axis are relative, with 4096 being the sampling frequency and the y-axis is in db.

parser and compiler were also developed [Wang88]. The output of the compiler is a description of the algorithm in RL, a procedural C-like language, which serves as the input to the C-to-Silicon environment.

Silage was also selected as the specification language for the CATHEDRAL-II environment, developed at the IMEC laboratory in Belgium [Deman86]. CATHEDRAL-II targets the synthesis of dedicated multi-processor architectures. Whereas in the C-to-Silicon environment the data-path architecture is manually defined by the user (or by a data-path template), the CATHEDRAL-II environment attempts to automatically derive the structure of the data paths as well. In order to help the compiler to converge towards an efficient hardware solution, CATHEDRAL-II makes extensive use of the Silage *pragmas*, which are pragmatic hints to the compiler. Pragmas can be used as structural specifications (where the rest of the Silage input only describes behavior). Here are two examples.

```
pragma(alloc, mult, 2);
pragma(assign, "a*_", mult1);
```

The first pragma tells the compiler to provide and use two multipliers, while the second causes all multiplications with the variable a to be performed on a multiplier called

Chapter 15   DSP Specification Using the Silage Language   219

mult1 (the "_" serves as a wild card, matching every possible identifier). Within CATHEDRAL-II, a Silage simulator, called s2c was developed as well. The simulator first translates the Silage code into a C program, which is compiled into an executable simulator. This approach is very efficient and allows for the debugging and optimization of large Silage programs (such as for singular value decomposition, adaptive interpolation for digital audio, and CELP-based speech coders) using extensive data files.

The CATHEDRAL-II compiler has been adapted and reworked by the Philips Corp. into the PIRAMID environment [Delaruelle88]. From this commercialization and others (EDC) a number of language adaptations and extensions have been introduced. In addition other compilers have been either interfaced to or developed around Silage. The CATHEDRAL-I environment targets bit-serial architectures [Jain86]. Silage compilers for general purpose DSP processors such as the TMS320C25 and TMS32030 have proven to be far more efficient than their C counterparts [Genin89].

Finally, Silage is also being used as the specification language for the HYPER system described in Chapter 15. HYPER provides a complete synthesis path from high-level behavioral description in Silage to chip layout within LAGER framework. It is particularly efficient for data-path intensive architectures such as those encountered in image, video and speech processing.

15.5. SUMMARY

Silage was specialized for use in DSP applications, and was not intended to be a general-purpose programming language. Even within DSP, certain applications can be awkward to write or difficult to compile efficiently. This has motivated some planned extensions to Silage, most notably a facility for controlled use of traditional imperative procedures whose code uses side-effects, which can be useful, for example, in matrix computations.

Recent progress in programming languages has been toward more declarative programming, that is, towards a style in which the programmer describes the desired effect, rather than the method of achieving it. In DSP, Silage has provided some progress in this direction, by automating some of the "data choreography" and allowing programmers to abstract away from messy implementation details. With the development of libraries of Silage functions representing increasingly complex components, as illustrated in the previous section, we come still closer to the ideal of a declarative DSP programming environment.

REFERENCES

[Delaruelle88] A. Delaruelle et al., "Design of a syndrome generator chip using the PIRAMID Design System", *Proc. ESSCIRC Conf.*, pp 256-259, Manchester, September 1988.

[Deman86] H. De Man, J. Rabaey, P. Six and L. Claesen, "CATHEDRAL-II: A Silicon Compiler for Digital Signal Processing Multiprocessor VLSI Systems", *Design & Test of Computers*, pp. 13-25, December 1986.

[Genin89] D. Genin, J. De Moortel, D. Desmet and E. Van de Velde, "System Design, Optimization, and Intelligent Code Generation for Standard DSP", *Proc. IEEE ISCAS Conf. 1989*, Portland, pp. 565-569, May 1989.

[Goering90] R. Goering, "DSP-Tools—Hammering Out New Solutions", in *High Performance Systems*, pp. 20–31, Feb. 1990.

[Jain86] R. Jain et al., "Custom Design of a VLSI PCM-FDM transmultiplexer from system specifications to circuit layout using a computer aided design system", *IEEE Journal on Solid State Circuits*, vol. SC-21, no. 1, pp. 73-85, February 1986.

.[Lee89] E. Lee, *et al.*, "Gabriel: A Design Environment for DSP", *IEEE Transactions on ASSP*, pp. 1751-1762, Nov. 1989.

[Wang88] E. Wang, "A Compiler for Silage", Master's Thesis, University of California at Berkeley, December, 1988.

16

Synthesis of Datapath Architectures

Jan Rabaey, Chi-min Chu, Phu Hoang and Miodrag Potkonjak

The computationally intensive parts of high performance real time systems, such as those encountered in the areas of telecommunications, speech, robotics, video and image processing, are usually implemented on clusters of heavily pipelined datapaths, controlled by a relatively simple finite state machine.

A typical example of such an architecture can be found in Figure 16.1, which shows the datapath section of a Viterbi Processor used in connected speech recognition. The extremely high throughput requirements exclude the use of a classical Von Neumann style architecture where all operations are multiplexed on a single data general purpose datapath and the functionality is determined primarily by the contents of the controller. The small ratio between the sampling frequency and the clock frequency prohibits extensive operation multiplexing. The emphasis is therefore shifted from the control to the datapath section. As can be seen in Figure 16.1, the datapath is an almost direct representation of the computational graph. A dedicated hardware unit has been assigned to each operator. Pipelining has been added to meet the clocking requirements. the controller of this processor is a simple finite state machine, containing only 20 states. Another important observation regarding this example is that a direct mapping of the computational graph into a hardware structure would not have resulted in a real-time performance. Extra flow-graph transformations, such as the overlapping of the iterations of the inner loop were necessary.

Designers usually adopt either a direct one-to-one mapping approach where a dedicated hardware unit is provided for each operation, or a multi-processor approach where the algorithm is partitioned over a number of concurrent processors, each consisting of a multi-function ALU with a simple controller. The latter approach is often used in image and video processing (e.g. [Roermund89]). However, in many cases the ideal solution is situated somewhere between the two extremes: the algorithm is best implemented on a cluster of dedicated datapaths with limited resource sharing, and the computation is controlled by a simple controller.

Synthesizing such an architecture is however a tough task for a human designer. For example, designers tend to over-emphasize pipelining, not realizing that the cost of a register is between 1/3 to 1/2 the cost of an adder. A major part of the design effort is usually devoted to optimizing the clock frequency, whereas optimizing the algorithm flow graph may result in solutions that are orders of

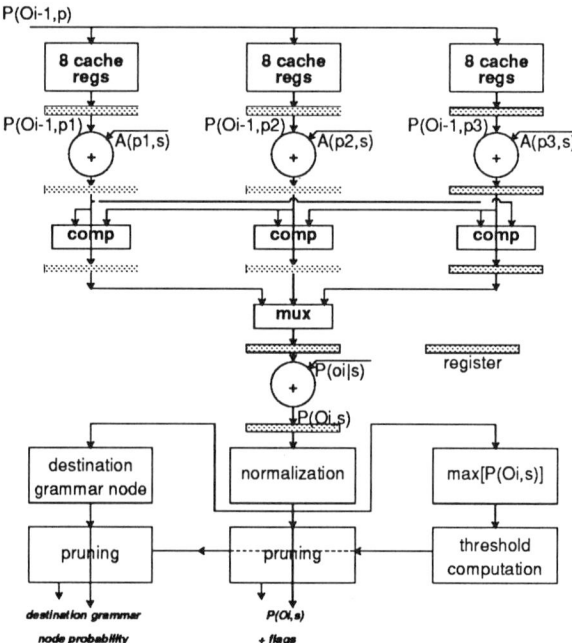

Figure 16.1: Datapath section of Viterbi processor

magnitude better. The HYPER set of tools were developed to automate the process of optimizing an algorithm flow graph and synthesizing a physical implementation within the LAGER framework.

16.1. HYPER OVERVIEW AND METHODOLOGY

The synthesis for real time applications can be defined as the following optimization process:

> Given: An input computational graph, a number of real-time constraints and a hardware cell library.

> Find: The hardware implementation with the minimal area.

This process requires the execution of a large number of operations, each of them having considerable complexity. This is demonstrated in Figure 16.2 which shows the overall composition of the HYPER system. The real time application is described using the SILAGE signal flow graph language. This description is parsed and compiled into an intermediate *Control/Data Flow Graph* database. This CDFG represents the algorithm as a data flow graph, extended with some macro control flow statements such as loops and if-else structures. During the parsing process, a number of standard, architecture independent, compiler transformations, such as the elimination of dead code, manifest expressions, common sub-expressions and algebraic identities, are executed.

The data flow serves as a central repository on which synthesis operations, such as complexity estimations, flow graph transformations, hardware allocation and scheduling, are executed. The results of these operations are back-annotated into the database. As a result, the HYPER system has a modular composition which allows new tools to be easily integrated. At every point in the design process a simulation model of the flow graph can be generated, which allows for a verification of the correctness of the executed operations and a checking of their effects on the performance parameters such as the signal-to-noise ratio.

The HYPER synthesis tool box will be discussed in detail in the subsequent sections. A brief description of each of class of operations is however instructive at this point.

- **Module Selection:** This is one of the first steps in the synthesis process. It selects an appropriate hardware library module for each flow graph operation. At the same time, groups of operations are clustered together into *primitive hardware nodes*, which can be executed within one clock-period, and are thus fully combinational.

- **Estimation:** Derives the minimum and maximum bounds on the required hardware resources. This information serves as an initial solution and helps in selecting the next synthesis operation to perform.
- **Transformations:** Manipulate the signal flow graph of the algorithm to improve the final implementation, without changing the input-output relation. Typical transformations are retiming, loop unrolling and software pipelining.

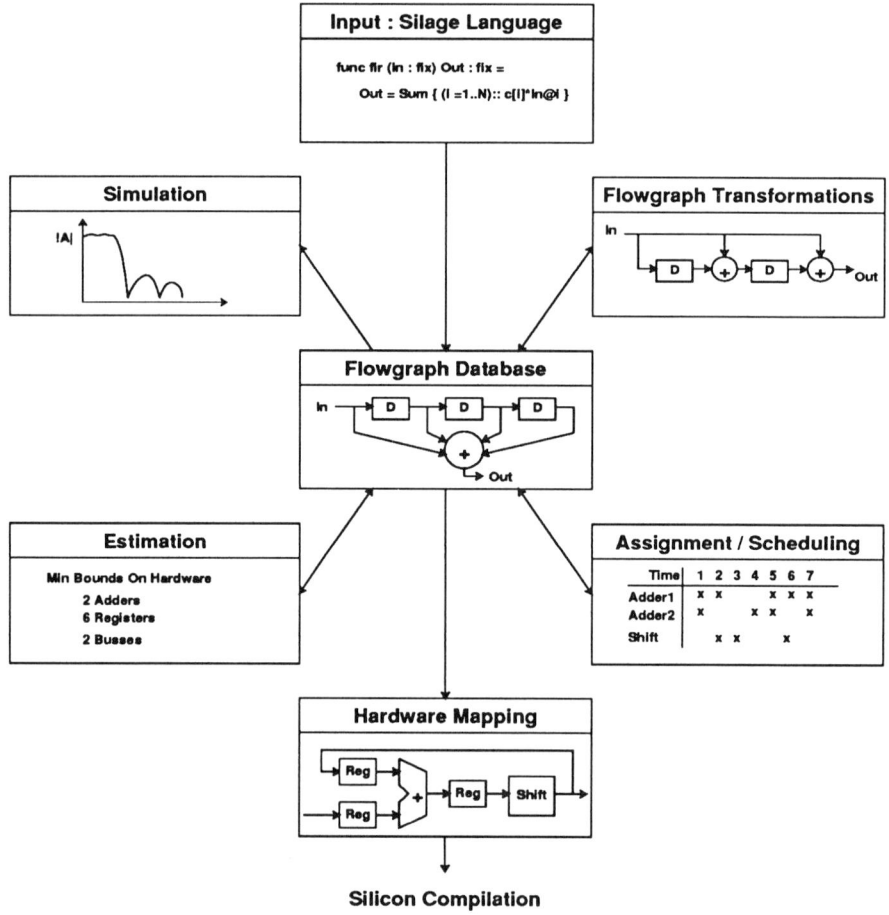

Figure 16.2: HYPER - general overview

- **Allocation, Assignment and Scheduling:** Select the amount of hardware resources (execution units, registers and interconnect), needed for the execution of the algorithm. Bind each flow graph operation to a particular hardware unit and time slot.

Finding an optimal solution for the hardware synthesis problem is a non-trivial task, due to the facts that most of the synthesis operations mentioned above are at least NP-hard. Furthermore, the ordering of the operations (such as the transformations) affects the quality of the final solution. We have opted in HYPER to implement the overall synthesis procedure as a search process: Starting from an initial solution, new solutions can be proposed by executing a number of basic moves such as adding or removing resources (in other words, changing the allocation), changing the time-allocation for the different sub-graphs of the algorithm and applying an optimizing graph transformation. The feasibility and the precise cost of the current solution is checked by the assignment-scheduling module. The overall search is managed by the **synthesis-manager**, which manages the overall synthesis process and decides (either interactively or automatically) what move to perform next. This decision is driven by the results of the estimation process and the feedback of the scheduling module on bottlenecks and problem areas. The user interface of the synthesis manager is shown in Figure 16.3.

Once an acceptable solution is obtained, the search is halted and the graph is mapped into hardware (Figure 16.2) and stored in the OCT database as a structure_instance view (Chapter 5) for use by DMoct and the layout generation tools.

The different modules mentioned above will now be described in detail. A single example (7th order biquadratic IIR filter) will be used throughout to demonstrate the effects of the various synthesis operations and the quality of the proposed algorithms.

16.2. BEHAVIORAL SPECIFICATION AND SIMULATION

A proper representation of the input algorithm is crucial to the performance of any synthesis environment. The input representation should allow for efficient synthesis of the algorithm independent of whether it is data flow oriented, control flow oriented, or both. Information on the data flow of the algorithm exposes all of the available parallelism in the algorithm, which allows for area/performance trade-off's. Information on the overall control flow of the algorithm results in a fast and area efficient control unit. For these reasons, a mixed *control/data flow graph* (CDFG) representation is used at the heart of the HYPER synthesis system.

16.2.1. Control/Data Flow Graph (CDFG)

The CDFG represents the algorithm as a flow graph with nodes, data edges, and control edges. The nodes represent data operations, while the data edges represent data precedences between those nodes. In addition, control edges can be introduced to enforce extra precedence rules between nodes (e.g. the execution time of operation X has to trail the execution of operation Y by at least N clock cycles).

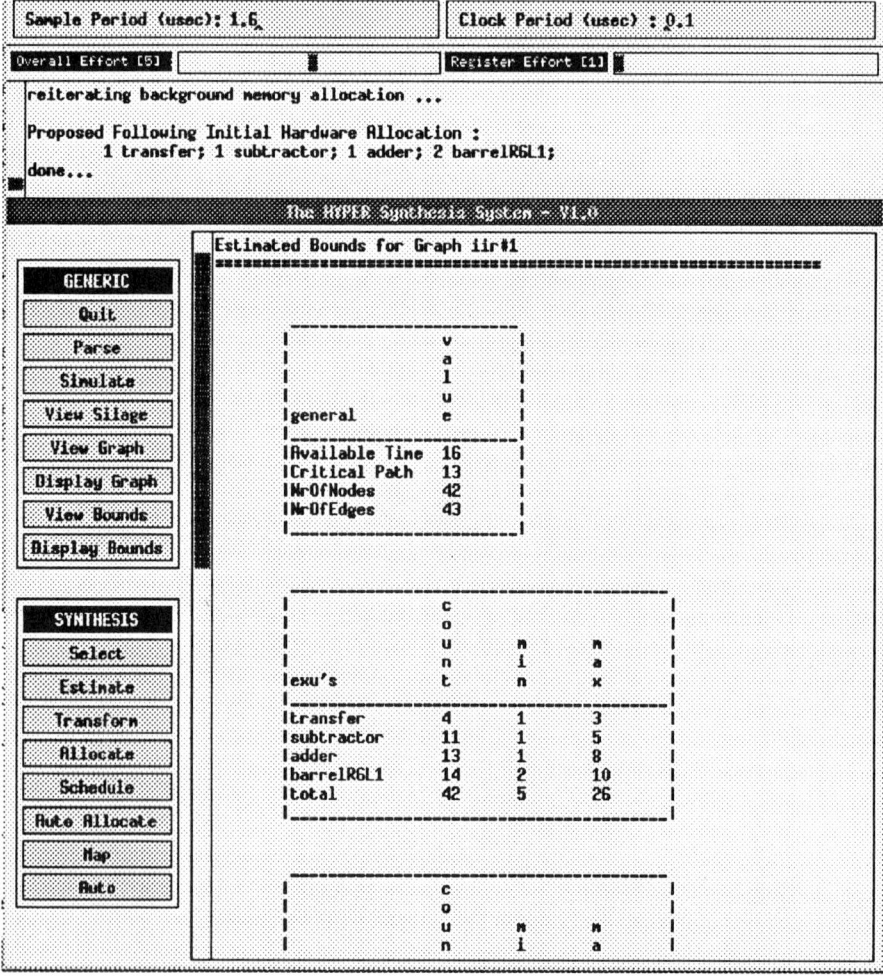

Figure 16.3: HYPER Synthesis Manager - User Interface

Chapter 16 Synthesis of Datapath Architectures

Besides the standard arithmetic operations, the CDFG allows a number of *macro control flow* operations such as loops and if-then-else blocks. The introduction of these control statements results in a hierarchical graph: The body of a loop or a conditional is represented by a sub-graph, which is contracted into a single node at the next higher level in the hierarchy. This hierarchical representation has the advantages of compactness and descriptiveness. The flow graphs are stored more efficiently. It also allows for a clean definition of the macro control-flow of the algorithm resulting in more efficient control structures. Finally, the hierarchical structure helps in preserving the structural hints from the designers.

The flow graph is stored in the OCT database (Chapter 2) using a policy detailed in [Rabaey89]. This policy is meant to capture the *structure* of the flow graph, not its behavior. It describes the interconnection of nodes and edges, as well as where specific information on the behavior of the node is found. A library of behavioral primitives (addition, multiplication, delay, decimation, etc.) is provided as a start. The user can easily define his own primitives. By storing only the structure, the same representation can be used to support a variety of front-ends, such as SILAGE (Chapter 15), GABRIEL [Lee89] and McDAS [Hoang92].

In addition, an ASCII format flow graph description language [Rabaey89] is available. This format (called AFL) has a one to one correspondence to the OCT policy, and serves as an easy readable interface to the OCT database. A tool to translate the AFL format to OCT, and vice versa, has been developed.

16.2.2. Silage To Flow Graph

Silage (Chapter 15) is a signal-flow language developed especially for DSP specification. A translator was developed to convert a Silage program to a CDFG

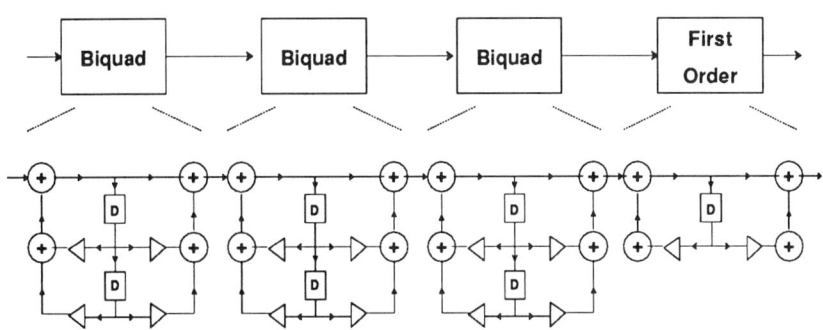

Figure 16.4: 7th order IIR: flow graph structure

with essentially the same hierarchical structure. A function call in Silage is represented by a *Func* node which has a pointer to a subgraph representing the function body. Similarly, an iteration is Silage is represented by an *Iter* node with a pointer to a subgraph representing the loop body. The translator generates the flow graph in OCT by default, and optionally also in the ascii AFL format.

```
#define num16 fix<32,8>
#define Coef0 0.001953125
#define Coef1_1 -1.3125
#define Coef1_2 0.625
#define Coef1_3 1
#define Coef1_4 1
#define Coef2_1 -1.25
#define Coef2_2 0.75
#define Coef2_3 0.0625
#define Coef2_4 1
#define Coef3_1 -1.125
#define Coef3_2 0.921875
#define Coef3_3 -0.25
#define Coef3_4 1
#define Coef4_1 -0.71875
#define Coef4_2 1
func main (In : num16) Out : num16 =
begin
      In1 = num16(In*Coef0);
      In2 = biquad(In1, Coef1_1, Coef1_2, Coef1_3, Coef1_4);
      In3 = biquad(In2, Coef2_1, Coef2_2, Coef2_3, Coef2_4);
      In4 = biquad(In3, Coef3_1, Coef3_2, Coef3_3, Coef3_4);
      Out = firstorder(In4, Coef4_1, Coef4_2);
end;
func biquad(in, a1, a2, b1, b2 : num16) : num16 =
begin
      state@@1 = 0.0;
      state@@2 = 0.0;
      state = in - (num16(a1*state@1) + num16(a2*state@2));
      return = state + num16(b1*state@1) + num16(b2*state@2)
end;
func firstorder(in, a1, b1: num16) : num16 =
begin
      state@@1 = 0.0;
      state = in - num16(a1*state@1);
      return = state + num16(b1*state@1);
end;
```

Figure 16.5: 7th order IIR: Silage description

Figure 16.5 gives the Silage program of our example 7th order IIR filter. The filter is composed of 3 cascaded biquadratic and 1 first-order section, as also shown in the block diagram of Figure 16.4. This hierarchical structure, defined by the user in the Silage program, is retained in the initial CDFG. Subsequent synthesis operations might however manipulate this hierarchy, either by flattening or by clustering nodes.

16.2.3. Behavioral Simulation

Simulation of the algorithm is an essential part of the synthesis process. It is needed not only to verify the functionality of the algorithm and the correctness of the transformations performed, but also to optimize and check the values of performance parameters like the signal to noise ratio, the effects of truncation on the transfer function, the distortion and the presence of small and large scale limit cycles. For instance, a simple flow graph transformation, which replaces a multiplication with a constant by a sequence of add/shift operations is known to change the effects of truncation and hence also the signal to noise ratio of the algorithm.

A compiler was therefore developed to translate the CDFG description into executable C-code. Two simulation models are generated: the first uses floating point data types, while the second models data as fixed point entities. The floating point mode offers (quasi)-infinite precision, while the fixed point mode uses the exact data type, defined in the flow graph and thus allows for the modeling of quantization effects. In this way, the distortion of the system response due to quantization effects can be accurately monitored.

16.3. MODULE SELECTION

Given the behavior description of an algorithm represented by a signal flow graph, the goal of the hardware selection process is to select the clock period (if not specified by the user), to choose proper hardware modules for all operations and to determine where to pipeline (or where to put registers), such that a minimal hardware cost is obtained under given timing and throughput constraints.

Most published datapath synthesis systems either consider only a fully pipelined architecture [Jain88] (which means that each intermediate result is stored in a register) or do not consider pipelining and resource sharing simultaneously [Note90]. These restrictions tend to result in inefficient solutions. If an algorithm is fully pipelined, the available clock period might not be completely exploited, due to mismatches between the execution times of the operators. Furthermore, performing operations in sequence without intermediate buffering can result in a reduction of the critical path. For instance, the delay of two carry-propagate additions in series is shorter than two times the delay of a single addition, since the carry propagation has to be accounted for only once. HYPER attempts to perform as many operations in sequence as allowed by the clock period, while still trying to maximize the resource sharing factor. This requires that when clustering operations into non-pipelined hardware modules, the reusability of these modules over the complete computational graph be maximized.

Both the hardware selection and the pipelining process are at least NP-hard problems. These problems become even more complicated when considering the timing constraints and possible multiple-function units, such as an ALU. Therefore we have opted for a heuristic approach based on operation clustering as shown in Figure 16.6.

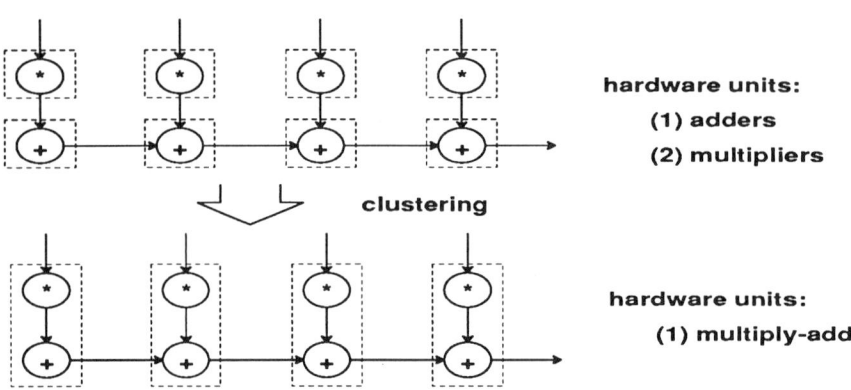

Figure 16.6: Operation clustering

The search starts from an initial solution with all operations implemented on the cheapest hardware and with full pipelining. HYPER then clusters operations in a way that favors structures with a high reusability factor. It simultaneously ensures that clustering does not violate timing constraints. During clustering, more expensive but faster hardware may be swapped in for operations on the critical path.

This approach requires the implementation of three basic tasks: a search strategy to determine the next move, a hardware cost estimation to evaluate the

Chapter 16   Synthesis of Datapath Architectures

effect of a move on the cost function and a timing estimation module to check if a proposed solution does not violate any timing constraints.

Currently, a simple greedy search mechanism is used. The basic moves in the search process are either the clustering of a set of nodes or the swapping in of more expensive hardware. The effect of a move on the overall hardware cost is evaluated by estimating the required hardware using a fast parallelism analysis (as discussed later in the section on estimation). The reusability of a cluster can further be determined by an analysis of the occurrence of that cluster in the computation graph.

During the clustering process, the delay of each proposed cluster has to be checked against the available clock-period. An accurate, yet easily computable timing model is therefore required. A fully expanded bit-level model was proposed in [Note90]. This model is accurate but inefficient to calculate. In our case the timing analysis has to be performed repeatedly during the optimization process. HYPER therefore uses a *ripple model* to simplify the timing estimation problem.

The ripple model characterizes a hardware block by three parameters: the ripple delay, the ripple direction and the one-bit delay. The propagation delay of an individual module is defined by the *ripple delay*, which is normally a function of the word length of the operator. This function can be any expression and can capture the characteristics of complex operators such as carry-select and carry-look-ahead adders. The *ripple direction* expresses the direction of the internal ripple (left, right or no-ripple) and is used to determine the propagation delay of cascaded modules. Finally, the one-bit delay is the delay of the one-bit operation. It can also be a function of some parameters such as the word length.

The ripple model estimates the critical path of a flow graph by tracing the graph and maintaining three parameter values for each edge - the longest ripple delay so far, the ripple direction, and the total accumulated delay. These values of an edge can be derived from the three ripple parameters of its input node and the input edges of the node. If a ripple operation has a smaller ripple-delay than the longest ripple-delay of its input edges and both ripples have the same direction, only the one-bit delay is added to the accumulated delay and thus to the critical path. The ripple delay of this operation is ignored since it overlaps with the longest ripple delay. Through simple rules such as this an algorithm using the ripple model efficiently and accurately calculates the critical path of a flow graph.

16.4. ESTIMATION

In the estimation phase, min and max bounds on the required resources are deduced [Rabaey90]. These bounds are important for several reasons: first of all, they delimit the design space, thus speeding up the design synthesis search process. The computed min-bounds can serve as an initial solution, which from our experience is often very close to the final solution. The bounds also serve as entries in a **resource utilization table** which helps to guide the transformation, assignment and scheduling operations. In order to be useful, it is essential that these bounds be as sharp as possible. To obtain this goal a technique of gradual refinement is used. Let us consider first a flat graph with a max-bound on the execution time t_{max}.

The estimation process starts with a topological ordering and leveling of the graph with respect to the input nodes and the output nodes, yielding a minimum and maximum execution time for each operation Oi (t_{min}^{Oi} and t_{max}^{Oi} respectively). In the literature these times are often called the *as soon as possible* (ASAP) and *as late as possible* (ALAP) execution times.

An upper bound on each resource is easily obtained from the ordered graph by computing for each clock cycle the maximal possible usage of that resource (in other words, the maximal parallelism available in the graph) and by finding the maximum of this value over the entire time period. Notice that a resource could be an execution unit, a register, an interconnection between execution units or an input/output bus. For the sake of brevity, we will concentrate our description on execution units.

This procedure is demonstrated for the example of a 7th order IIR low pass filter, whose signal flow graph was given in Figure 5. Since a hardware multiplier is too expensive, it was decided to expand the multiplication operation into shifts and adds. The parallelism graph for this expanded flow graph is shown in Figure 16.7. The graphs shown display the maximum available parallelism (here in terms of the number of additions, subtractions and shifts), plotted over time. It is assumed here that a maximum of 16 clock cycles is available and that each operation takes exactly one clock cycle. It is clear from this figure that the maximum parallelism in additions, subtractions and shifts equals respectively 6, 4 and 8.

Deriving a precise lower bound is somewhat more complicated. A crude guess (called the *naive lower bound*) is first obtained by observing that given a number of resources of class Ri (N_{Ri}), at most $N_{Ri}*t_{max}/d_{Ri}$ operations can be

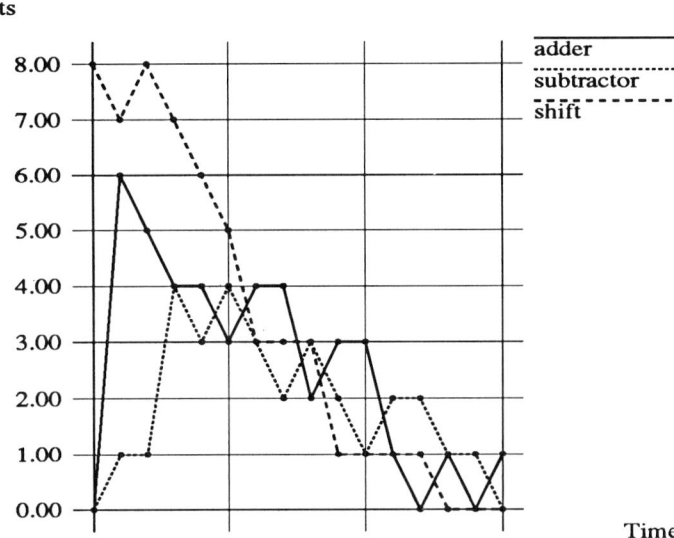

Figure 16.7: Parallelism graphs (multiply, shift, add) for 7th order filter

performed on those resources (where d_{Ri} is the duration of a single operation). The required number of operations (O_{Ri}) can be easily derived from the computational graph, resulting in the following lower bound on N_{Ri}:

$$N_{Ri} \geq \frac{O_{Ri} \times d_{Ri}}{t_{max}} \tag{16.1}$$

This bound is however too optimistic: it assumes that the flow graph contains sufficient parallelism to support a 100% utilization of the resource, which is obviously rarely the case. This is demonstrated in Figure 16.7, where the available number of shift operations drops below 3 after cycle 8. This observation results in a more precise min bound:

$$N_{Ri} \geq \frac{O_{Ri} \times d_{Ri} + \text{UnusedTime}}{t_{max}} \tag{16.2}$$

Unused Time actually depends on N_{Ri} and can be deduced from the parallelism graph. Therefore, (EQ 2) has to be solved iteratively starting from the initial solution obtained using (EQ 1).

A sharper lower bound can be determined using a technique called *discrete relaxation*. Given N_{Ri}, the minimal execution time t_{min} is determined using a slack based list scheduling, considering only operations of class i. This too is an iterative procedure: starting from the solution obtained in (EQ 2), N_{Ri} is increased till $t_{min} \leq t_{max}$. The scheduling operation is of complexity $N_{Ri} \log(N_{Ri})$ [Simons81], if $d_{Ri} = 1$ (an approximate solution can be found with the same complexity when $d_{Ri} \neq 1$).

The min and max bounds obtained by the above algorithm for the example of Figure 16.4 are shown in Table 16.1.

Resource	No. of Operations	MinBound	MaxBound
add	15	2	6
shift	12	2	8
subtract	9	1	4
add (registers)	30	8	30
shift (regsiters)	12	8	12
subtract (registers)	18	4	8
add-add	4	1	3
add-shift	4	1	4
add-subtract	6	1	3
add-io	1	1	1
shift-add	8	1	7
...			

Table 16.1: Min and max bounds on execution units, registers and interconnect for 7th order biquadratic filter (Figure 16.4)

The situation becomes more complex when considering hierarchical graphs (containing loops and if-else constructs). Since the only time constraint is the total execution time for the complete graph, the time allotted to each subgraph is unknown and is a subject of optimization itself. Once again *discrete relaxation* offers the solution. For each sub-graph and each resource a resource-time graph is

Chapter 16 Synthesis of Datapath Architectures 235

constructed. It plots the minimum number of resources it takes to execute the sub-graph as a function of the available time. Figure 16.8 shows the resource-time graphs for the adder and shifter resources in the filter example.

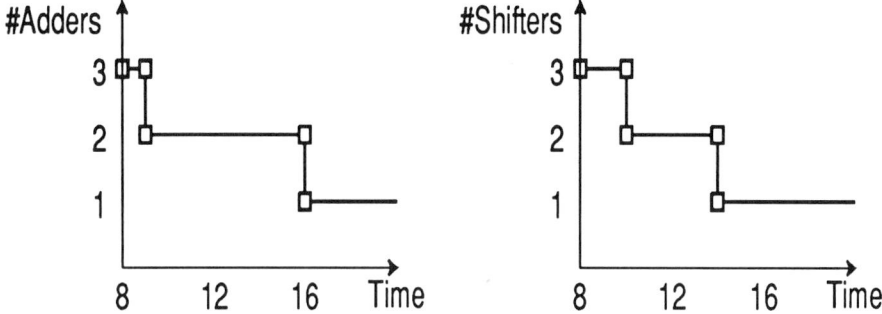

Figure 16.8: Area-time trade-off graphs for the 7th order filter

Hierarchical estimation is now straightforward: the resource-time graph of a hierarchical graph is constructed by combining the results of its sub-graphs. This is demonstrated in Figure 16.9 for a function *Total* which is composed of two sub-graphs (*Func1* and *Func2*). From the combined resource-time graphs it can be deduced that at least two adders are needed when the total available time to execute *Total* equals 34 clock cycles.

The results of the estimation process are summarized in the **resource utilization table** which tabulates for each sub-graph the min-bounds on hardware and time resources. The entries in this table serve as an initial seed as well as selection measure in the allocation and transformation process described next. A sample table is shown in Table 16.2.

16.5. EXPLORING THE DESIGN SPACE

The goal of the design space exploration and resource allocation process is to find the minimal area solution, which still complies with the timing constraints.

Before this process starts, it has to determine whether a feasible solution exists. By checking the critical paths, it can be determined if the proposed graph violates the timing constraints. If so, performance optimizing transformations such as retiming for critical path [Leiserson83], pipelining and tree height reduction can be applied. After an acceptable graph is obtained, the resource allocation

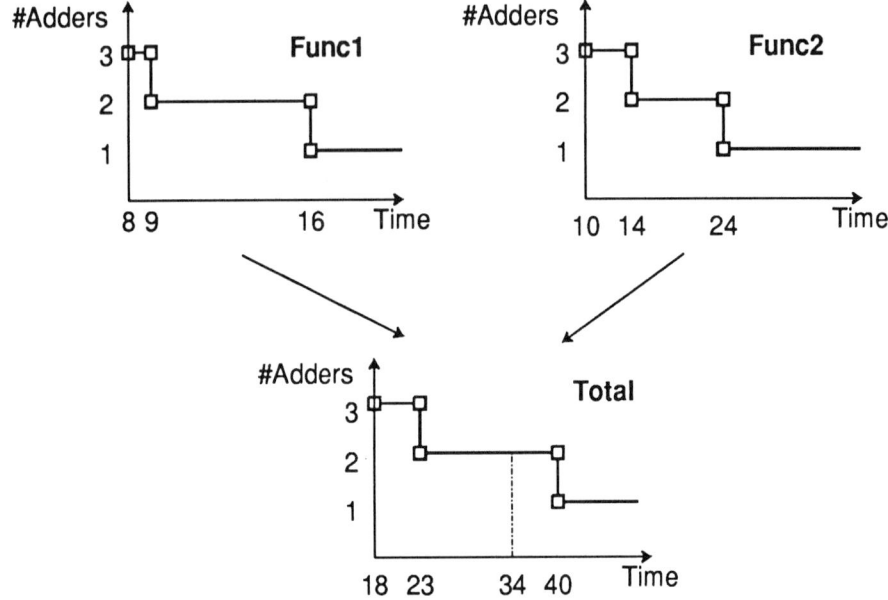

Figure 16.9: Hierarchical composition of resource-time graphs.

Block	Critical Path	Cycles	IO	*	+	Registers
Graph1	c1	t1	1	0	1	12
Graph2	c2	t2	0	1	4	36
Graph3	c3	t3	1	2	1	18
Total	$c = \sum c_i$	$t = \sum t_i$	1	2	4	36

Table 16.2: Sample resource utilization table

process is initiated. As we explained earlier, the design space exploration in HYPER is organized as a search process [Rabaey90]. The search is organized as an iterative process, where new solutions are proposed by applying basic moves. Those moves can be categorized in three classes:

Chapter 16 Synthesis of Datapath Architectures 237

- Changing the available hardware (also called the **hardware allocation**). The hardware allocation includes the number of execution units, registers and busses.
- Redistributing the time allocation over the sub-graphs.
- Transforming the graph to reduce the hardware requirements. Possible transformations here are pipelining and the application of arithmetic laws such as associativity and commutativity.

Since many different moves are feasible at any point during the search, we need an accurate, yet easily computable measure to rank the candidate moves according to their effect on the overall cost function. This information is given by the **resource utilization table**, resulting from the estimation process. Consider for instance Table 16.2. The number of adders required for sub-graph 2 is disproportionate to the required adders in the other sub-graphs. One move would be to extend the time allocated to sub-graph 2 or to select a transformation (such as retiming) which reduces the min-bounds on the additions in this sub-graph. In effect, the latter transformation would also reduce the overall min-bound, as the bottom row of the table shows. In this sense, the utilization table serves both as a global measure of the quality of a proposed solution and as a guide for selecting moves.

Further move ranking can be obtained from the assignment and scheduling process discussed later. Each time a promising solution is proposed, the assignment and scheduling module is applied to determine the feasibility of this solution. During its execution the module gathers a number of statistics regarding the ease of scheduling - such as which resources are in short supply (hence hampering the chances for successful scheduling) and which resources were over-supplied. This feedback information is extremely useful in helping to select the next move.

Since the search mechanism has to simultaneously address resource allocation and transformations, it is obvious that the optimization strategy should be flexible enough to handle the variety of constraints imposed by these problems. This suggest a probabilistic iterative improvement algorithm such as simulated annealing. However, the application of transformations (as well as the scheduling and the assignment process) is computationally expensive. We have therefore adopted a rejectionless probabilistic iterative search technique, where moves are always accepted, once executed [Welch84]. Our approach gives faster convergence and reduces computational complexity. Our initial experiments have demonstrated that convergence is obtained with a fairly small number of steps - three to six allocation moves appear to be sufficient for most benchmarks.

16.6. TRANSFORMATIONS

A behavioral transformation reorganizes the signal flow graph of an algorithm to improve the quality of its final implementation without altering the input-output relationships. Most of those transformations have been introduced in the field of software compilers. They include the elimination of constant arithmetic and common sub-expressions, dead-code elimination and the application of algebraic laws such as commutativity, distributivity and associativity.

Most of the recent attention in this area has been focused on the transformation of loops, since most of the parallelism in an algorithm is embodied in the loops. Thus loops are likely to have the most dramatic effects on the quality and performance of a solution.The most important loop transformations are loop jamming, partial and complete loop unroling, strength reduction and the more recent loop retiming and software pipelining [Lam85, Goossens89]. An interesting observation is that loop transformations are even more effective in real time systems, where most programs contain an infinite loop over time. This feature has been used extensively in the signal processing literature, for instance as a means to implement very fast recursive filters [Messerschmitt88].

Optimizing transformations are extremely important in the hardware synthesis process and have far more effect on the quality of the final solution than for instance the assignment and scheduling. As mentioned earlier, in HYPER the transformation process is an integral part of the design space exploration. The majority of the above discussed transformations have been implemented. In addition, a number of novel transformations specifically geared towards the needs of a hardware compiler have also been developed. An example of such a transformation is the **retiming for resource utilization** which is described in detail below.

Retiming is a powerful and conceptually simple transformation which has been applied successfully in several areas of design synthesis and automation. The goal of the retiming transformation is to move delays (which can be either clocked delays in a circuit or algorithmic delays in a behavioral flow graph, depending upon application) such that a certain objective function is optimized. Until recently, the objective function has been exclusively the critical path or the number of delays in a graph or a circuit [Leiserson83]. However, the potential of retiming is significantly higher. We have developed a new formulation, this time

Chapter 16 Synthesis of Datapath Architectures

targeted towards behavioral synthesis: given a signal flow graph, retime it in a such a way that the resulting signal flow graph will have a minimum hardware cost, while still satisfying all timing constraints. Since the implementation with minimum cost has the most efficient resource utilization, we call this transformation: retiming for efficient resource utilization.

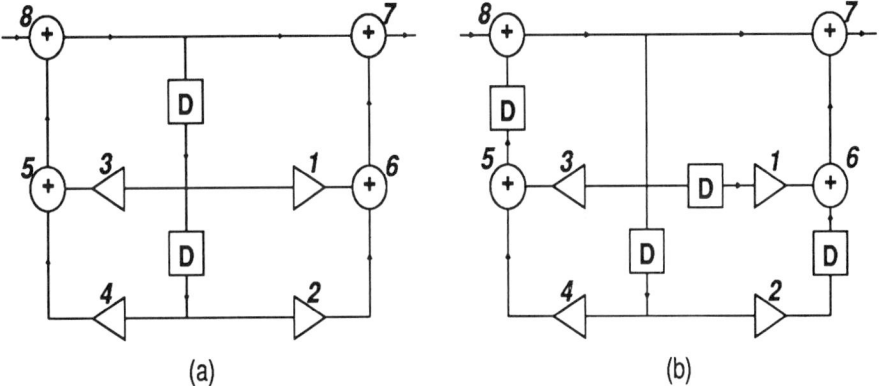

Figure 16.10: Biquadratic filter (a) before and (b) after retiming for scheduling

Consider, for instance, the biquadratic second order filter shown in Figure 16.10. Assume that at most 4 clock cycles are available for the execution of this flow graph and that both multiplication and addition take a single clock cycle. The critical path of this computational graph equals four clock cycles as well. It is obvious that at least two multipliers are required to implement this flow graph. This is due to the fact that all multiplications are clustered in the earlier stages of the program, while the additions can only be performed in the final cycles. For example, no multiplication can be executed in the fourth cycle, and no addition can be executed in the first cycle. The resource utilization is obviously not equally balanced over time. If we define the resource utilization as the ratio of the number of cycles a resource is used versus the total number of available cycles, then the resource utilization for adders and multipliers in this example is 50%, which is an indication of a relatively low quality solution. Table 16.3a shows a possible schedule for this filter.

Consider now an equivalent flow graph, shown in Figure 16.10b, obtained by moving the delays (retiming) in the original flow graph. A solution with one

	BEFORE		AFTER	
Cycle	Multiplier	Adder	Multiplier	Adder
1	3,4	-	1	8
2	1,2	5	3	6
3	-	6	4	7
4	-	7,8	2	5

Table 16.3: (a) Possible filter schedule (before retiming)
(b) Filter Schedule after Retiming

multiplier and one adder can be achieved as shown in Table 16.3b. The resource utilization for the execution units now equals 100%.

While the traditional retiming problem is of a polynomial complexity, we have proven that retiming for efficient resource utilization is an NP complete problem [Potkonjak90]. A probabilistic algorithm has been developed and implemented. The proposed algorithm has the advantage that other transformations such as associativity and pipelining can be easily combined with the retiming operation.

Application of the transformation on a large number of examples has demonstrated its distinctive advantage over the traditional retiming for minimal critical path in the high level synthesis arena. The benefits of the **retiming for resource utilization** transformation are demonstrated for the example of the 7th order filter. The parallelism graphs before retiming have already been shown in Figure 16.7. It can be noticed that most of the parallelism in the original graphs is present in the first 5 cycles. Resource utilization will be low in the later phases of the algorithm. The parallelism graphs after the retiming operation are plotted in Figure 16.11 from which it is clear that due to an overall increase in the available parallelism, the resource utilization is improved and the min-bounds on the number of required resources drops.

16.7. SCHEDULING AND ASSIGNMENT

The scheduling task selects the control step in which a given operation will happen. The assignment operation determines on which particular execution unit a given operation will be realized, from which register it will request data and

Chapter 16 Synthesis of Datapath Architectures

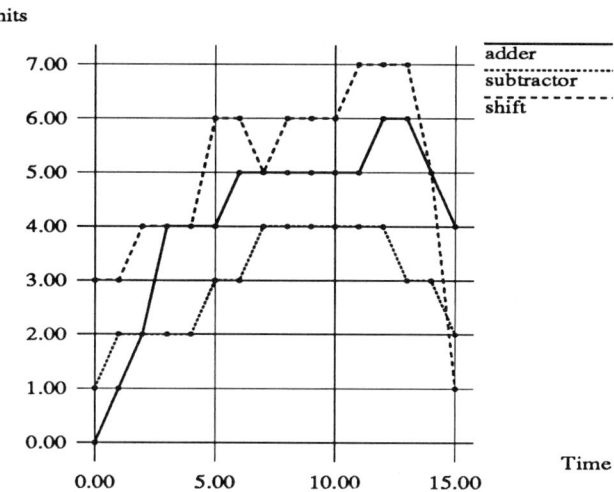

Figure 16.11: Parallelism graphs for 7th order IIR filter (after retiming)

where it will send the result using which connection. The resource allocation is closely related to the above tasks: it reserves the amount of hardware (in terms of execution units, memory registers and interconnect) necessary for realization. Obviously, those three tasks are interdependent.

Almost all scheduling and assignment problems, even when posed in a highly restricted form, are at least NP-complete. Numerous approaches have been proposed in the recent literature. They can be categorized in several groups: brute force approaches, heuristics techniques, using as soon as possible (ASAP) and as late as possible (ALAP) scheduling to obtain a global picture of the solution space, integer programming, probabilistic approaches. such as simulated annealing and neural nets and continuous relaxation techniques, including linear programming and gradient methods.

Despite the intense activity in solving scheduling and assignment problem, some aspects of the problem have not been adequately addressed. First, in VLSI technology it is essential to simultaneously address all three components of the cost function (the number of execution units, memory registers and interconnect). Very few scheduling and assignment algorithms do this. Second, none of the published approaches describe how to cope with hierarchical graphs (containing loops or if-else constructs) such that a global optimum can be pursued. Furthermore, it is necessary that the scheduling process considers not only the structure of the SFG but also the available hardware and its properties. It is obvious that schedules for two different technologies, where for instance the relative hardware costs of the functional units are different, could be radically different.

The key characteristics of our approach can be summarized as follows:

- In contrast to most published approaches, we perform assignment before scheduling. Assignments are produced using an iterative, probabilistic approach. We succeeded in characterizing a proposed assignment with a simple quality measure, which predicts the chances to find a successful schedule for this assignment.

- Once an assignment is accepted, scheduling is performed using resource utilization as the priority function. Operations, which relax the constraints on the critical resources (execution units, interconnect or registers) are given a higher scheduling priority. A critical resource is a resource which is in large demand and short supply. A resource can, for instance, be critical if due to precedence constraints it is not usable during some control steps. The resource utilization can be measured using the *discrete relaxation* technique discussed in Section 16.4.

- The graph is scheduled hierarchically in a bottom-up fashion.

The algorithms have been tested on a wide variety of examples, and performed better or at least as good as other algorithms using only a fraction of the time required by those algorithms. A detailed description of the algorithms can be found in [Potkonjak89].

16.8. HARDWARE MAPPING

The last step in the synthesis process maps the allocated, assigned and scheduled flow graph (called the decorated flow graph) onto the available hardware blocks. The result of this process is a structural description of the processor architecture in the SDL language (Chapter 3) which serves as the input to DMoct (Chapter 5) and the layout-generation tools in LAGER. The mapping process transforms the decorated flow graph into three structural sub-graphs: the datapath structure graph, the controller state machine graph, and the interface graph. The interface graph determines the relationship between the datapath control inputs and the controller output signals. Three dedicated mapping tools then translate those graphs into the corresponding structure_instance views. An overview of the basic features of each of these tools is given below. A more complete description of the hardware mapping process can be found in [Chu89].

Chapter 16 Synthesis of Datapath Architectures

16.8.1. Datapath Generation

The hardware mapping process of datapaths consists of a set of transformation steps, applied on the datapath structure graph. The key steps will be discussed briefly.

During the **register file recognition** and the **multiplexer reduction** steps, individual registers are merged as much as possible into register files. This process reduces the number of bus multiplexers, the overall number of busses (since all registers in a file share the input and output busses) and the number of control signals (since a register file uses a local decoder). The idea of the multiplexer reduction transformation is explained in Figure 16.12.

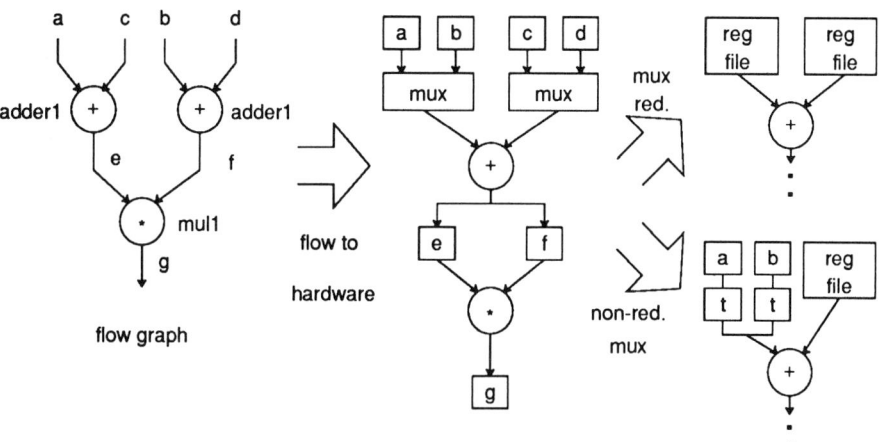

Figure 16.12: Multiplexer reduction

Our approach addresses the multiplexer and I/O bus minimization problems simultaneously. In the literature, it has been shown that the I/O bus minimization can be modeled as a clique partitioning problem, which is NP-complete. Adding the multiplexer minimization to it makes the problem even harder and also excludes a clique-partitioning formulation. Experiments with a heuristic approach on a number of benchmark examples did not yield acceptable results. Therefore, a simulated annealing based algorithm [Kirkpatrick83] was selected.

The **datapath partitioning** step optimizes the processor floorplan. Non--partitioned datapaths tend to be long and thin with long local interconnect wires and lost space due to cell stretching (Chapter 10). Three criteria are used to decide when and where to partition: First, sets of disjointed blocks are put in different partitions. Each group is further divided if different word lengths are found within the group. If the number of blocks in a partition is still too large after the above processes, further partitioning is applied using a a probabilistic approach [Greene86] based on clustering and rejectionless annealing.

In addition to the above algorithmic transformations, the hardware mapping process performs a number of translation steps (such as final hardware module selection and operator expansion into basic library cells) which require an accurate knowledge of the available cell library in terms of functionality, speed, area, and black box views. This is provided by a rule-based library database. A set of access routines are provided to support an efficient search of the database, using either the operator functionality or the cell name as the key. The search can be constrained by timing or area requirements. A rule editor allows for an easy introduction of new cells into the library.

16.8.2. Control Path Generation

The control path of a processor can also be derived from the decorated flow graph. First, a state transition diagram is generated based on the scheduling information. This is a recursive procedure due to the hierarchical nature of the flow graph. The transition diagram is then optimized by removing the dummy states such as the ending states of the if-then-else constructs.

Next, hardware is allocated for the status registers, the interface logic and the finite state machines. Notice that the resource allocation, described in previous sections, did not include these three parts (also called the control path). The interface logic between the datapaths and the finite state machine is allocated based on a demand driven algorithm so that no redundant logic is allocated. This algorithm traces the flow graph recursively and uses a set of heuristic rules to decide if a logic operation is to be performed in the interface logic or in the finite state machine. The interface logic is then partitioned in correspondence with the datapath partitioning.

From the transition diagram and the interface logic, a finite state machine description is generated. To reduce the size of the finite state machine and also to simplify the wiring between the control path and the datapath, further optimization steps are performed before the final control structure is generated.

- The first pass recognizes control signals that are independent of the control states and replaces them by a local control in the interface logic. This optimization is especially useful for the control of pipeline regis-

ters and multiplexers: the load and output-enable signals of pipeline registers can be driven directly by the clock signals, while a multiplexer can often be controlled locally without needing any central control (for instance in a max/min operation).
- The second transformation merges equivalent or complementary signals. The boolean value DON'T CARE is assigned as much as possible to facilitate the merging (for instance when an execution unit is not in use). This gives more freedom to the logic synthesis system (in this case MIS-II [Brayton87]) for the final synthesis of the control logic.
- Decoders for register files are allocated to reduce the wiring and the minimize of the number of control signals.
- Status registers are generated when the production and consumption times of a status signal are different and a temporary storage is required. These registers will be part of the state of the FSM.

The output of the control synthesis process is a number of BDS files (Section 8.2.), describing the contents of both the controller FSM and the interface logic. These BDS-files serve as input to the MIS-II logic synthesizer [Brayton87].

The hardware mapping process discussed above resulted from the analysis and critique of a number of layouts produced by the system. The data and control path partitioning heuristics were particularly influenced. This has resulted in a dramatic improvement in the area-efficiency of the layouts generated by the hardware mapper. Some of the important conclusions of the layout analysis are given below:

- The interface logic should be properly partitioned for area and timing reasons. The standard cell implementation should be implemented as a single row and match the pitch of the datapath as closely as possible. Local control slices abutting the datapath are preferred.
- Datapaths should be partitioned into approximately equal size. Irregular size datapaths usually produce inefficient layouts.
- In the digital signal processing area, datapaths tend to occupy far more area than the control paths. This implies datapath optimizations have more significant effects than the control path optimizations.
- Wiring is still one of the dominant area consumers. We are currently studying alternative layout strategies (e.g. using standard cell datapaths) as well as more accurate estimation wiring estimation techniques to cope with this issue.

16.9. COMPARING IMPLEMENTATIONS OF THE IIR FILTER

Three versions of the IIR filter of Figure 16.5, each with different timing constraints, were generated (using a 2 μm CMOS library). All the implementations went through the retiming process to achieve a high utilization rate for each hardware unit. Each implementation is partitioned into three datapaths and three control slices. A Finite State Machine (FSM) is used as the central control for each filter. The functional correctness of all produced layouts has been analyzed using the THOR functional simulator (Section 13.1.) and the simulation results have been checked against the simulation results at the SILAGE level. Although this is by no means a complete test, it has assured us about the correctness of the applied transformations and synthesis operations.

Number of	Implem. 1	Implem. 2	Implem. 3
Clock Cycles	20	16	10
Adders	1	2	2
Sub1tractors	1	1	2
Shifters	11	1	2
Registers	36	37	46
Buffers	15	17	25
Total Area	13mm^2	18.9 mm^2	27.9 mm^2

Table 16.4: Comparison of 3 Filter Implementations

Table 16.4 shows the results of the synthesis with three solutions, indicating the tradeoff between area and speed. Implementing the same filter on a general purpose signal processor, such as the Motorola 56000 - a 24-bit 20 Mhz fixed point signal processor with concurrent multiply-accumulate and address generation unit - would take at least 27 cycles of manually optimized assembler code.

The layouts of three implementations are shown in Figure 16.13, properly scaled to show the relative sizes. The area grows almost linearly, when the com-

Chapter 16   Synthesis of Datapath Architectures   247

putation time goes down. Table 16.5 summarizes the distribution of the CPU time over the synthesis modules for this particular example.

Flow graph generation	0.8 sec
Module Selection	1.8 sec.
Estimation/Allocation	1.7 sec.
Retiming	7.4 sec.
Assignment/Scheduling	1.9 sec.
Hardware Mapping	~ 1 min.
Layout Generation	~1 hour

Table 16.5: CPU time for IIR Synthesis operations (on Sun 4/100)

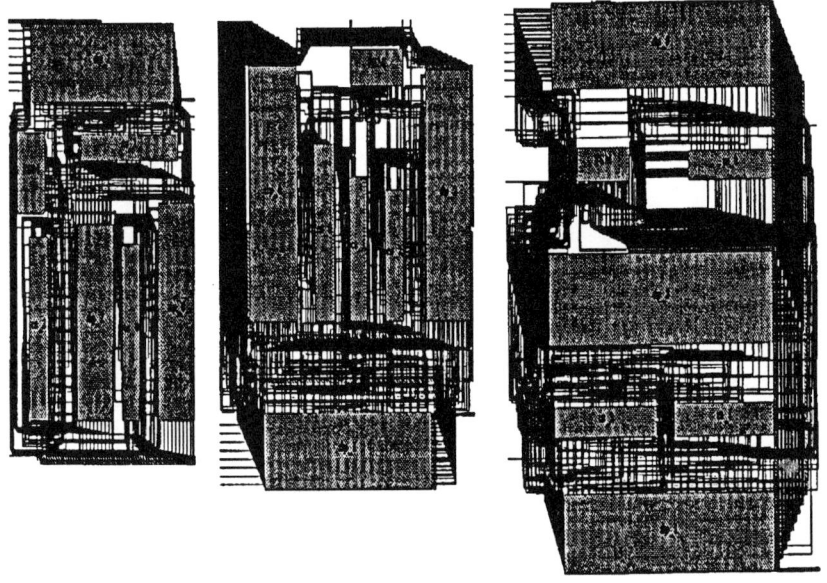

Implementation 1   Implementation 2   Implementation 3
20 clock-cycles   16 clock-cycles   10 clock-cycles

Figure 16.13:Layouts of the 3 IIR Filters

16.10. SUMMARY

The overall composition of the HYPER, a synthesis system for arithmetic intensive processors, has been described. The most important characteristics of the system are:

- a single, global quality measure, called the **resource utilization**, is used throughout the system to drive the design space exploration process. This approach effectively merges the hardware allocation, transformation application and hierarchy handling in a consistent way.
- The modular organization of HYPER allows for an easy introduction of new software modules, such as transformations, schedulers and user feedback tools.

HYPER allows the designer to quickly make area-time trade-offs and thus shortens the design cycle dramatically for this portion of the design task.

REFERENCES

[Brayton87] R. Brayton, et al., "MIS: A Multiple-Level Logic Optimization System," *IEEE Transactions on CAD*, vol. CAD-6, no. 6, pp. 1062-1081, November 1987.

[Chu89] C. Chu, et al., "*Hyper*: An Interactive Synthesis Environment for High Performance Real Time Applications", *Proc. IEEE ICCD Conf., Nov. 1989*.

[Goossens89] G. Goossens, et al., "Loop Optimization in Register Transfer Scheduling for DSP Systems", *Proc. Design Automation Conf.*, pp. 827-831, June 1989.

[Greene86] J. W. Greene and K. J. Supowit, "Simulated Annealing Without Rejected Moves," *IEEE Transactions on CAD*, vol. CAD-5, no. 1, pp. 221-228, January 1986.

[Hoang92] P. Hoang and J. Rabaey, "A Compiler for Multiprocessor DSP Implementation," *Proceedings of ICASSP*, March 1992.

[Jain88] R. Jain, et al., "Module Selection for Pipelined Synthesis," *Proceedings 25th ACM/IEEE Design Automation Conference*, Anaheim, June 1988.

[Kirkpatrick83] S. Kirkpatrik, et al., "Optimization by Simulated Annealing," *Science*, vol. 220, no.4598, pp. 671-680, May 1983.

[Lam85] M. Lam, "A Transformational Model of VLSI Systolic Design", *Computer*, pp. 42-52, Feb. 1985.

[Lee89] E. A. Lee, et al., "Gabriel: A Design Environment for DSP," *IEEE Trans. on ASSP*, November 1989.

[Leiserson83] C. Leiserson and F. Rose, "Optimizing Synchronous Circuitry by Retiming", Third Caltech Conf. On VLSI, March 1983.

[Messerschmitt88] D. Messerschmitt, "Breaking The Recursive Bottleneck", in *Performance Limits in Communication Theory and Practice*, Kluwer Academic Publishers, 1988.

[Note90] S. Note, et al., "Combined Hardware Selection and Pipelining in High Performance Data-Path Design," *Proceedings IEEE ICCD '90*, Cambridge, September 1990.

[Potkonjak89] M. Potkonjak and J. Rabaey, "A Scheduling and Resource Allocation Algorithm for Hierarchical Signal Flow Graphs", *Proc. Design Automation Conf.*, pp. 7-12, June 1989.

[Potkonjak90] M. Potkonjak and J. Rabaey, "Retiming for scheduling," *Proceedings of 1990 IEEE Workshop on VLSI Signal Processing*, November 1990.

[Rabaey89] J. Rabaey and P. Hoang, "OCT Flowgraph Policy," *U.C. Berkeley Internal Document*, September 1989.

[Rabaey90] J. Rabaey and M. Potkonjak, "Resource Driven Synthesis in the *Hyper* system," *ISCAS-90*, vol. 4, pp. 2592-2595, New Orleans, LA, May 1990.

[Roermund89] A.H. Van Roermund, P.J. Snijder, H. Dijkstra, C.G. Hemeryck, C.M. Huzier, J.M.P. Schmitz and R.J. Sluijter, "A General Purpose Programmable Video Signal Processor", in *IEEE Transactions on Consumer Electronics*, pp 249-258, August 1989.

[Simons81] B. Simons, "On Scheduling with Release Times and Deadlines", in Deterministic and Stochastic Scheduling, D. Reidel Publishing Co., pp. 75-88, 1981.

[Tseng86] C. Tseng and D. Siewiorek, "Automated Synthesis of Data Paths in Digital Systems," *IEEE Transactions on CAD*, vol. 5, no. 3, pp. 379-395, July 1986.

[Welch84] Welsh, D.J.A., "Correlated percolation and repulsive particle systems," Tautu, P: Stochastic Spatial Processes, Springer Lecture Notes 1212, pp. 300-311, 1984.

17

From C to Silicon

Lars E. Thon, Ken Rimey and Lars Svensson

A major difficulty in the design of application-specific programmable signal processors is the program development for the processor [Pope84, Ruetz86]. Common software development tools are typically poorly suited for the task of generating code that must be targeted to varying architectures. This results in the program development being done directly in the machine language of the processor, discouraging architecture exploration because of the large time investment needed for machine code development. Determining whether a malfunction in the functional simulation of the processor is caused by an error in the code or by an error in the hardware design is very difficult. To ease the effort required in the software development and verification, a compiler for RL, a C-like language, has been developed.

A critical issue is whether (and how easily) the compiler can be retargeted from one architecture to another, and also the range of architectures for which it can be applied. The RL compiler covers a fairly general class of architectures, that can be retargeted by the architecture designer (as opposed to a compiler expert) through a *machine description* file.

17.1. ARCHITECTURE EXAMPLES

The processor architecture Kappa, which was used as the initial target, was originally developed for a robot-control application [Azim88]. The Kappa datapath is shown in Figure 17.1. A few details are not shown explicitly in the figure:

- The parallel-to-serial converter rcoef at the bottom of the picture is used in shift-and-add multiplication and in long division.
- The barrel shifter shifts by an amount that is specified in the instruction (an immediate constant).
- Status signals tell the controller if the output of either adder is negative. When overflow occurs in the main adder (on the left), saturation may be applied before the result is saved in the accumulator.

Examples of other architectures which have used the C-to-Silicon code generation and simulation tools in their design include one optimized for a trigonometrically intensive robotics calculation and another for a communications application.

The PUMA processor performs a computationally intensive task normally assigned to the host processor of a robot arm: solving for sets of joint angles that correspond to a desired position and orientation of the manipulator. PUMA incorporates a shifter that can shift by data-dependent amounts. A parallel multiplier was also considered during the architecture exploration phase, but was found to be too expensive in terms of silicon area. A detailed discussion of the PUMA design can be found in Chapter 19.

Another application was a channel equalizer for digital mobile radio [Svensson90]. Again a number of architecture and algorithm tradeoffs were investigated. To achieve the necessary throughput, it was essential to incorporate a parallel multiplier. This in turn made it necessary to rebalance the architecture. To keep the multiplier supplied with operands, the memory bank was split. To keep the dual memory banks supplied with addresses, the address-arithmetic unit was modified. The resulting datapaths are shown in Figure 17.2 and Figure 17.3.

The basic restrictions on the architectures are that they are Harvard machines (instructions and data are stored separately), and that they use a horizontal control word with no or little restrictive encoding. A basic structure is assumed for the microcode control unit, based on finite state machines, as shown in Figure 17.4. The detailed specifications are generated by the compiler.

Chapter 17 From C to Silicon 253

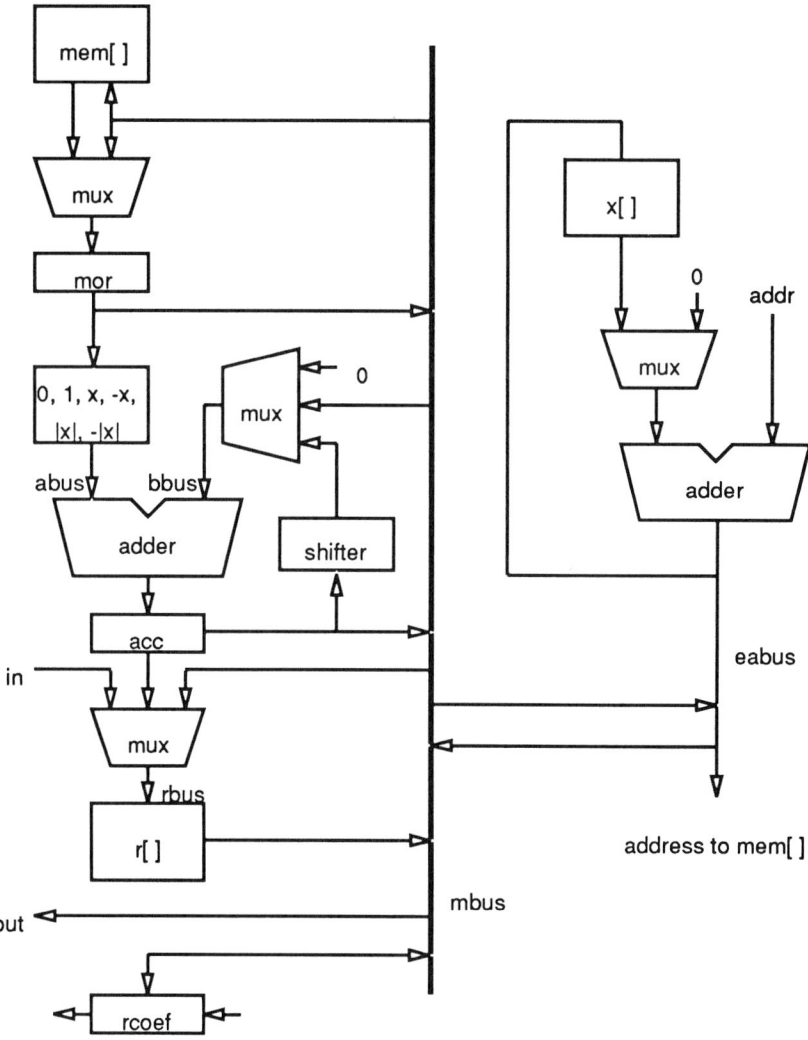

Figure 17.1: The Kappa datapath

The flexibility lies in the design of the datapaths. They may differ in the functional units and registers provided, and in topology. However, they also must retain some common features: The microoperations that they implement must belong to a standard set of microoperation types supporting the operations of a C-like language. The register structure and datapath topology must meet some

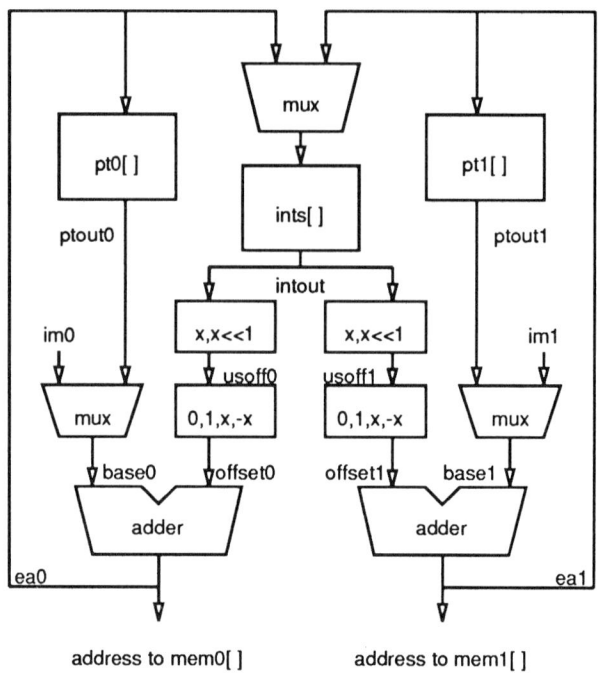

Figure 17.2: Integer datapath for channel equalizer

basic connectedness requirements imposed by the compiler. Finally, the overall design style should ideally be similar to the style demonstrated by Kappa. The compiler is tuned to irregular datapaths with moderate amounts of parallelism, and will certainly produce substandard code for radically different datapaths.

17.2. THE RL LANGUAGE

The RL language is an approximate subset of C. To keep the compiler simple, RL includes only those features of C that correspond closely to the capabilities of Kappa and related architectures—recursion, for example, is not supported. RL includes two major extensions: fixed-point types and register type modifiers. It is therefore not strictly compatible with C. For behavior simulation, we provide a translator that converts RL into standard C acceptable to other compilers.

Fixed-point types are a convenience for the programmer. The underlying integer arithmetic is inconvenient to write by hand, partly because simple fixed--point constants correspond to huge integers, and partly because the natural multi-

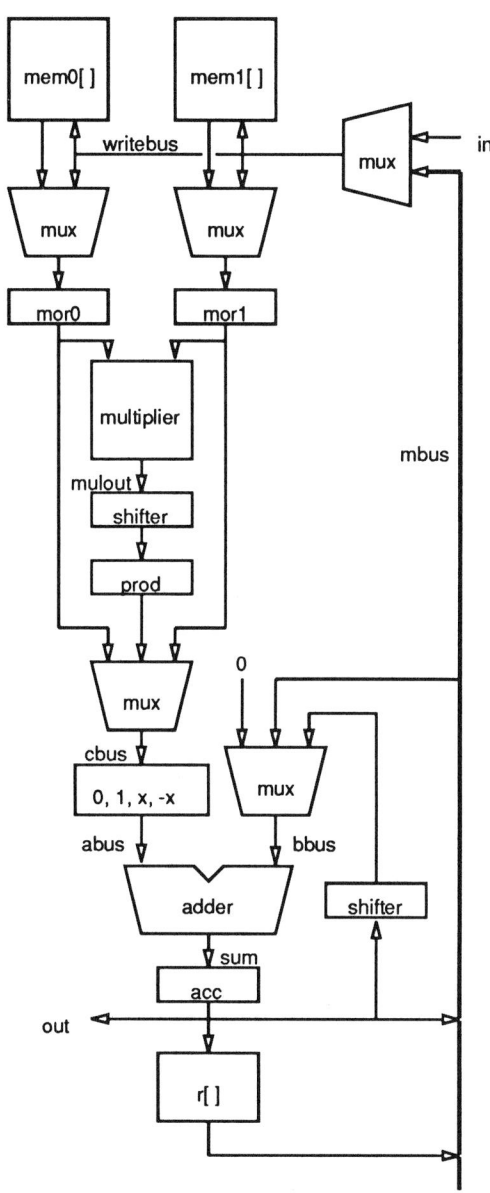

Figure 17.3: Fixed point datapath for channel equalizer

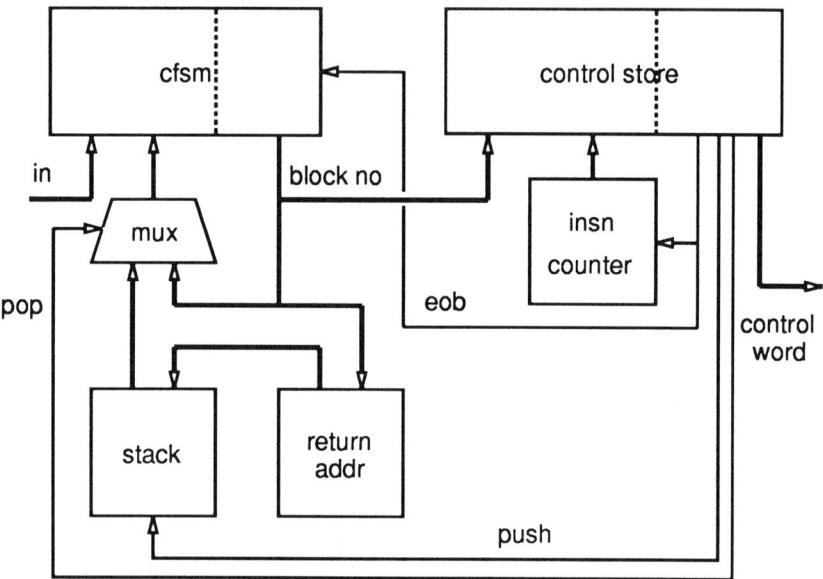

Figure 17.4: The control sequencer

plication for fixed-point numbers is not integer multiplication. In adding a new numerical type to a programming language, finding an elegant notation for the new constants can be difficult. In RL, all constants are typeless real numbers that take on appropriate types from context. In declarations and type casts, the fixed-point type of range $-2^n \leq x < 2^n$ is denoted by fix:n; or if $n = 0$ then by just fix.

Register type modifiers, which are generalized C register declarations, let the programmer suggest storage locations for critical variables. For example,

```
register "r" fix y;
```

declares the variable y to be a fixed-point number to be stored in the register bank r. A reasonable default is chosen if the name of the register bank is omitted (as in Figure 17.5). Register type modifiers are also helpful with multiple memories, and they can be applied to pointers. For example,

```
"mem" fix * "mem2" p;
```

declares p to reside in mem2 and point into mem.

17.2.1. Limitations

Many parts of C have been left out of RL for the sake of simplicity:

- There is no separate compilation.
- There are no explicit or implicit function declarations; functions must be defined before they are used.
- Initial values may only be specified in declarations of variables that are to be stored in read-only memory.
- There are no `struct`, `union`, or `enum` types; no `char`, `float`, or `double` types; and no `short`, `long`, or `unsigned` modifiers. This leaves only `void`, `int`, pointer types, array types, and the RL-specific types, `bool` and `fix`.
- There are no `goto`, `switch`, `continue`, or `break` statements.
- There is no `typedef`, no `sizeof`, and there are no octal or hexadecimal constants.

Because the target processors do not provide a stack for local variables, it is also necessary to prohibit recursive function calls. For the same reason, the programmer has to be aware that doing a function call within the scope of a register declaration will force the compiler to produce rather poor code.

17.2.2. Type modifiers

In RL, the `const` type modifier is used mainly in declaring variables that are to be stored in read-only memory. The `volatile` type modifier is used mainly to identify boolean variables that represent signals on external pins. A `volatile bool` variable represents an output pin which is set by the processor. A `const volatile bool` variable represents an input pin which is sensed by the processor.

17.2.3. Pragmas

In RL, pragmas have the same form as the `#define` preprocessor command, but start with `#pragma` instead. Pragmas define flags and parameters that control the RL compiler and other software, as in these examples:

`arch_file` gives the name of the machine description file to use;

`word_length` determines the number of bits in a processor word;

x`_capacity` sets a limit on the number of registers that the compiler may assume for the register bank x.

17.2.4. Register declarations and register type modifiers

The RL compiler assigns a variable to a specific memory or register bank depending on

- whether or not it is a `register` variable,
- its base type, and
- if the base type is a pointer type, the bank that it points into.

The defaults for a given architecture are specified by pragmas in the machine description, but can be overridden by pragmas in the RL program. For example, to override the usual defaults for Kappa and store non-`register` integer variables, and pointers into bank `mem`, in bank `x` instead of in bank `mem`, the programmer would put the following pragmas into the RL program:

```
#pragma int_memory "x"
#pragma mem_pointer_memory "x"
```

Assigning all variables to default memory and register banks is sometimes too crude. For such cases, RL has register type modifiers. A register type modifier is written as the name of a memory or register bank in double quotes. It is a type modifier, like `const` and `volatile`, that can appear wherever `const` and `volatile` can appear. For example, an integer variable `x` stored in the bank `foo` would be declared like this:

```
"foo" int x;
```

A more complex example is a pointer to `int`, residing in the bank `bar` and pointing into the bank `foo`:

```
"foo" int * "bar" p;
```

17.2.5. The boolean type

In C, boolean values (true and false) are represented as integers, which is convenient for typical general-purpose computers. In contrast, our application--specific target processors perform boolean operations on (and store) individual bits. This is the reason for having a distinct boolean type, `bool`, in RL.

In RL, there are no implicit conversions to or from `bool`, except in certain cases involving literal numbers. True can be written as `(bool) 1`; false, as `(bool) 0`; and in most cases, the casts can be omitted.

The operations that return booleans as results are the relationals (`<`, `>`, `<=`, `>=`, `==`, `!=`) and the boolean operations (`&&`, `||`, `!`). The operations that require boolean operators are the three boolean operations, and the conditional expression (*condition* ? *then-part* : *else-part*). The tests in if, while, do-while, and for statements are also required to be boolean.

17.2.6. Fixed-point numbers

RL has a set of fixed-point types. Arithmetic on fixed-point numbers is saturating, except in shift operations. This is in contrast to integer arithmetic, which is always non-saturating.

The fixed-point types have names of the form fix:*n*, where *n* is a possibly negative integer. The form fix is a shorthand for fix:0. Values of type fix:*n* have a machine-dependent precision (controlled by the pragma word_length) and lie in the range $-2^n \leq x < 2^n$. Casts may be used to convert between the different fixed-point types, but conversions between fixed-point and integer types are not allowed. A cast of a fixed-point datum to another fixed-point type is typically implemented with an arithmetic shift operation.

All of C's floating-point arithmetic operators are available in RL for fixed point arithmetic. With the exception of multiplication and division, the arguments of a binary fixed-point operator must have the same type, as must the second and third arguments in a conditional expression. Casts are commonly used to accomplish this.

In addition, fixed-point values may be explicitly shifted with the arithmetic shift operators << and >>.

17.2.7. Predefined functions

RL has three predefined functions: abs(), in(), and out(). These functions are overloaded to take arguments of type int as well as type fix:*n*. The value returned by in() may be considered to be of type number, that is, the resulting type (after implicit conversion) depends on a limited amount of context. In ambiguous cases, casts must be used.

17.2.8. User-defined operations

Hardware-supported operations that are not predefined in RL can be specified in the machine description file. An operation is defined and given a name, and one or several implementations of the operation are specified in the same way as for the predefined operations. An operation defined in this way is available in RL in the form of a "function call", where the function has the same name as the operation. This is useful for hardware lookup tables and in general for handwritten, idiomatic instruction sequences. For example, a multiplication step with some particular behavior on overflow might be implemented as a user-defined operation because it would not be compiled into efficient code if written in pure RL.

17.2.9. Preprocessor commands

There are four new preprocessor commands in addition to those of standard C. They are useful for unrolling and partially unrolling loops: #repeat, #endrepeat, #rrepeat, and #endrrepeat. The form

```
#repeat id N
...text...
#endrepeat
is roughly equivalent to
#define id 0
...text...
#undef id
#define id 1
...text...
#undef id
...
#define id N-1
...text...
#undef id
```

#rrepeat and #endrrepeat are similar, except that they count backwards.

17.2.10. Program structure

The last difference between RL and C is that the RL programmer may (and often will) leave main() undefined. In its place, the code should define loop() and optionally init(). The compiler then supplies an implicit main(), where init() is called once (if it has been defined), and then loop() is called indefinitely. This is an appropriate form for a program which reads a indefinite input stream.

```
#pragma word_length 16
#pragma arch_file "kappa"
register fix y;
init() {y = 0;}
loop() {
        y = (3/4) * y + (1/4) * (fix) in();
        out(y);
}
/* Provided by the compiler */
main() {
        init();
        for (;;) loop();
}
```

Figure 17.5: A simple filter program in RL.

17.3. FIR FILTER EXAMPLE

The sample program in Figure 17.5 is a trivial low-pass filter. Using the recurrence

$$y_n = (3/4)*y_{n-1} + (1/4)*x_n \qquad (17.1)$$

the filter smooths the input sequence $x_1, x_2, x_3, \ldots$ to produce the output sequence $y_1, y_2, y_3, \ldots$. Instead of main, RL programs define the functions init and loop. This implicit form for the outermost loop will facilitate, for example, automatic insertion of memory-refresh code. The *pragma* construct has recently been introduced into C by the ANSI draft standard [Kernighan88]. A pragma at the head of the sample program is used to specify the word width of the processor.

17.3.1. The Register-Transfer Notation

The compiler output for the low-pass filter program is shown in Figure 17.6. The code consists of two straight-line segments, each of which is a sequence of instructions terminated by semicolons[1]. Each instruction is a sequence of microoperations separated by commas. We have abbreviated "*a* = *b*, *b* = *c*" by "*a* = *b* = *c*" to make the code easier to read.

---

[1] Chapter 19, "*The PUMA processor*", describes a more realistic example in which the RL program is several hundred lines long

```
init:   acc=0;
        r[0]=acc;
        GOTO loop;
loop:   acc=bbus=mbus=r[0], mor=mbus, r[1]=iodata=in();
        mor=mbus=r[1],abus=mor,bbus=acc>>1,
        bbus.1=(abus+bbus)>>1;
        acc=bbus;
        acc=abus=mor, mor=mbus=acc;
        abus=mor, bbus=acc>>2, acc=abus+bbus;
        iodata=mbus=acc, out(iodata), r[0]=acc;
        GOTO loop;
```

Figure 17.6: Compiler output (microcode) for the FIR filter

The microoperations are written in a register-transfer notation reminiscent of C or RL. Subscripting (e.g., r[0]) is used to refer to an element of a register (or memory) bank. Member selection (e.g., bbus.1) is used to refer to the value of a register or bus at a given time: *register*.0 refers to the value of *register* in the current instruction, *register*.-1, to the value in the previous instruction, and so on.

Reasonable defaults reduce the need for the latter notation. An unqualified name means *register*.0, except on the left hand side of an assignment. There, it means *register*.*delay*, where *delay* is defined in the machine description. For example, in the microoperation

```
bbus.1 = (abus.0 + bbus.0) >> 1
```

the qualifiers on the right can be omitted, but the one on the left must remain because the *delay* attribute of bbus is 0. This register transfer designates a single microoperation, even though its right-hand side appears to be a compound expression. This microoperation is equivalent to "acc = abus + bbus", followed in the next instruction by "bbus = acc >> 1", except that the compound operation allows correction of overflow by shifting the overflowing bit back into the word.

17.3.2. The machine description

As mentioned earlier, the RL compiler needs a description (*md*-file) of the architecture for which to compile. Figure 17.7 shows a portion of the md-file for

Chapter 17　　　　　From C to Silicon　　　　　　　　　263

```
#define bus node :delay = 0
#define reg node :delay = 1 :static
#define file reg :bank

field       :signed x_addr_imm[]
field       output_enable_x[], load_x[],select_immed

file        x
bus         addr, xbus, xsum
bus         xsign /* this "bus" carries a boolean value*/

micro       addr = Immediate
            { x_addr_imm = Immediate }
micro       xbus = x[N]
            { output_enable_x[N] = 1, select_immed = 0 }
micro       xsum = addr
            { select_immed = 1 }
micro       xsum = xbus
            { x_addr_imm = 0 }
micro       xsum = addr + xbus
micro       xsum = xbus + addr
micro       xsign = xsum.-1 < 0
micro       x[N] = xsum
            { load_x[N] = 1 }
```

Figure 17.7: Machine description for the Kappa address arithmetic unit

Kappa. The portion showed corresponds to the address datapath, which is the right half of Figure 17.1.

The *md*-file contains declarations of control word fields (field), resources (bus, file) and microoperations(micro). The microoperations may optionally specify their implementation in terms of control signals. Information about control signals is used later on by the assembler.

17.3.3. Simulation and profiling

The RL compiler is accompanied by a translation tool which can translate RL programs into real C code for execution on a regular computer. Table 17.1 shows the names and some options for the various programs. The float translator replaces fix variables with floating point variables. The fix translator replaces `fix` variables with integer variables and replaces operators `+, -, *, /` with calls to fixed point subroutines from a library.

There is also an option which inserts profiling code, using the same mechanism as the UNIX C compiler `-pg` option. This allows us to collect a *dynamic instruction count* (cycle count) for a given program executed on a given target architecture and with given input data. The cycle count is for the actual target chip architecture executing the RL program (*not* for the computer which executes

Program name	Program function	Comment
kc	RL to microcode compiler	c=compiler
kt -float	RL to C translator (floating point)	t=translator
kt -fix	RL to C translator (fixed point)	
kt -fix-p	RL to C translator (for profiling)	p=profiler
kprof	Profiling postprocessor	

Table 17.1: The RL compiler and translator programs

the C program). The program `kprof` analyzes the profiling data and prints out the cycle count.

17.3.4. Microcode assembly and layout generation

The actual layout of the processor which executes a RL program is performed by the layout-generation tools in LAGER, as presented in the earlier chapters. The interface between the RL compiler and the layout-generators is a microcode assembler which translates the microcode and associated information into *layout parameters* making use of the parameterization facility provided by the SDL language (Chapter 3). Examples of parameters are PLA content specifications and bus width specifications. Chapter 19 shows the results of such a layout process.

17.4. IMPLEMENTATION OF THE RL COMPILER

Ideally, the chip designer should not have to know how the high-level-language compiler works. He should be able to treat it as a black box that takes descriptions of an algorithm and of a machine and produces symbolic microcode according to these specifications. The languages in which the two inputs and the output are expressed will be described in the following sections. All these languages are preprocessed by the standard C preprocessor. Comments and macro definitions are therefore allowed just as in C.

The compiler consists of two parts. The *front end* translates the program into successive straight-line segments of code expressed in an intermediate language. For each straight-line segment, the *back end* selects microoperations and packs them into instruction words. Only the back end uses the machine description.

17.4.1. The Front End

The front end performs a variety of routine tasks and simple optimizations, including parsing, constant folding, building the symbol table, and type analysis. Because they reflect our target machines, two additional optimizations should be noted:

The first is the reduction of multiplications by constants into minimal sequences of shifts, additions, and subtractions. For example, a fixed-point multiplication by 7/8 is reduced to a three-bit right-shift and a subtraction.

The second is control-flow optimization, intended to take advantage of multi-way jump/call/return operations. The compiler uses a structured dataflow algorithm to move control-flow operations upward within the program so that they can be coalesced. All jumps to jumps, to calls, and to returns are eliminated in the process.

For each straight-line segment, the front end passes to the back end a dataflow graph whose nodes are intermediate-language operations such as "add" and "shift-right." The front end eliminates all variables that are not live across branches by adding edges to the graph. The back end assigns the remaining variables to register banks using the register declarations from the RL program together with reasonable defaults. These variables are accessed by special "read" and "write" nodes in the dataflow graph, which itself makes no reference to particular busses or registers.

17.4.2. The Back End

Most of the effort in developing the RL compiler has gone into developing the algorithms used in the back end [Rimey88]. Here, we describe the relationship of our basic approach to the approaches that have been tried elsewhere in the compiler-construction community.

The usual approach to generating horizontal code has been to separate the process into two phases. First loose sequences of microoperations are generated. Then these are packed tightly into a small number of instructions in the *compaction* phase. Researchers have studied compaction in various forms:

> *Local compaction* is the packing of straight-line code segments one at a time. Good heuristics for this have been developed and thoroughly studied.
>
> *Global compaction* generalizes local compaction to include movement of microoperations across branches. The best known method, *trace scheduling*, has been developed in conjunction with very-long-instruction-word (VLIW) supercomputers [Colwell87]. Unfortunately, trace scheduling is unsuitable for signal processing applications because it improves average running time only at the expense of worst-case time.
>
> *Software pipelining* is a specialized technique for the compaction of loops. It rearranges the body of a loop to overlap the execution of many successive iterations. Touzeau describes the use of software pipelining in a compiler for the FPS-164 array processor [Touzeau84].

Fisher, Landskov, and Shriver's paper on microcode compaction [Fisher81] is a good general introduction to these techniques.

Vegdahl critically examines this separation of code generation into two phases [Vegdahl82]. He concludes that some coupling of them is useful: feedback from the compaction phase is needed for effective microoperation selection. We find this issue to be particularly important in generating transfer microoperations for Kappa-like architectures. However, rather than couple the two phases, we have chosen to completely integrate them.

The back end of the RL compiler *schedules* microoperations as they are selected. This approach creates numerous opportunities for code improvement—even when it is applied just to one straight-line program segment at a time. Hence we have limited ourselves to *local scheduling*.

Our scheduler is similar to the "operation scheduler" developed by Ruttenberg and described in a paper written with Fisher et al [Fisher84]. It does *greedy* scheduling of function microoperations and *lazy* scheduling of transfer microoperations. For each node of the directed acyclic graph (DAG) of operations, it finds the function microoperation that can be inserted into the earliest possible instruction, and inserts it. It also finds transfer microoperations that deliver the arguments of the function microoperation, and inserts them into the appropriate instructions. This subtask is *lazy data routing*.

How best to route an intermediate result between functional units depends on the time interval between the definition and the use, as well as on what resources in the datapath are free during that interval. By postponing the selection of the route until the use is scheduled (and more of the schedule is known), we get more constraints to guide the selection. We also get a complication: the possibility that all feasible routes for the value will be unwittingly closed off by an unfortunate scheduling decision in the meantime.

We detect and avoid such disaster by maintaining a feasible route to the indefinite future, a *spill path*, for each value with as-yet-unscheduled uses. These spill paths do not represent scheduling commitments; indeed, they are continuously adjusted as scheduling proceeds. Their purpose is to identify scheduling decisions for which no accommodating adjustment exists.

Spill-path adjustment can be performed efficiently; any number of spill paths can be rerouted in a time linear in the size of the *space-time graph*. This is a directed graph whose nodes are ordered pairs $[r, t]$ such that r is a node of the datapath and $t \in \{1, 2, 3, ..., t_\infty\}$ is an instruction number. The quantity t_∞ is the number of instructions in the current schedule, and then some. There is an edge from $[r_1, t_1]$ to $[r_2, t_2]$ if and only if there is a transfer microoperation that copies a value from r_1 to r_2 in $t_2 - t_1$ instruction times. Spill paths respect the node capacities of the space-time graph: if the *capacity* attribute of the datapath node r is n, and if r is currently committed to holding k values at t, then no more than $n - k$ spill paths may pass through the space-time node $[r, t]$. By a standard construction that splits each node of the graph into two, the space-time graph with its node capacities can be converted into a space-time *network* with edge capacities. On this the spill paths form a *network flow*, to which well-known algorithms apply. In particular, if the flow can be adjusted to accommodate a unit reduction in the capacity of an edge, this can be done in a time linear in the size of the network.

17.5. SUMMARY

We have described a class of programmable processors and a user-retargetable compiler that form the basis for a practical ASIC development strategy. The use of irregular horizontal-instruction-word architectures facilitates the tuning of the processor datapath but limits the applicability of standard code-generation techniques. To generate efficient code for diverse datapath topologies, we have developed the technique of lazy data routing.

REFERENCES

[Azim88] Syed Khalid Azim, "Application of silicon compilation techniques to a robot controller design," *PhD thesis, UC Berkeley UCB/ERL memo M88/35*, May 1988.

[Colwell87] Robert P. Colwell, Robert P. Nix, John J. O'Donnell, David B. Papworth, and Paul K. Rodman, "A VLIW architecture for a trace scheduling compiler," *Second International Conference on Architectural Support for Programming Languages and Operating Systems*, pages 180–192, 1987.

[Fisher81] Joseph A. Fisher, David Landskov, and Bruce D. Shriver, "Microcode compaction: Looking backward and looking forward," *Proceedings of the National Computer Conference*, pages 95–102. AFIPS, 1981.

[Fisher84] Joseph A. Fisher, John R. Ellis, John C. Ruttenberg, and Alexandru Nicolau, "Parallel processing: A smart compiler and a dumb machine," *Proceedings of the ACM SIGPLAN '84 Symposium on Compiler Construction*, pages 37–47, June 1984.

[Kernighan88] Brian W. Kernighan and Dennis M. Ritchie, "The C Programming Language," Prentice-Hall, Englewood Cliffs, New Jersey, second edition, 1988. This edition reflects the draft ANSI C standard.

[Pope84] Stephen S. Pope, "Automated Generation of Signal Processing Circuits," *PhD thesis, UC Berkeley*, 1984.

[Rimey88] Ken Rimey and Paul N. Hilfinger, "Lazy data routing and greedy scheduling for application-specific signal processors," *The 21st Annual Workshop on Microprogramming and Microarchitecture*, pages 111–115, November 1988.

[Rimey89] Kenneth Edward Rimey, "A compiler for Application-Specific Signal Processors," *PhD thesis, UC Berkeley*, September 1989.

[Ruetz86] Peter A. Ruetz, "Architectures and design techniques for real-time image processing IC's," *PhD thesis, UC Berkeley UCB/ERL memo M86/37*, May 1986.

[Svensson90] Lars Svensson, "Implementation aspects of decision-feedback equalizers for digital mobile telephones," *PhD thesis, Lund Institute of Technology*, May 1990.

[Touzeau84] Roy F. Touzeau, "A Fortran compiler for the FPS-164 scientific computer," *Proceedings of the ACM SIGPLAN '84 Symposium on Compiler Construction*, pages 48–57, June 1984.

[Vegdahl82] Steven R. Vegdahl, "Phase coupling and constant generation in an optimizing microcode compiler," *The 15th Annual Workshop on Microprogramming*, pages 125–133, 1982.

18

An FIR Filter Generator

Paul Yang and Rajeev Jain

High performance FIR filters have applications in several video processing [Privat86] and digital communications systems [Samueli90]. While compiler tools exist for low sample rate applications such as audio and telecommunication, techniques for automating the design of high sample rate FIR filters have only recently been emerging [Hartley89, Reutz89]. These techniques have been shown useful for sample rates of 10-30 MHz. However, methods for higher rates such as required in high-speed digital data communications [Samueli90] or high rate video transmission systems [Isnardi88] have not yet been reported. In this chapter a silicon compiler tool, `Firgen`, is presented that has been successfully used to generate compact FIR filter circuits operating at sample rates up to 112 MHz in relatively mature technologies.

`Firgen` allows the user to generate the chip layout starting from filter specifications or from coefficient values. It consists of programs for (a) filter synthesis, optimization and simulation; (b) architecture generation; and (c) floorplan generation. The architecture generation techniques are based on ideas proposed in [Reutz89, Ulbrich85, Lin89], and the floorplanning techniques are based on results reported in [Lin89, Hatamian87, Noll86]. The architecture and floorplan generation tools are specifically aimed at achieving very high sample rates with compact layout and minimal pipeline latency. The final layout generation is accomplished by the LAGER silicon assembly system.

Relative to existing filter compiler tools Firgen provides two advanced features: (a) high performance with relatively compact layouts and (b) integration of filter synthesis, simulation, architecture generation and layout generation tools.

18.1. ARCHITECTURAL REVIEW

The bit-serial compilers FIRST [Denyer84], Inpact [Yassa87], and Cathedral I [Jain86] are applicable to both FIR and IIR filters but are mostly useful for low sample rate filters such as in speech, audio and telecommunication applications [Ginderdeuren86]. Microcoded bit-parallel compilers [Rabaey85, Ruetz86] have also been used for these low sample rate applications. Several techniques have been proposed for achieving higher sample rates in FIR filter circuits. These require varying amounts of hardware complexity and in many cases rely on heavily pipelining the architecture since pipeline latency can be introduced in FIR structures without affecting the input/output transfer function. However, the use of indiscriminate pipelining leads to substantial increases in power dissipation as well as chip area due to the pipeline registers [Reutz89, Jain87]. The increase in power dissipation is of special concern at high speeds. The aim in developing Firgen has been to achieve high sample rates with minimum possible use of pipelining.

The FIR filter compiler Parsifal [Hartley89] offers the user the choice of combining multiple digit-serial datapaths to achieve performance that is higher than that of pure bit-serial structures but lower than that of full bit-parallel architectures. The idea is to offer a spectrum of trade-offs between bit-serial and bit--parallel architectures. In contrast Firgen is optimized for hardwired bit-parallel architectures only to get maximum performance for a given technology.

The FIR filter compiler proposed by Privat [Privat86] is also targeted at hardwired bit-parallel architecture but uses carry-ripple addition which offsets the advantage of bit-parallel processing. Ruetz [Reutz89] also uses bit-parallel architecture but uses carry-save addition and additionally employs a tree structure for accumulating the tap products to reduce the critical path delay through the carry-save adders. A further reduction in critical path can be achieved by using carry-save bit-parallel processing together with the transpose form instead of the direct form. Ulbrich and Noll [Ulbrich85] have proposed such an architecture with both carry-save addition and a fast vector-merge adder to get full advantage

Chapter 18 An FIR Filter Generator 271

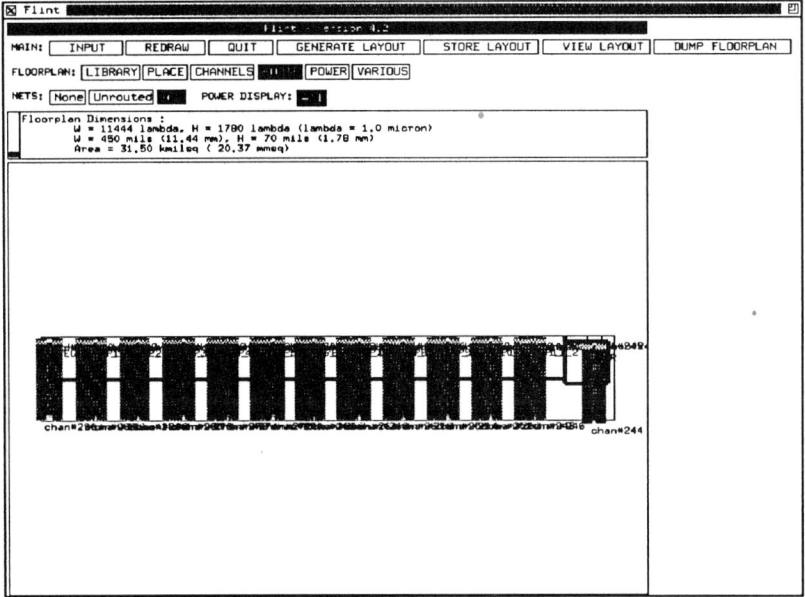

Figure 18.1: `Flint` place and route without floorplan constraints

of bit-parallel processing, and Lin and Samueli [Lin89] have shown a bit-level pipelined version of such an architecture. This is the approach used in `Firgen`.

18.1.1. High performance layout requirements

To achieve the high-performance requirement, special attention is paid to the construction of power, clock and signal cables. In particular, it was found that if `Flint` were allowed to perform automatic place and route, the resultant layout (Figure 18.1) not only consumes more area but also causes excessive signal delays and clock skews that would result in incorrect high-speed operation due to sub-optimal routing solution. To remedy this, `Firgen` generates both the netlist describing the structure of the filter and a floorplan file specifying the placement and routing of the macrocells and cables. The floorplan file directs `Flint` in the place & route operation. The resultant layout, shown in Figure 18.2, reduces the size of the routing channel between macrocells and produces regular routing for the clock and signal bus. The area of this floorplan-directed layout is less than half of that produced by `Flint` without a floorplan. The details of the floorplan generation process will be discussed in a later section.

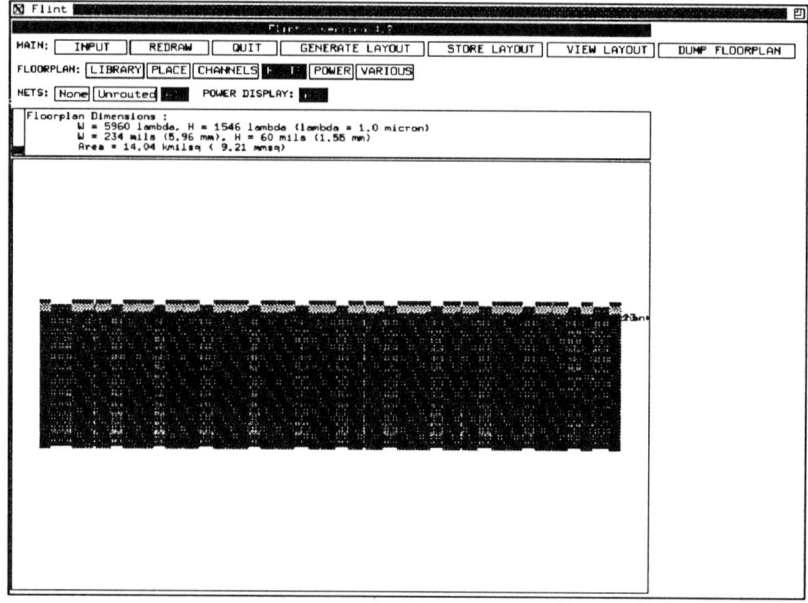

Figure 18.2: `Flint` place and route with `Firgen` generated floorplan

18.2. FIRGEN FILTER DESIGN SYSTEM

The filter design system `Firgen` consists of three major subsystems: (a) filter synthesis and coefficient quantization; (b) filter architecture and floorplan generation; (c) chip layout generation. Details of each subsystem are given below. Figure 18.3 provides an overview of the entire filter generation process.

In the first step the user determines the filter coefficient values using the filter synthesis, simulation and optimization software. The coefficient values are then entered in the architecture and floorplan generation programs. Both programs obtain information about the circuit cells from the cell library database using a mapping function. The architecture generator provides a netlist of the library cells to implement the filter. The floorplan generator creates a floorplan describing placement of the cells in the layout. The netlist and floorplan information is then used by `DMoct` to obtain the final layout.

Frequency domain design specifications can be entered for lowpass, highpass, bandpass or arbitrary multi-band filters. The specifications are entered interactively through a command window (Figure 18.4). The user can then choose to

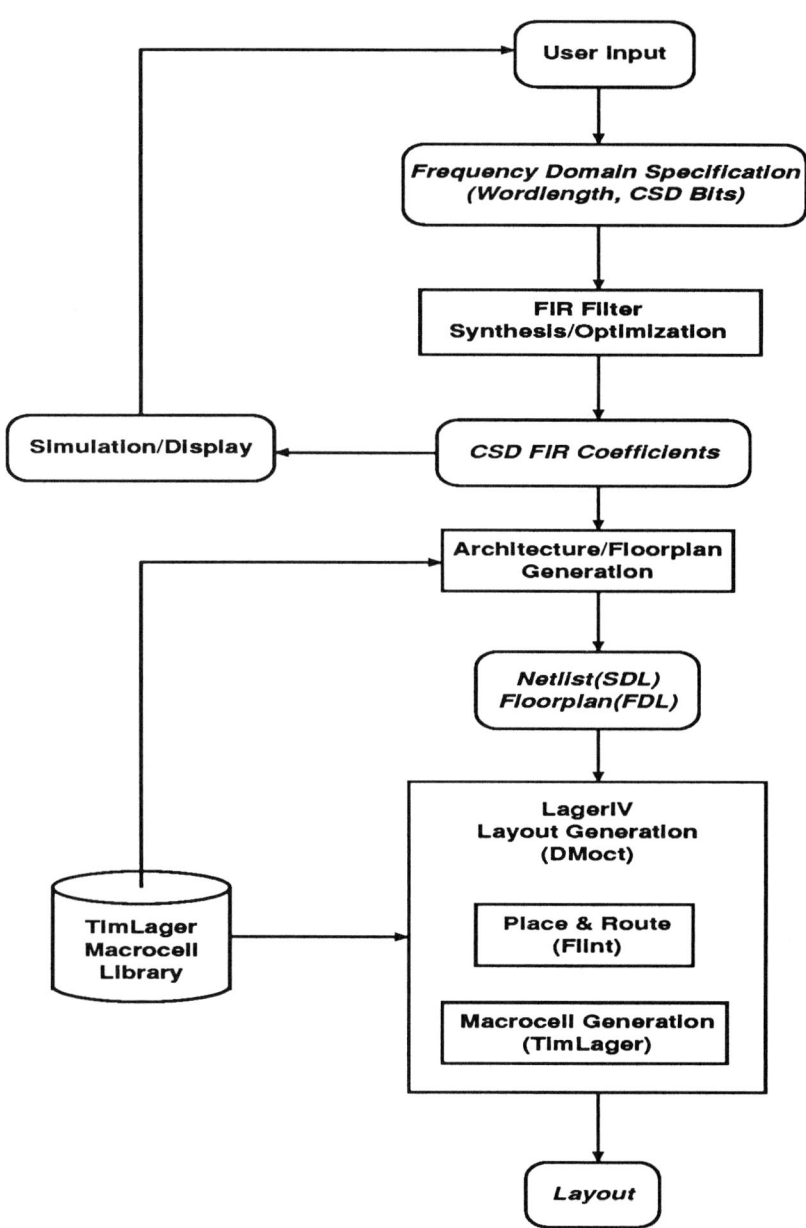

Figure 18.3: Firgen overview

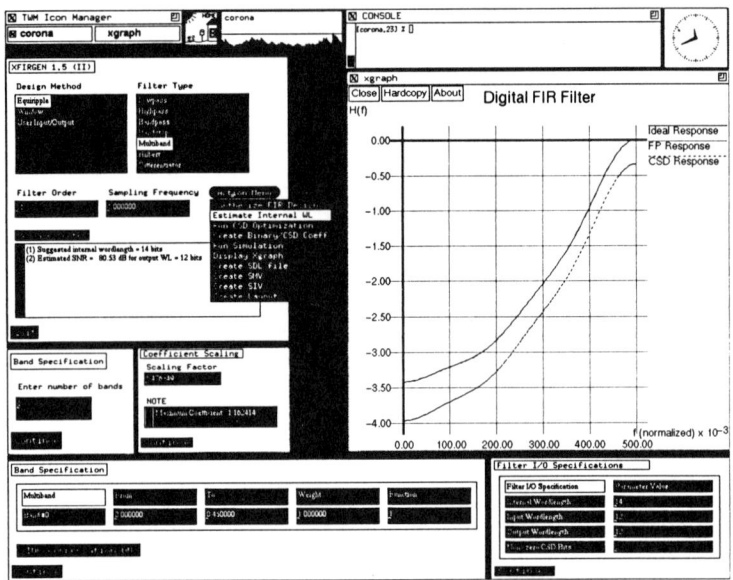

Figure 18.4: Snap shot of Firgen user interface

use either the synthesis tool FIR [Parks85] or WINDOW [Rabiner85] to obtain floating point values for the filter tap coefficients. Firgen then processes these values through a coefficient quantization routine MKCSD to convert the coefficients to canonical signed digit form [Reitwiesner66]. Next, the user specifies the number of CSD digits and the internal wordlength to be used. The CSD values are optionally optimized to obtain the best possible frequency response with the given number of non-zero canonical signed digits using the CSDOPT program [Samueli87]. Firgen then displays a plot of the frequency response obtained with the floating-point coefficient values as well as with the CSD quantized values (Figure 18.4). At this point the designer can iterate on the quantization until the number of CSD digits used produces an acceptable response.

18.3. ARCHITECTURE GENERATION

Once the CSD coefficient values have been decided on, Firgen can proceed with the architecture and floorplan generation (Figure 16.2). Note that the user can also enter the filter coefficients directly.

The architecture used in Firgen has been described in [Jain91,Yang90] and is shown in Figure 18.5 (in this figure, two non-zero CSD bits are assumed

Chapter 18 An FIR Filter Generator

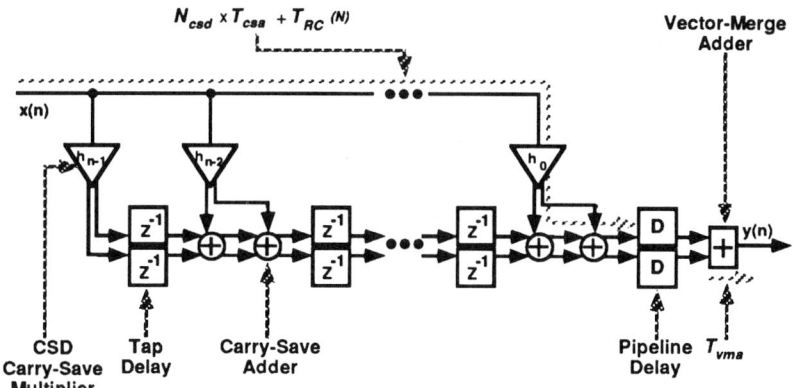

Figure 18.5: Filter architecture

for the CSD multiplier). It basically consists of a transposed form FIR filter structure with carry-save adders and registers implementing the filter core. A vector-merge adder is used to reduce the carry-save outputs to conventional two's complement final output. A pipeline register is inserted between the final tap output and the vector-merge adder to isolate the final tap delay and the vector-merge adder delay. The input signal is broadcasted to all filter taps, and the multiplication is performed by hardwired shift-and-add circuits using canonical signed digit multiplication. The critical path of the architecture is either : (1) the delay through the data broadcast $T_{RC}(N)$ and the delay through one filter tap, $N_{csd} \times T_{csa}$, where $T_{RC}(N)$ is the RC delay for broadcasting data to N filter taps, N_{csd} is the number of non-zero CSD bits used, and T_{csa} is the delay through a carry-save adder; or (2) the delay through the vector-merge adder, T_{vma}, which is a function of the internal wordlength of the filter. The two delay paths are indicated in Figure 18.5 and the actual dominant delay depends on the complexity of the filter as well as the process parameters for circuit parasitics. This target architecture provides high-speed operation with minimal latency and reduced hardware complexity.

Based on this target architecture, Firgen generates an architectural schematic that consists of integrated circuit macrocells for the basic hardware functions addition, subtraction, and sample delay. The mapping flow from signal flow representation of the filter to the actual hardware implementation is illustrated for one tap of a filter in Figure 18.6. Additionally, Firgen generates the netlist for

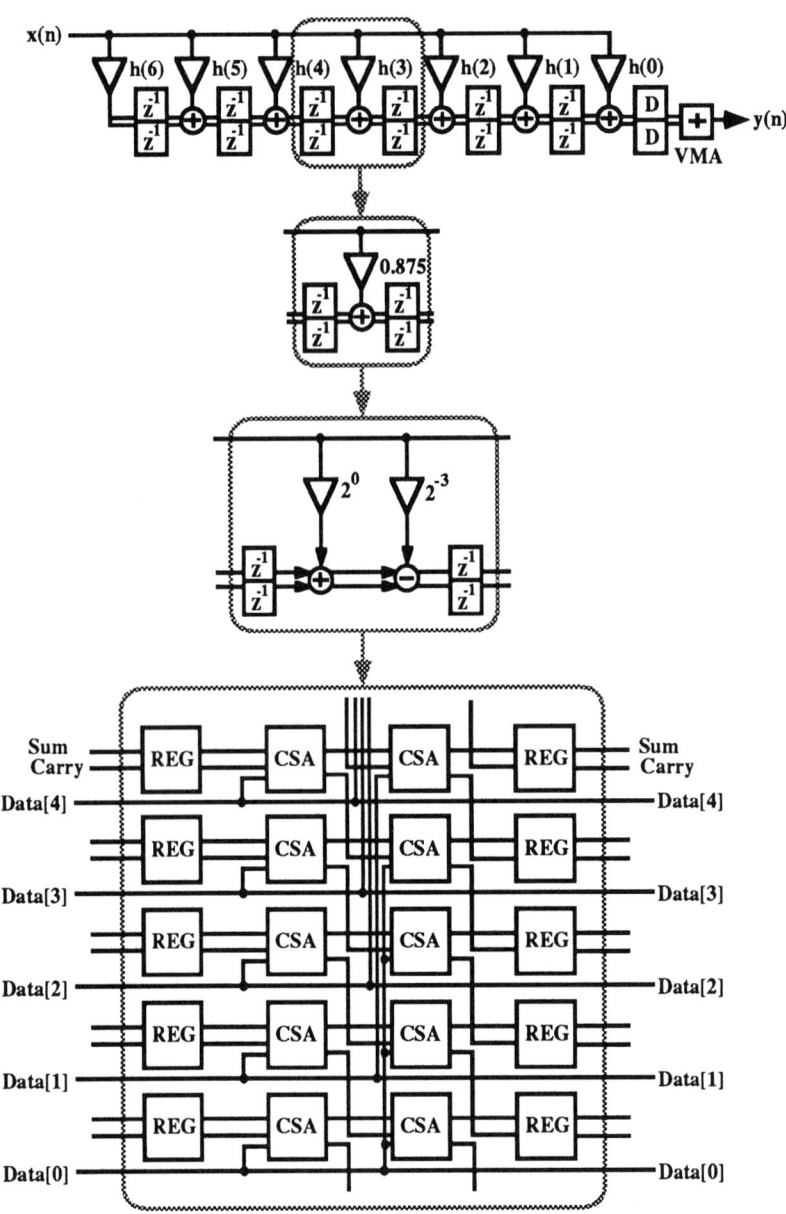

Figure 18.6: Filter architecture showing CSD multiply/accumulation for one tap

the clock generation and distribution network, and for the data distribution network, using various buffer macrocells.

The netlist is created as a text file using SDL and then entered into the normal layout generation process using `DMoct`.

18.3.1. Floorplan generation

The floorplan provides placement constraints to automatic place and route tool, `Flint`. In the floorplan, the placement of all macrocells, including those of clock and data buffers, in the schematic, as well as the definition of routing channels are specified. The actual optimized floorplan is derived using a multi-step user-guided process which is described in [Jain91, Yang90].

The basic floorplan strategy can be arrived at by looking at the architecture (Figure 18.5) and surmising that there are two kinds of nets that have to be routed:

1. Global nets: the input data bus distributed to all of the carry-save adder macrocells, and the clock signals distributed to all of the registers.
2. Local nets: the carry/sum signals are connected between neighboring adder or register macrocells.

To make the layout compact, all adder and register macrocells are placed in a row in the same order that they appear in the schematic (Figure 18.5). Thus, all of the local nets can be routed in channels between the macrocells. The global databus net has to be shifted at each carry-save adder input to implement the coefficient scaling (Figure 18.6). To avoid congestion in routing this net, it is routed by a combination of local channels between the macrocells and feedthroughs provided in the macrocells.

As noted earlier, particular attention has been paid to the data and clock distribution in the floorplan since this is crucial in high-speed circuits. A tree distribution of the clock and data network with multiple level buffering scheme as shown in Figure 18.7 is used to minimize clock skew and reduce signal propagation delays.

The floorplan for `Flint` is stored using FDL. In the FDL description, macrocells are clustered into modules that also contain routing channels. Associated with each of these modules (which can be hierarchical) are descriptions of (1) neighboring modules; (2) signal and power routing entering and/or exiting the module; and (3) preferred routing layer and routing direction. Thus, (1) specifies the relative (rather than the absolute) macrocell placement, while (2) and (3) serve to give routing hints to the place and route tools.

The place and route tool Flint takes this floorplan file and perform physical cell placement and generate physical connections for signal and power routing. To obtain a compact layout, feedthroughs are provided in the macrocells. These are exploited to divide the global routing into local channel routing problems in a manner similar to dpp [Srivastava87]. Note, dpp is not employed because of the restriction in dpp that the power and data signals be orthogonal to each other; in the FIR filter architecture, the optimal layout for the adder which makes up the core of the filter is to have the power and signal buses run along the same direction. In addition, bit slices are not identical (as dpp assumes) due to shifting of buses to realize scaling by power of two.

18.3.2. Layout generation

The final layout generation can be performed using the automatic macro-cell generation and place and route programs of LAGER under the control of DMoct. Flint performs the automatic place and routing of the macrocells using the netlist and floorplan generated and TimLager generates the actual macro-cells with parameters.

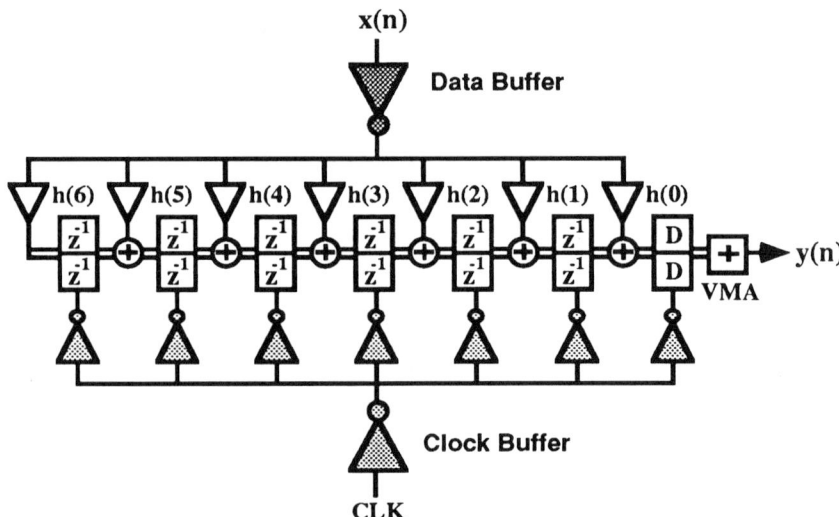

Figure 18.7: Data and clock distribution tree network

Chapter 18 An FIR Filter Generator 279

18.4. DESIGN EXAMPLE AND TEST CIRCUIT

18.4.1. Example design of a FIR filter

In this section, the design of an 11-tap FIR pre-distortion filter for sinc compensation in a D/A converter [Samueli88] will be described.

When the Firgen system is first started, an X-window based graphical-user-interface appears on the user's workstation (the XFIRGEN window in Figure 18.4). The user can then initiate the filter design by selecting the appropriate options (in this case, selecting the "Equiripple" design methodology) and the correct filter order and sampling frequency. Next, the multiband filter type is selected, after entering a "1" for a single band specification (as shown in the "Band Specification" submenu beneath the XFIRGEN window in Figure 18.4), a band specification menu appears and the appropriate values are entered (see the bottom "Band Specification" window in Figure 18.4). For the sinc filter, the "SINC compensation" button is set to "ON" to signal to the filter design/synthesis program to pre-distort the frequency specification in order to achieve the X/SIN(X) compensation desired; in normal filter design, the SINC compensation option would have been turn "OFF". Upon the completion of the filter specification, various options in the "Action Menu" can be invoked (as seen in XFIRGEN window in Figure 18.4), these include options to run programs for filter synthesis and design (Synthesize FIR Design), CSD coefficient synthesis (Create Binary CSD Coeff, where a popup window appears to prompt the user for hardware--structure related parameters, see the "Filter I/O Specifications" window in Figure 18.4), CSD optimization (Run CSD Optimization), filter simulation (Run Simulation), display simulation results (Display Xgraph), netlist and floorplan generation (Create SDL file), and the final layout generation (Create SMV, Create SIV, Create Layout).The filter specifications are also stored in OCT database and is automatically restored in the next Firgen session.

Feedback to the user are provided in the "Message Center" window, these includes warning and error messages generated by the programs as well as providing information to assist the filter design. For example, by selecting the "Estimate Internal WL" option from the "Action Menu", a popup menu will appear requesting the user to enter the desired output wordlength (not shown in Figure 18.4); based on this required output wordlength and the filter order, a suggested internal wordlength as well as the estimated SNR at the output is generated and displayed in the "Message Center" window as is shown in the XFIRGEN window in Figure 18.4.

Additional feedback on the filter performance is provided by first simulating the design using the built-in simulation program and then display the simulation results via the "Display Xgraph" option as is shown in the xgraph window in Figure 18.4. Note, in this particular example, the CSD implementation and hence the CSD response has been scaled to prevent overflow in the filter and is thus offset from the ideal frequency response of the filter. The scaling of the coefficients are done when the "Synthesize FIR Design" option is exercised. A scaling factor is computed and presented to the user for possible alteration in a popup window (see the "Coefficient Scaling" window in Figure 18.4).

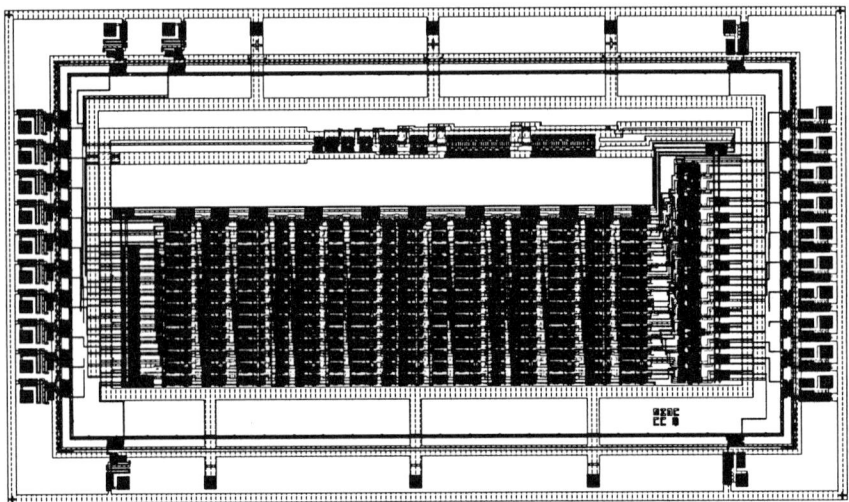

Figure 18.8: CIF plot of the 1.2-micron CMOS chip

18.4.2. An 11-tap FIR filter chip

A CMOS chip implementing the example described in the previous section has been fabricated and tested. The CMOS chip is fabricated in 1.2-μm n-well CMOS technology through MOSIS. The critical path of the test chip is the delay through a 14-bit vector-merge adder, T_{vma} (Figure 18.5); and in order to speed up the operation of the vector-merge adder, a carry-select adder is used in place of the conventional carry-ripple adder. The chip has been tested functionally up to

112 MHz and dissipates 332 mW at the maximum operating frequency. A plot of the chip is shown in Figure 18.8.

18.5. SUMMARY

Firgen provides a means to take high level filter specifications and automatically synthesizes filter coefficients, generates the architecture annd provides the information necessary for the silicon assembly tools to generate an optimized layout. The critical characteristic which allows such high level operation is the limitation to a particular application area.

REFERENCES:

[Denyer84] P. Denyer, A. Murray, D. Renshaw, "FIRST : Prospect and Retrospect," in VLSI Signal Processing (P. Cappello, ed.), pp. 252-263, IEEE Press, 1984.

[Ginderdeuren86] J. Ginderdeuren, H. DeMan, A. Delaruelle, H. Wyngaert, "A Digital Audio Filter using Semi-Automated Design," in ISSCC Digest of Technical Papers, pp. 88-89, 1986.

[Hartley89] R. Hartley, P. Corbett, P. Jacob, S. Karr, "A High Speed FIR Filter Designed by Compiler," in Proc. of the Custom Integrated Circuits Conference, pp. 20.2.1-20.2.4, 1989.

[Hatamian87] M. Hatamian, G. Cash, "Parallel Bit-Level Pipelined VLSI Designs for High-Speed Signal Processing," Proc. of the IEEE, vol. 75, pp. 1192-1202, September 1987.

[Isnardi88] M. Isnardi, T. Smith, B. Roeder, "Decoding Issues in the ACTV System," Trans. Consumer Electronics, vol. 34, pp. 111-120, February 1988.

[Jain86] R. Jain, F. Catthoor, J. Vanhoof, D. Loore, G. Goossens, L. Claesen, J. Ginderdeuren, J. Vandewalle, H. DeMan, "Custom Design of a VLSI PCM-FDM Transmultiplexer from System Specifications to Circuit Layout Using a Computer-Aided Design System," IEEE J. Solid-State Circuits, vol. 21, pp. 73-85, February 1986.

[Jain87] R. Jain, P. Ruetz, R. Brodersen, "Architectural Strategies for Digital Signal Processing Circuits," in VLSI Signal Processing II (S. Kung, R. Owen, and J. Nash, eds.), pp. 361-372, IEEE Press, 1987.

[Jain91] R. Jain, P. Yang, T. Yoshino, "*Firgen* : A Computer-Aided Design System for High Performance FIR Filter Integrated Circuits," IEEE Trans. on Acoustics, Speech and Signal Processing, July 1991.

[Lin89] T. Lin, H. Samueli, "A CMOS Bit-Level Pipelined Implementation of an 11-Tap FIR X/SIN(X) Predistortion Digital Filter," in Proc. IEEE Int. Symp. Circuits and Systems, pp. 351-354, May 1989.

[Noll86] T. Noll, S. Meier, "A 40 MHz Programmable Semi-Systolic Transversal Filter," in ISSCC Digest of Technical Papers, pp. 180-181, 1986.

[Parks85] T. Parks, J. McClellan, L. Rabiner, "Optimal Equiripple FIR Program," in Programs for Digital Signal Processing, pp. 1078-1105, IEEE Press, 1985.

[Privat86] G. Privat, L. Paris, "Design of Digital Filters for Video Circuits," IEEE J. Solid State Circuits, vol. 21, pp. 441-445, June 1986.

[Rabaey85] J. Rabaey, S. Pope, R. Brodersen, "An Integrated Automated Layout Generation System for DSP Circuits," IEEE Trans. on CAD, vol. 4, pp. 285-296, July 1985.

[Rabiner85] L. Rabiner, "WINDOW - FIR Filter Design Program," in Programs for Digital Signal Processing, pp. 1192-1202, IEEE Press, 1985.

[Reitwiesner66] G. Reitwiesner, Binary Arithmetic, vol. 1 of Advances in Computers, pp. 231-308. Academic Press, 1966.

[Reutz86] P. Ruetz, S. Pope, and R. Brodersen, "Computer Generation of Digital Filter Banks," IEEE Trans. on CAD, vol. 5, pp. 256-265, April 1986.

[Reutz89] P. Reutz, "The Architectures and Design of a 20-MHz Real-Time DSP Chip Set," IEEE J. Solid-State Circuits, vol. 24, pp. 338-348, April 1989.

[Samueli87] H. Samueli, "An Improved Search Algorithm for the Optimization of the FIR Filter Coefficients Represented by a Canonic Signed-Digit Code," IEEE Trans. on Circuits and Systems, vol. 75, pp. 1192-1202, September 1987.

[Samueli88] H. Samueli, "The Design of Multiplierless FIR Filters for Compensating D/A Converter Frequency Response," IEEE Trans. on Circuits and Systems, vol. 35, pp. 1064-1066, August 1988.

[Samueli90] H. Samueli, B. Wong, "A VLSI Architecture for a High-Speed All-Digital Quadrature Modulator and Demodulator for Digital Radio Applications," IEEE J. in Selected Areas in Communication, October 1990.

[Srivastava88] M. Srivastava, "Automatic Generation of CMOS Datapaths in the Lager Framework," Master's thesis, University of California at Berkeley, May 1987.

[Ulbrich85] W. Ulbrich, T. Noll, "Design of Dedicated MOS Digital Filters for High-Speed Applications," in Proc. IEEE Int. Symp. Circuits and Systems, pp. 225-228, 1985.

[Yang90] P. Yang, "A Functional Silicon Compiler for High-Speed FIR Digital Filter," Master's thesis, University of California at Los Angeles, 1990.

[Yassa87] F. Yassa, J. Jasica, R. Hartley, S. Noujaim, "A Silicon Compiler for Digital Signal Processing : Methodology, Implementation and Applications," Proc. of the IEEE, vol. 75, pp. 1272-1282, September 1987.

Part V

Applications

19

The PUMA Processor

Lars E. Thon

The PUMA processor application and algorithm will demonstrate many of the features of the C-to-Silicon system described in Chapter 17. It also demonstrates architecture exploration and system simulation. This chapter is arranged as follows: we first present an overview of the design problem and the computation involved. Then we examine the computation task in more detail, and extract its primary characteristics. The next step is to apply this knowledge to select algorithms that will lead to an efficient integrated circuit implementation. Simulating the algorithm is important especially with respect to its implementation in fixed-point arithmetic. Of critical importance is the development of an efficient architecture, supported by the C-to-Silicon tools, for implementing the algorithms. The finished chip developed in this process, and its physical characteristics, is presented.

19.1. THE INVERSE POSITION PROBLEM

The problem chosen comes from the field of robot control and path planning, and is known as the inverse position-orientation problem (IPOP), or the Inverse Kinematics problem. The most advanced industrial robots have six revo-

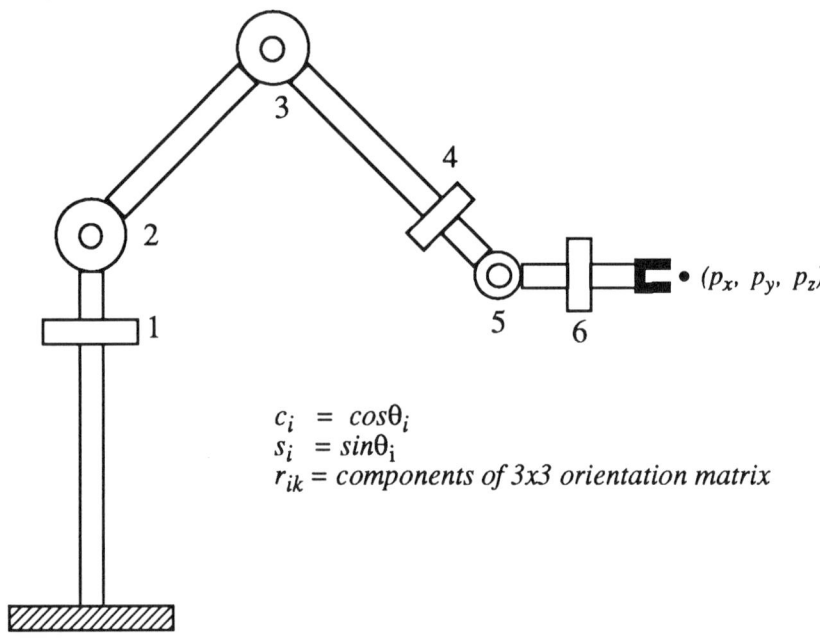

Figure 19.1: A fully articulated robotic arm

lute joints driven by independent actuators. A typical robot is shown in Figure 19.1.

The robot is typically controlled by executing separate position/speed/acceleration profiles (over time) for each joint, and employing feedback to correct deviations from the given profile. However, the robot task is more naturally described in cartesian space. Hence, we need to be able to compute a set of joint angles that correspond to a given position and orientation of the robot hand in the cartesian space. Our task is to compute the solutions of the IPOP for the Puma 560 industrial robot.

19.1.1. The Computation Task

The method for computing IPOP for a 6-R robot is well known for the case where the three last axes intersect in a point [Craig88]. Not coincidentally, most, if not all, industrial robots are designed so that such is the case. The input to the algorithm is the mechanical (Denavit-Hartenberg) parameters of the robot, and the desired position/orientation of the hand, given as the T matrix. The following shows the essentials of the computation. The actual code has to cover four different cases, so most of the code is in effect iterated four times.

$$\theta_1 = \text{atan2}(p_y, p_x) - \text{atan2}(d_3, \pm\sqrt{p_x^2 + p_y^2 - d_3^2}) \tag{19.1}$$

$$K = (p_x^2 + p_y^2 + p_z^2 - a_2^2 - a_3^2 - d_3^2 - d_4^2)/(2 a_2) \tag{19.2}$$

$$\theta_3 = \text{atan2}(a_3, d_4) - \text{atan2}(K, \pm\sqrt{a_3^2 + d_4^2 - K^2}) \tag{19.3}$$

$$\theta_{23} = \text{atan2}\,((-a_3 - a_2 c_3)\, p_z - (c_1 p_x + s_1 p_y)(d_4 - a_2 s_3),\, (a_2 s_3 - d_4)\, p_z + (a_3 + a_2 c_3)(c_1 p_x + s_1 p_y)) \tag{19.4}$$

$$c_4 s_5 = -r_{13} c_1 c_{23} - r_{23} s_1 c_{23} + r_{33} s_{23} \tag{19.5}$$

$$s_4 s_5 = -r_{13} s_1 + r_{23} c_1 \tag{19.6}$$

$$\theta_4 = (c_4 s_5 c_4 s_5 + s_4 s_5 s_4 s_5 < \varepsilon) \,?\, \theta_4^{old} : \text{atan2}(s_4 s_5, c_4 s_5) \tag{19.7}$$

$$s_5 = -r_{13}(c_1 c_{23} c_4 + s_1 s_4) - r_{23}(s_1 c_{23} c_4 - c_1 s_4) + r_{33} s_{23} c_4 \tag{19.8}$$

$$c_5 = -r_{13} c_1 s_{23} - r_{23} s_1 s_{23} - r_{33} c_{23} \tag{19.9}$$

$$\theta_5 = \text{atan2}(s_5, c_5) \tag{19.10}$$

$$s_6 = -r_{11}(c_1 c_{23} s_4 - s_1 c_4) - r_{21}(s_1 c_{23} s_4 + c_1 c_4) + r_{31} s_{23} s_4 \tag{19.11}$$

$$c_6 = r_{11}((c_1 c_{23} c_4 + s_1 s_4) c_5 - c_1 s_{23} s_5) + r_{21}((s_1 c_{23} c_4 - c_1 s_4) c_5 - s_1 s_{23} s_5) - r_{31}(s_{23} c_4 c_5 + c_{23} s_5) \tag{19.12}$$

$$\theta_6 = \text{atan2}(s_6, c_6) \tag{19.13}$$

The notation a ? b : c used above has the same meaning as in the C programming language. The number of solutions is doubled from four to eight by the following symmetry:

$$\theta_4 = \theta_4 + \pi \qquad \theta_5 = -\theta_5 \qquad \theta_6 = \theta_6 + \pi \tag{19.14}$$

19.1.2. Characteristics of the Task

The IPOP algorithm is intensive in multiplication and trigonometric computations. Table 19.1 gives a summary of the operations involved for computing all eight solutions. The table makes it clear that there is strong need for using efficient algorithms for the sin/cos/atan2/sqrt operations. The standard method used in general purpose computer systems is *rational approximation*. That is, using an approximation which is the ratio of two polynomials [Cody80]. Rational approxi-

	Operation count		
Mult (var)	208	atan2	22
Mult (const)	24	cos	14
Add	108	sin	14
Sub	104	sqrt	2
Divide(var)	1	Divide(const)	1

Table 19.1: The IPOP algorithm is intensive in multiplication and trigonometric functions

function	shift and add	add	mult
atan2	34	17	0
cos and sin	36	19	0
sqrt	32	18	1

Table 19.2: CORDIC functions consist mostly of shift/add operations

mations usually involve polynomials of degree 3 and 4 each. This adds up to eight multiplications and one division as a minimum, assuming the polynomials are evaluated with Horner's scheme.

To minimize area requirements we will employ an architecture without an array multiplier, which would make the rational approximation approach very time consuming. In fact, just evaluating atan2 twenty-two times would cost more than all the remaining multiply and add operations in the algorithm. We will therefore use alternative algorithms for sin/cos/atan2/sqrt, based on CORDIC operations.

19.2. ALGORITHM SELECTION

CORDIC [Blahut85,Volder59,Walther71] is a family of algorithms that meets our computational requirements. It can compute all the functions we need, and in the absence of an array multiplier, it is also much more efficient than rational approximation. For a 20 bit word length (and full accuracy), the operation count is given in Table 19.2

Note that cos and sin are computed at the same time at no extra expense. This is quite handy in our case. As we shall see later, CORDIC can be efficiently implemented on a datapath which has an adder and a variable preshifter for one of the inputs. An overview of the CORDIC algorithm is given below.

19.2.1. An RL program for atan2

This is the program text of the CORDIC atan2 (catan2) function, as programmed in the RL language:

```
fix       catan2 (sin, cos)
fix       sin, cos;
{
    register int    k;
    register fix    x, y;
    fix             theta;

    /* Start Cordic. The first step takes care of quadrants
       2 and 3. */
    if (cos < 0) {
        if (sin >= 0) {
            theta = FIXPIHALF; x = sin; y = -cos;
        } else {
            theta = -FIXPIHALF; x = -sin; y = cos;
        }
    } else {
        theta = 0; x = cos; y = sin;
    }

    /* Scale x,y so they don't overflow when amplified. */
    x= (x>>1); y= (y>>1);

    /* The Cordic iterations work in quadrants 1 and 4. */
    for (k = 0; k <= NUMIT; k++) {
        fix     xnew, ynew;
        if (y > 0) {
            theta += ctable[k];
            xnew = x + (y >> k); ynew = y - (x >> k);
            x = xnew; y = ynew;
        } else {
            theta -= ctable[k];
            xnew = x - (y >> k); ynew = y + (x >> k);
            x = xnew; y = ynew;
        }
    }
    return theta;
}
```

In the above program, the operations that have the most impact on the architecture turn out to be the ones of the type $x\!>\!>\!k$, where k is the loop index. We will refer to this later as the *variable shift* operation.

The complete RL program for the IPOP algorithm consists of five functions and a main program. Total code size is 658 lines of text, of which 263 lines are actual RL statements (i.e., not counting comments, blank lines, etc.). It is clear that the IPOP algorithm is nontrivial both in size and complexity, and therefore constitutes a good test case for the `C-to-Silicon` tools.

Alternative 1	Alternative 2
R6L1 shifter	R16L1 logarithmic shifter
inline multiplication code	subroutine call
constant shifter (r>>I)	variable shifter (r>>x[I])
array multiplier (possibly pipelined)	iterative shift/add multiplier

Table 19.3: Design tradeoffs affect layout area, static instruction count and dynamic instruction count

19.3. ARCHITECTURE DESIGN AND EXPLORATION

The starting target architecture is known as Kappa. It was originally developed for speech processing applications [Pope84], and later modified for use in a PID robot joint controller [Azim88]. See Figure 17.1 for a description of the Kappa datapath. Starting with this datapath, we went through architecture design iterations, each time making inexpensive modifications that would improve the efficiency in executing the algorithm. It is emphasized that most of these changes only had to be done on paper or in the machine description file, as explained in Chapter 17. Hence we could quickly evaluate a number of alternatives without expensive investment in circuit layout generation. The tradeoffs we considered are shown in Table 19.3. Some of the tradeoffs are interrelated.

For each of the alternatives, we investigated the effect on the static instruction count (program size) and dynamic instruction count (running time in cycles). The cost of the hardware involved, if any, was also considered. The results are shown in Table 19.4

Chapter 19　　　　　　The PUMA Processor　　　　　　　　　　291

Let us first discuss the shifter type: we use R<n>L<m> to denote a shifter that can shift up to n places to the right or m places to the left in one cycle. Since the CORDIC routines need shifts almost as large as the word length W=20, we decided to eliminate the original R6L1 shifter and replace it by a R15L1 shifter. This allows rapid CORDIC iterations (the improvement is not shown in the table, but the hardware cost was small).

Case	Shifter	Mult type	Num blocks	Code size	Cycles
0	constant	inline code	201	2924	18156
1	constant	subroutine	255	1920	18156
2	variable	inline code	66	1720	18156
3	variable	subroutine	120	717	18156
4	variable	array (delay 1)	66	683	9192
5	variable	array (delay 1*)	66	642	9028
6	variable	array (delay 3*)	66	723	9352

Table 19.4: Effect of design decisions on code size (static instruction count) and code execution time (dynamic instruction count)

Entries 0-1 in the table reflect the use of inline code versus a subroutine call for multiplication. Using the subroutine means an increase in the number of basic program blocks, but a large decrease in code size since one piece of code is shared by all the multiplication operations. Since the architecture has a low-overhead subroutine call, there was essentially no difference in execution time. Entries 0-1 versus 2-3 reflect the result of introducing variable shift capability. The original architecture could only shift by an immediate constant which is coded into the instruction at compile time. Looking at the catan2 RL program, this means that the loop over k would have to be unrolled and the code repeated 17 times (NUMIT=16), each time with a different value for k. This is very expensive in terms of static instruction count, as evident from the table. It was therefore decided to introduce an extra instruction bit which selects between the immediate constant and the lower 4 bits of the index registers X0-X2 as source for the shift amount. This was a very inexpensive addition to the hardware, but the benefit was very significant, reducing program (and hence ROM) size without changing the execution speed.

Entries 4-6 show what happens if we introduce an array multiplier unit into the datapath. First of all, the number of blocks is reduced because the multiplication subroutine calls go away. More impressive is that the execution time is cut in half. A reduction was expected, considered that the program has a large amount of multiplications. The code size, however, shows little improvement. Considering the cost and design time for an array multiplier, we decided against using one. A previous layout indicated that a 20x20 array would be at least 2.54x2.60mm in 2μ technology, plus a substantial overhead in hooking up the busses between the multiplier and the datapath.

The three different cases 4-6 were done as an experiment to see whether the introduction of pipeline delay and/or input multiplexers would make a large difference in the performance. Case 4 assumes that each multiplier input can only come from one particular source, for example, the left input from mbus and the right input from the RAM. Cases 5-6 assumes that either input can come from either source (marked with a * in the table). We observe that neither the pipeline delay nor the input routing had much of an impact on either static or dynamic instruction count. This is positive evidence that the compiler is doing a good job at both scheduling and data routing. In summary, we decided to use the architecture with the R16L1 variable shifter, and leave out the multiplier. Assuming that the chip can run at a 10MHz clock rate, this means we can solve the IPOP equations at a rate of 10e+07/18156=551 times per second. This will be sufficient for most purposes (most robots have a control loop that runs at less than 100Hz, and the IPOP typically is run at a slower rate than the control loop). The resulting datapath is shown in Figure 19.2

As mentioned in Chapter 17, the RL compiler requires that the program sequencer for the chip has a certain minimum instruction set, but the implementation can vary. The PUMA chip uses the sequencer shown in Figure 19.3 [Azim88].

19.4. FIXED POINT COMPUTATION

Since our target processor does not support the floating point data type, it is important to perform a careful analysis of how to implement the algorithm efficiently in fixed-point arithmetic. The goal is to minimize the word length w. The parameters determining word length are the *precision* and *range* requirements of

Chapter 19 The PUMA Processor 293

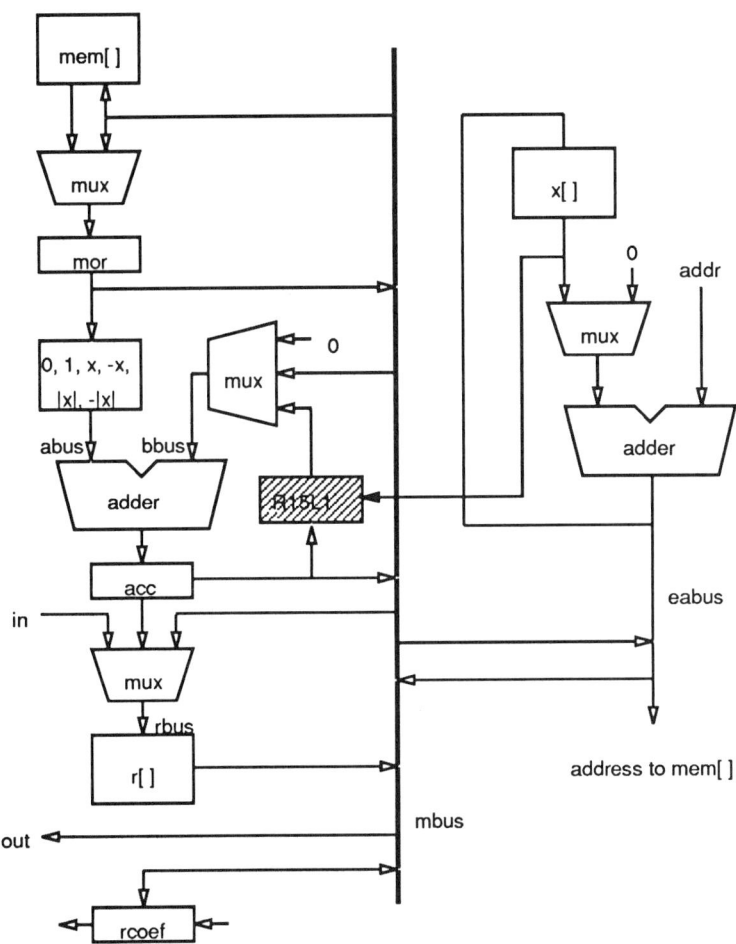

Figure 19.2: The PUMA Datapath

the variables in the program. The basic concepts of fixed-point computation are reviewed briefly in Table 19.5. If abs(x)<2^s we can use the scale $S=2^s$. The basic tradeoff is to choose 2^s large enough to give sufficient range and small enough to provide sufficient accuracy.

Scaling by powers of 2 is convenient, because the processor can easily convert between numbers of different scale by using shift operations. Increasing the scale will, however, lead to loss of significant bits as the bits are shifted out. It is

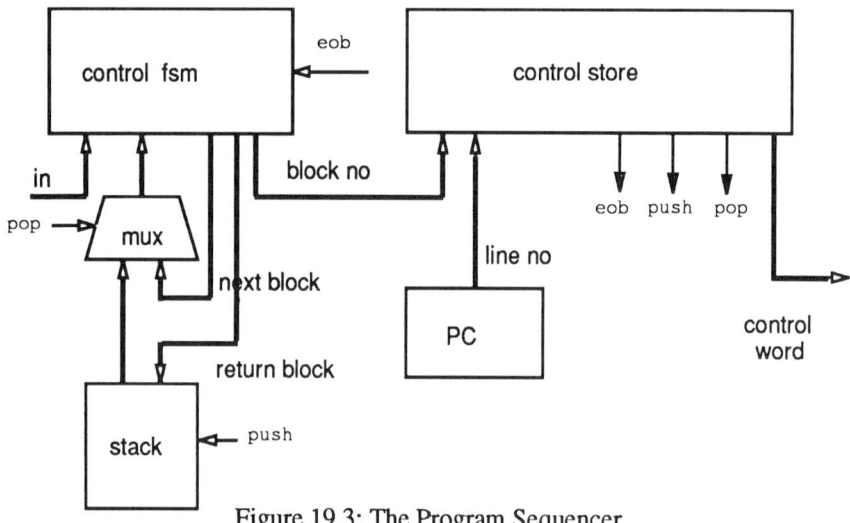

Figure 19.3: The Program Sequencer

Real number	Requirement	Representation (wordlength w)	
x	$abs(x) < 2^s$	$rep(x,w) = integer(x*2^{w-1} / 2^s)$	
operation	scales	requirement	result scale
$x_1 \; x_2$	s_1, s_2	$s_1 = s_2$	$s = s_1 = s_2$
$x_1 * x_2$	s_1, s_2	-	$s = s_1 + s_2$
x_1 / x_2	s_1, s_2	-	$s = s_1 - s_2$
class	range	actual scale	binary point position
lengths (p_x, a_i, d_i)	±2048	2048	12.8
lengths²	±2048²	2048²	23.(-3)
angles (q_i)	π	π	3.17 (approx)
units (s_i, c_i, r_{ik})	±1	2	2.18

Table 19.5: Rules for fixed-point representation and computation. Scaling classes for IPOP variables.

sometimes handy to use scale values other than powers of two, for example π as the scale value for angles, as seen below.

Since RL only allows power-of-two scales, we must simulate other scales by doing the appropriate scaling outside the chip and declare the angle variables to be of type *fix:0*. In fact, it was easier in our case to declare ALL variables in the program puma.k to be of the type *fix:0*. However, constants and input data are assumed to be prescaled according to Table 19.5

In Table 19.5, the scale 2048 for lengths is chosen because the maximum reach of the robot is about 900mm. We cover this with a safety factor of two. Products of lengths get the scale 2048^2 for consistency. The reason for scaling angles to π is the following: The formulas for θ_1 and θ_3 involve the subtraction of two angles. Since each of the two angles may be in $[-\pi,+\pi]$, the result can in principle be anywhere in $[-2\pi,+2\pi]$. Hence there will be a need to reduce the value modulo π so that it comes inside $[-\pi,+\pi]$. When we use π as the scale of the angles, the modulo reduction comes for free during the subtraction (due to the modulo arithmetic of the processor). It would seem reasonable to use scale 1 for the c_i and the s_i. We know that a sine/cosine will always be between -1 and 1, so a scale of 1 should be sufficient. This seems appealing, but consider the effect of inaccuracy: If cos=0.999 becomes cos=1.001, the value will wrap around and become cos=-0.999. These values are not at all "close", because they correspond to very different angles. (It is not analogous to the situation with angle values, where +179.99° and -179.99° describes essentially the same angle.) Hence, we decided to use a scale of 2.

The word length used was w=20. It was chosen as follows: our target was to compute $\theta_1...\theta_6$ with an error of less than 0.05°, or 4.5 decimal digits. To achieve this, we determined that we need about 5.5 decimal digits precision in the intermediate calculations. This corresponds to roughly 19 bits. Adding one bit to represent the negative numbers we end up with w=20.

19.5. HIGH-LEVEL SIMULATION

Using the above scaling scheme, the IPOP computation was programmed in RL, using CORDIC subroutines for the elementary functions. To make sure that the program and the scaling were sound, we used the KT tools to perform first floating point and then fixed point simulation.The simulations showed that the program works well unless the specified goal frame is close to a singularity ([Craig88], p146). It should be noted that a floating point program will also pro-

Characteristic	Value	Comment
wordlength	20	size of all the datapaths
cstore PLA	13 x 649 x 77	49973 bits (microcode ROM)
lgu PLA	16 x 32 x 8	inputs • minterms • outputs
cfsm PLA	21 x 171 x 26	inputs • minterms • outputs
data RAM	172 x 20	3440 bits
technology	2 micron	scalable CMOS (nwell)
width x height	9864 x 9608	lambda^2
transistors	44802	54582 inactive transistors in PLA/ROM
pads	126	
package	208 pin PGA	

Table 19.6: Physical design characteristics of the PUMA chip

Block	Speed	Comment
chip (IRSIM)	6.2 MHz	without resistance modelling
chip	4.6 MHz	limited by RAM speed
RAM	4.6 MHz	long poly lines (area optimized)
datapath	8.2 MHz	
control store	8.4 MHz	
control fsm	8.4 MHz	
program counter	>10 MHz	

Table 19.7: Measurements on the PUMA chip

duce inaccurate results in this case. Moreover, the loss of accuracy is often accompanied by the fact that the position/orientation is only weakly dependent on the value of the particular inaccurate angle. It is also possible to detect during the computation that we are close to a singularity, and issue an error signal. The typical case had an angle error of less than 0.02^o for each one of the 48 angles when simulated using target positions/orientations generated with a random number generator.

19.6. CHIP DESIGN AND VERIFICATION

The IPOP algorithm is quite complex compared to algorithms employed in DSP applications. In particular, the resulting microprogram is large (about 670 lines after compression), yielding a chip of 9.8 x 9.6 mm^2 in 2µ technology. The PUMA chip core consists of 221 macrocells (six at the top level) and seven levels of hierarchy. Table 19.6 summarizes the key aspects of the physical chip design.

19.6.1. Chip Simulation

The PUMA chip has been simulated extensively at both the logical level with THOR [Thor88] and at the switch level, from extracted layout, using IRSIM [Salz89,Salz90]. The execution of the microcode, including data input and output, has been simulated in its entirety, and the results have been verified against those computed in the high-level (RL) simulations.

19.6.2. Physical design results

The completed chip layout is shown in Figure 19.4. There was complete functional agreement between measurements on the chip and the THOR/IRSIM simulation results. IRSIM is usually a conservative predictor of chip speed. For the PUMA chip, the simulation worked up to 6.5 MHz. Measurements on the chip showed that it was fully functional only up to 4.6MHz. The first block to fail was the RAM. The datapaths, the program ROM and the block sequencer were all functional up to 8.2MHz at 5V. The discrepancy is due to the fact that the circuit extraction did not include wire resistances. Resistance extraction is important when there are long polysilicon lines in the layout. This was the case in the RAM and PLA modules. These modules had been optimized with respect to area by using polysilicon lines instead of metal lines in certain key circuits. It is clear that the speed could be increased substantially by spending more area on metal lines.

19.7. SUMMARY

The design of the PUMA chip demonstrates the feasibility of the C-to-Silicon system, and served as the main driving force behind its development. The system allows architecture-level experimentation, that is, finding the best architectural tradeoffs without having to go through a detailed design process. The detailed netlist design and the physical layout is deferred until the

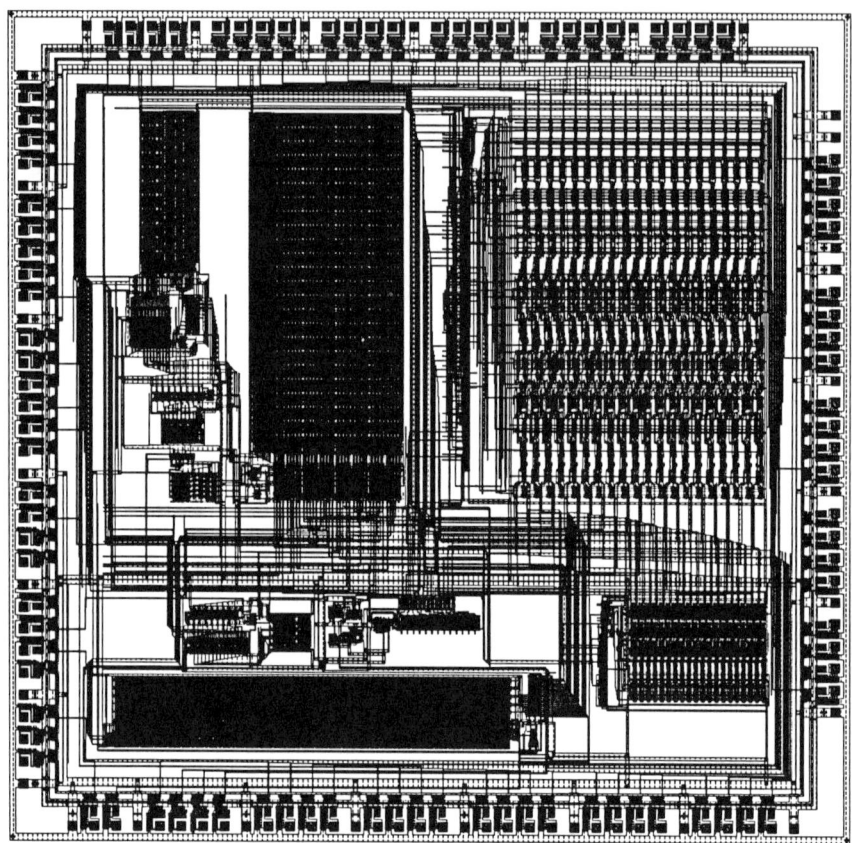

Figure 19.4: CIF plot of the PUMA chip

desired architecture has been chosen. The system also provides facilities for fixed-point simulation, so that the numerical correctness of the user's algorithm can be verified.

REFERENCES:

[Azim88] Syed Khalid Azim, *Application of silicon compilation techniques to a robot controller design*, PhD thesis, UC Berkeley, May 1988. UCB/ERL memo M88/35.

[Blahut85] Richard Blahut, *Fast Algorithms for Digital Signal Processing*, Addison-Wesley, 1985.

[Cody90] William J Cody and William Waite, *Software Manual for the Elementary Functions*, Prentice-Hall, 1980.

[Craig88] John J Craig, *Introduction to Robotics*, Addison-Wesley, 1986.

[Pope84] Stephen S. Pope, *Automated Generation of Signal Processing Circuits*, PhD thesis, UC Berkeley, 1984.

[Salz89] Arturo Salz and Mark Horowitz, "IRSIM: An incremental MOS switch-level simulator," Proceedings of the 26th ACM/IEEE Design Automation Conference, June 1989, pp. 173-178.

[Salz90] Arturo Salz, Stanford University, IRSIM m*anual*, 1990.

[Thor88] CAD Group, Stanford University, *Thor tutorial*, 1988.

[Volder59] J E Volder, "The cordic trigonometric computing technique,"*IRE Transactions on Electronic Computers*, pages 330–334, 1959.

[Walther71] J S Walther, "A unified algorithm for elementary functions," *Proceedings of the 1971 Spring Joint Computer Conference*, pages 379–385, IEEE, IEEE, 1971.

20

Radon Transform Using the PPPE

William B. Baringer

The Radon and inverse Radon transforms offer attractive opportunities for image processing and analysis. These transforms involve computing and analyzing the "projections" of a digitized image along lines at various angles (for the forward Radon transform) and reconstructing an image from a set of projections (for the inverse Radon transform). The inverse Radon transform is used in medical computer aided tomography (CAT) imaging devices, and in image reconstruction for geophysics, for example. The use of the Radon transform enables efficient computation in image processing/analysis applications. This is accomplished by analyzing the image in its transformed space, thus reducing two-dimensional image processing to a set of independent and parallel one-dimensional processing steps.

Because of the autonomous relationship between these independent one-dimensional transform vectors, they can then be processed independently by the current breed of powerful commercial DSPs. Thus, many different image processing algorithms can be realized by simply programming the DSPs appropriately. This results in a highly flexible machine vision and image synthesis environment.

This chapter presents the custom ASICs that implement the forward and inverse Radon transform in real-time, the associated DSP micro-computers that process the transformed data, and the custom printed circuit boards that support them. Some of the image processing algorithms that benefit from operating in the Radon domain are also discussed.

20.1. MACHINE VISION ALGORITHMS

The Radon transform can be efficiently used to realize a wide variety of machine vision algorithms. Image processing algorithms employing the Radon transform can be divided into two general classes. One class produces certain numeric or symbolic information from the transformed data, while the other produces an image as output. These classes are referred to in general as "analysis" and "synthesis" algorithms.

20.1.1. Examples of General Analysis Algorithms

Examples of Radon transform-based image analysis algorithms that benefit from operating in projection, rather than spatial or frequency image representation, include robust high-accuracy line detection and measurement [Petk89], model-based shape analysis for image segmentation, image recognition, and geometric feature extraction such as centroids and moments [Sanz88]. All of these algorithms can be applied to either gray-level or binary input images.

The general computation involved in image analysis tasks involves pre-computing the desired angles through which projections are to be taken, for the specific analysis task at hand. Then, during the active image period, the projections of the image are computed along the given angles θ_i. These projections are then stored in local buffers.

At the end of the active image frame, the necessary transform-based image processing is performed on each of the projections to extract the desired parameters from the original image. Some examples of required transform-based operations on the projections include peak finding, to determine the location and orientation of lines or edges in the image; one-dimensional centroid or higher moment computation, to determine the two-dimensional centroid or higher moments of an object; averaging, to determine the total average grey-level of the image; one-dimensional discrete Fourier transform, to determine the spatial frequency content along the particular angle of projection, for filtering in the Fourier domain, and for Fourier phase analysis to determine shifts in the projection domain; removal of means and cross-correlation, to compare different projections against one another for object identification and tracking, etc.

20.1.2. Examples of General Synthesis Algorithms

Examples of Radon transform-based image synthesis algorithms include general image reconstruction from projections (using the inverse Radon transform) [Herm80, Brac79], very large kernel two-dimensional linear filtering, convex hull approximation, multi-color mask generation, mathematical morphology, and image compression / decompression [Shap80].

For this second class of algorithms, the necessary computation becomes more elaborate. In addition to collection and subsequent processing of the projections as in the analysis discussion above, a third phase is required.

Synthesis task specific computations are first performed on the projection data. The final image is reconstructed using this processed projection data. The projection bins that contribute to the appropriate output pixels in the reconstructed image are determined. Arithmetic or logic operations are performed to combine the appropriate projection bins to form a final processed or synthesized image. This output image is then sent to the system CRT monitor or output frame buffer.

Depending on the particular synthesis application, processing operations on the projections include one-dimensional convolutions, one-dimensional discrete Fourier and inverse-Fourier transforms with multiplications by Fourier kernels, decimation, etc. Some non-linear image operations, such as convex hull approximations, are achieved by treating the local projection buffers as general-purpose look-up tables (LUTs). After performing projection data analysis, the LUT is filled with "synthesized" data for reconstruction.

Thus, there is a wide class of image processing and machine vision algorithms that greatly benefit from operating in the Radon space, both in image analysis and image synthesis tasks. This wide class of algorithms would typically need many different architectures to support the diversity of operations required. A single architecture that achieves this wide range of functionality is presented in the following section.

20.2. A REAL-TIME RADON TRANSFORM ARCHITECTURE

This section presents an architecture that provides the forward Radon transform *and* the inverse Radon transform at video rates. The discrete Radon transform is given by Equation 22. The calculation of the Radon transform is comprised of the following nested loops: For each orientation angle θ; and for each offset ρ for a given θ; find all the pixels (x_s, y_s) that are contained in the digital line $L_d(\theta, \rho)$; and for each such pixel accumulate the pixel's value $f(x_s, y_s)$ into the register $P\theta(\rho)$.

$$P_\theta(\rho) \cong \sum_{(x_s, y_s) \in L_d(\theta, \rho)} f(x_s, y_s) \Delta s \qquad (20.1)$$

For 180 different angles of θ, and 1024 offsets ρ, and 512 pixels on a given digital line of projection, all at 30 frames per second, there is a total of approximately *three billion* inner loop executions required per second. Each inner loop includes a random access to a frame buffer, a read and write cycle to the appropriate Radon (ρ, θ) storage elements, and associated computation.

An architectural concept called the "Parallel Pipeline Projection Engine", or "PPPE" was developed [Sanz88]. The concept was extensively simulated, and was shown to efficiently perform raster-mode processing. This architecture has an "MISD" arrangement of its instruction and data streams. This is provided by a single wide pipelined video bus connected through all of the homogeneous (identical) processing elements, and separate control and instruction busses connected in parallel to each of the processing elements. This arrangement is shown in Figure 20.1.

The architecture is composed of a pipeline connection of "M" processor element stages. Each stage is responsible for one projection orientation in the Radon space. Thus, the architecture is "output-partitioned" such that each stage computes all of the output ρ values for a given fixed angle θ. This arrangement of identical processing stages provides a high degree of modularity and application specific flexibility. The pipelined video bus running through the processors operates at a 10 MHz (synchronous) rate. There is little pipeline delay in each stage, so the overall latency of the architecture is small. There are no explicit communication or control channels between the stages, except the video data bus described. Thus, there is no message passing, and no global memory contention between stages. This enables a highly efficient parallel implementation in the architecture. The stages essentially operate independently, transforming a two-dimensional input image to a set of "M" one-dimensional independent and parallel data vectors. These vectors are approximately 2K bytes in size. Thus, the transformation has effectively reduced the 10 MHz bandwidth of the input image to an amount that a commercial DSP chip can process in one image frame period.

Each processing element has several modes of operation allowing algorithm flexibility and computational power to be realized in a decoupled and modular fashion. Due to the highly computational and I/O intensive nature of the tasks performed in each processor stage, the functions enclosed in the dotted box in Figure 20.1 are implemented in a custom VLSI application specific IC. Extensive use of parallelism and pipelining is employed at all levels of the hierarchy within this architecture.

Each individual processor element consists of an address generator, a projection collector, a line-length counter, internal dynamic RAM banks, an arithmetic logic unit, and the necessary control and interface logic. Eight of these processor elements share access to a general-purpose digital signal processing (DSP) micro-computer, in a time-shared fashion. A local host CPU has direct access to each DSP and indirect access to the contents of each projection memory of each stage. The host can download program control to each stage's DSP chip, and obtain results of the locally executed algorithms from each stage for further high-level processing.

In summary, this "parallel pipelined projection engine" has many desirable features:

- It provides the forward *and* inverse Radon transforms.
- It operates in a raster scan fashion.
- It supports a wide range of projection-based image processing and image analysis tasks.
- The building blocks are simple and modular; the algorithms are completely decoupled with no global intercommunication connection required.
- The architecture is amenable to VLSI implementation.
- The architecture scales with the desired number of projections and stages.

20.3. THE RECONFIGURABLE RADON TRANSFORM SUBSYSTEM

This section describes the application specific printed circuit board that supports the implementation of the forward and inverse Radon transforms. This board also supports a wide range of real-time transform-based image processing, due to its high degree of parallelism and pipelining.

This custom printed circuit board is based around the VME bus. It is constructed as a 9U, 400mm VME board, using six signal layers and two planes for power and ground. A block diagram of the entire board is shown in Figure 20.2,

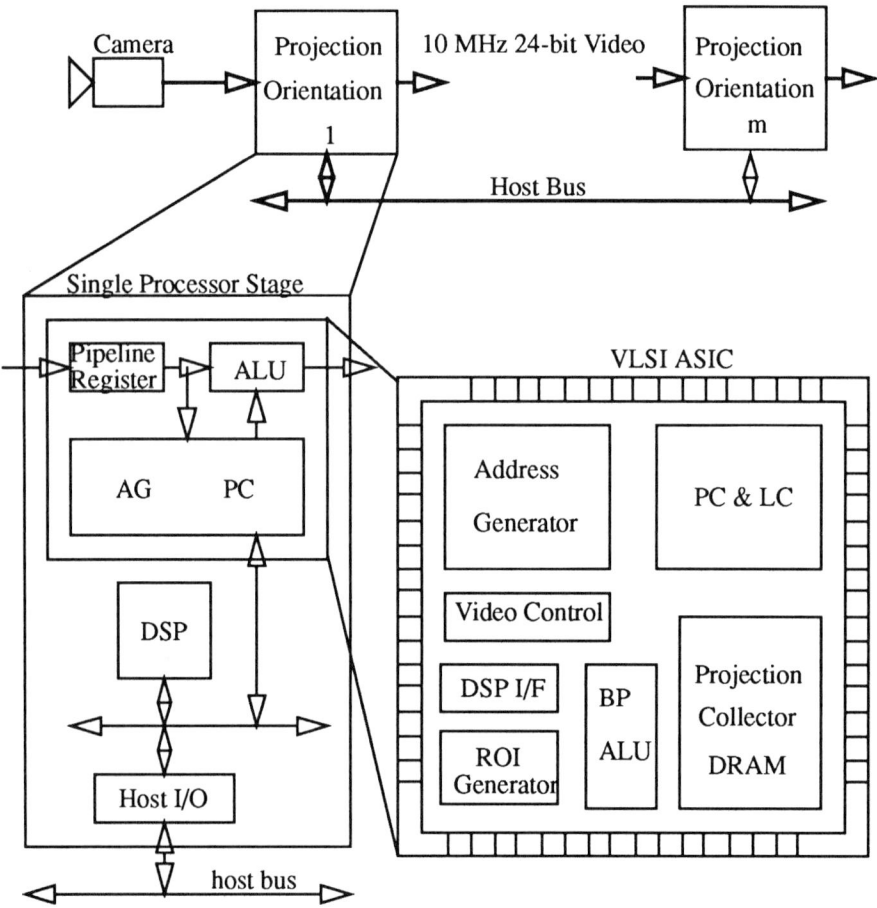

Figure 20.1: Architecture of the Parallel Pipelined Projection Engine

and a photograph of the fabricated and tested board is shown in Figure 20.3. Because of the hierarchical and modular nature of the design of the board, there is a close correspondence between its block diagram and its layout.

At its highest level of hierarchy, the board is composed of three internal functional circuit blocks, or "modules". These are the VME interface module, the video signal interface module, and the DSP-ASIC processor module. Referring to Figure 20.2, the DSP-ASIC processor module is instantiated four times on the board. In this way, placement of all the components contained in each of the four modules can be specified with a single placement offset coordinate in LAGER.

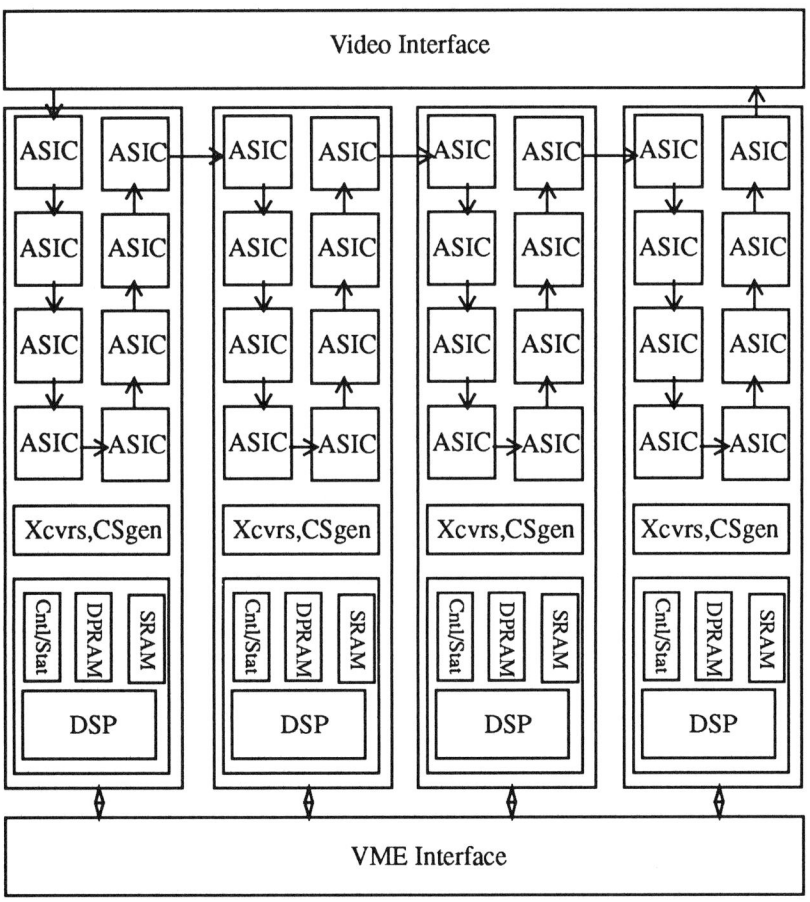

Figure 20.2: Block Diagram of Radon Transform Subsystem

Each of these DSP-ASIC processor modules is in turn made up of a DSP micro-computer module and an ASIC array module. Each of the DSP micro-computer modules is a complete, self-contained, micro-computer based on the AT&T DSP32C floating-point processor. This compact and very powerful module can be used in other board-level designs by simply instantiating its design files via LAGER. This module represents yet another layer of hierarchy, by encompassing the necessary modules containing the circuitry that comprises a DSP microcomputer. This includes the local high-speed static RAM module, a high-speed dual-port RAM module for communication to the VME interface module, and other modules for the necessary control, status, and interrupt registers, address buffering and decoding, reset circuitry and clock.

The ASIC array module consists of support for eight custom ASIC processors, as well as the required data, address, and control interface circuitry between the ASICs and the DSP micro-computer, and between the ASICs and the video interface circuitry. Referring again to Figure 20.2, the video pipeline flows from the video interface circuitry to the first ASIC in the far left DSP-ASIC module. After passing through the pipeline connection of the eight ASICs in the first ASIC array module, the video stream continues to the eight ASICs of the second ASIC array module. This continues until the image data has passed through all 32 ASICs on the board and finally returns to the output stage of the video interface module.

The video interface circuitry accommodates both Imaging Technology and Datacube commercial video signal protocols. This provides the flexibility to use the printed circuit board in image systems that employ either style of commercial image acquisition and processing boards. The video busses of multiple printed circuit boards are connected together in a daisy-chain fashion by plugging in ribbon cables between the video output port of one board's video interface circuitry with the video input port of the next board.

The VME interface circuitry includes the necessary address, data, and control lines to provide communication between the four DSP micro-computer modules and the VME bus (and thus the VME master on the back-plane). There are no communication channels made available between adjacent DSP micro-computer modules directly. The VME bus master communicates individually with each micro-computer module on each of these boards in the card-cage, and the analysis results from all of these processor modules are combined by the 68020 CPU in the bus master. This communication from micro-computer module to VME host is

Chapter 20 Radon Transform Using the PPPE 309

typically one of low bandwidth, as most of the processor-intensive computations are performed first by the ASICs, and then by the DSPs themselves.

The total board is comprised of approximately 300 integrated circuits, both commercial and custom. Forty-four of these ICs are in PGA packages, of size 84, 108, and 132 pins.

Figure 20.3: Thirty-Two Stage Radon Transform PCB

20.4. THE RECONFIGURABLE RADON TRANSFORM ASIC

This section describes the development and design of a VLSI ASIC that implements the forward and inverse Radon transforms at video rates. Transforming a 512 x 512 image into Radon space at video rates is a computational and I/O intensive operation. For the forward Radon transform, a pixel value must be read, the address of the correct projection bin must be computed, the previous value of the projection bin at this address must be read, its value updated, and written back to the projection memory. For the inverse Radon transform, the address of the correct projection bin must be computed, the previous value of the projection bin at this address must be read, a partial back-projection value must be read, the two data values must be combined, and the result must be sent to the next processing stage. In parallel with these operations, the desired region-of-interest must be computed, and the lengths of the digital lines must be updated. This must be done for each projection angle throughout the image, and each angle has 1K projection bins. Each bin must be at least 17 bits wide for 8-bit gray-level image pixels. All of these operations must be accomplished within a 100ns period, requiring tens of billions of operations per second. For this reason, the design and fabrication of a custom, application specific, VLSI circuit was necessary.

The data and I/O busses and control signals of the ASIC are created with a "generic" interface in mind. This means that the connections to an external DSP are made such that the ASIC appears as a memory-mapped device to the DSP. This simplifies the connections between the DSP and the ASIC, reducing the amount of required external components to a minimum. The only additional component required on the printed circuit board is an address decoder chip to produce a chip-select signal "cs" to choose which one of the ASICs is currently being addressed.

A "generic" treatment of the video signal interface is also built into the ASIC. Independent video input and video output busses are provided, as well as pins for "standard" RS-170 video control input signals.

The ASIC is composed of twelve internal functional circuit blocks or "modules". Most of these modules are parameterizable in size, number of words, number of bits, etc., as supported by LAGER. In this way, the ASIC can be redesigned for different image formats or for different internal data representations for accuracy requirements, for example.

Chapter 20   Radon Transform Using the PPPE

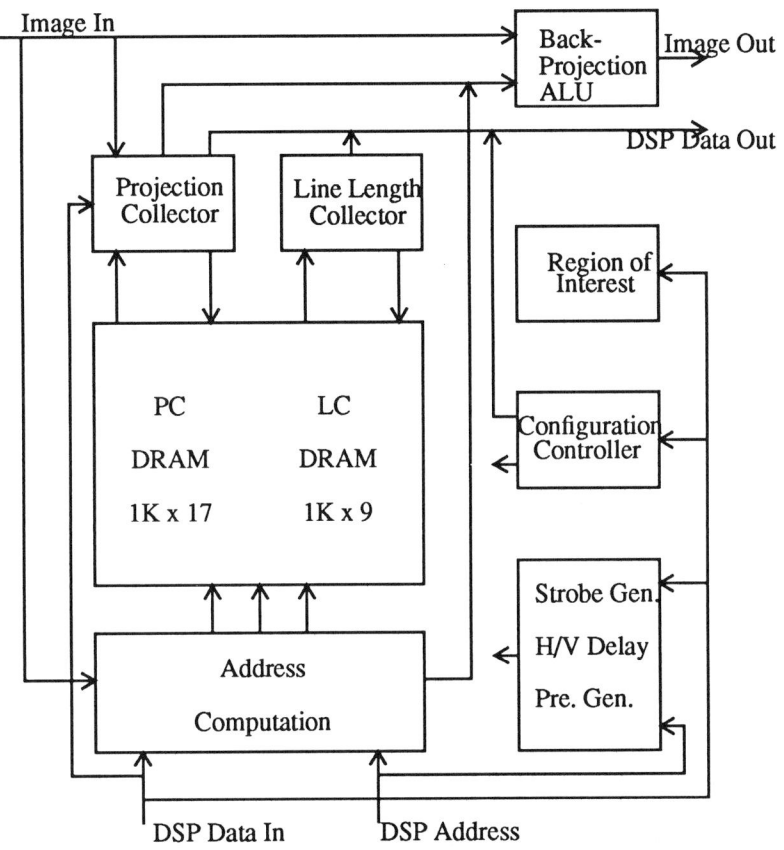

Figure 20.4: Block Diagram of ASIC

A block diagram of the ASIC, indicating the different modules and the address and data busses interconnecting them, is shown in Figure 20.4.

The "projection collector RAM" module, or "PCRAM", accumulates the projections of an image in raster scan mode. It is a self-timed read-modify-write dynamic RAM of size 1024 words by 17 bits per word. The maximum number of memory words arises when projections are taken at 45 degrees through an image, requiring 1024 projection bins in which to accumulate the slice of the Radon transform. The maximum length (actually the "area") of the projection lines is 512, so with eight-bit per pixel input images, the largest accumulated projection value will be 512 x 256, or 2^{17}, thus requiring 17 bits per word in the PCRAM.

The "line counter RAM" module, or "LCRAM", accumulates the length of the lines over which an image is projected. It is also a self-timed read-modify-write dynamic RAM. This module has a size of 1024 words by 9 bits per word, to accommodate up to 1024 projection lines with individual lengths, and with each line no longer than 512 units.

The modules that access the data busses and control lines of each of the PCRAM and LCRAM are designed so that the PCRAM and LCRAM can share the same address bus. The two memory modules could be specified to LAGER to be created as two different memory blocks by simply giving different specifications of the required data word widths. However, substantial area of the chip was saved by combining both modules into one memory block of size 1024 words by 26 bits per word. Routing area was also saved by this scheme. The PCRAM then consists of the common ten-bit address bus and the lower 17 bits of memory I/O busses, while the LCRAM makes up the upper 9 bits of memory I/O busses while sharing the address bus. The final layout actually uses two banks of 512 words by 26 bits per word to implement the combined 1024 words by 26 bits required by the PCRAM and LCRAM modules. Splitting the total memory into two banks provides a more optimal overall chip layout and decreases the address decoder delay time for the memories.

The "projection collector" module, or "PC", is tightly coupled with the PCRAM in order to provide the computational updates of the stored projection values depending on the external incoming data and the modes of operation chosen. In its "projection" mode of operation it reads values from the PCRAM, adds a new input image pixel value, and writes the result back into the same address location in the PCRAM. Other modes of operation include auto-clearing of the PCRAM, external data access of the PCRAM, gray-level histogram accumulation, and region-of-interest conditioning of the projection operation. To achieve all of this functionality, the PC module has input from the buffered DSP data bus, and input from a portion of the image input bus. It also has separate input and output data busses connected to the PCRAM module. An output bus from the PC can be read off of the chip by the DSP to support DSP read cycles of the PCRAM.

The "line counter" module, or "LC", is tightly coupled with the LCRAM in order to provide the computational updates of the stored projection line-lengths in the LCRAM. It also provides auto-clearing and conditional operation based on the calculated region-of-interest. This module also has separate input and output data busses connected to the LCRAM module. Likewise, an output bus from the

Chapter 20   Radon Transform Using the PPPE   313

LC can be read off of the chip by the DSP to support DSP read cycles of the LCRAM.

The "address computation" module, or "AC", provides the addresses to the LCRAM and PCRAM that correspond to the raster scan generation of Bresenham lines [Bres65]. It is programmable to generate projection lines at angles from -90 to 90 degrees, in increments of arctan(1/512). It is also highly reconfigurable, allowing the internal RAM modules to be addressed by the DSP's address bus, or by an internal refresh counter, or by the external image input bus for gray-level histogram operation. This flexibility requires a portion of the DSP's buffered data and address busses as inputs, as well as a portion of the input image bus.

The "back-projection arithmetic logic unit" module, or "BPALU", is used for image synthesis applications, such as image reconstruction, image filtering, and synthesized graphics. It performs the computational tasks necessary to combine previously stored projections or arbitrary functions into a raster scan output image, for further processing, storing, or viewing. It is highly programmable, to provide many arithmetic and logical functions on its input data streams. It can be configured to divide the internal data results by powers of two to rescale the data for viewing on a CRT monitor, or for processing on a subsequent stage. The BPALU's inputs are the external image input bus, an output bus from the AC module, and an output bus from the PC module.

The "region-of-interest" module, or "ROI", generates the appropriate clock signals to realize an effective region-of-interest over which computations by other modules in the ASIC take place. This programmable module is typically configured by an interactive X-window program that uses the workstation's mouse to select an area of interest. Automatic window selection is also commonly used. It uses a portion of the DSP's data bus as its input.

The "strobe generation", "horizontal blank programmable delay", "vertical blank programmable delay", "strobe generation", "programmable precharge generation", and "configuration control register" modules provide various levels of control of the operations of the ASIC from the external DSP. This design greatly reduces required external interface logic on the printed circuit board, as well as reducing unnecessary I/O pins on the ASIC.

The modules of the ASIC can be interconnected in different configurations to support a wide variety of different operating modes.The ASIC is fabricated in a

108 pin PGA package. A die photograph of the chip is shown in Figure 20.5. The chip consumes 7934 x 9031 microns in a 1.6 micron scalable CMOS process.

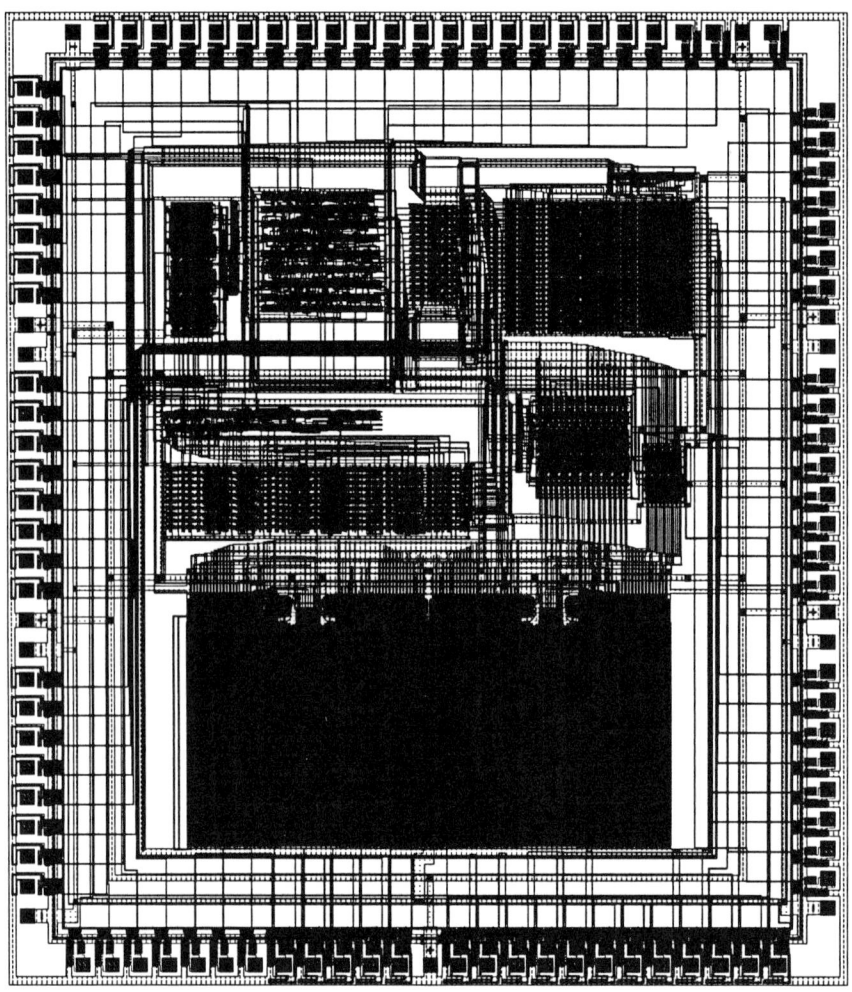

Figure 20.5: CIF plot of Radon Transform ASIC

20.5. DESIGN PROCEDURE

The development of an architecture to realize the Radon transform involves many trade-offs in hardware and software complexity. This is particularly true for the system implemented, as achieving video data rate throughput was a strict requirement. Careful analysis of the computational critical paths, data and address bus bandwidths, and all required control circuitry, was important to meet the specifications using the targeted ASIC and board-level technologies. These studies resulted in appropriated partitioning between commercially available components and custom ASIC circuits, as well as a hierarchical arrangement of processing power in the system.

The Radon transform ASICs can be considered as high-speed co-processors to the DSP micro-computers. The ASICs handle constant data stream 10 MHz video channels, while the DSPs have a lower data I/O bandwidth to process. The DSP micro-computers can be considered as high-speed floating point co-processors to the 68020/68882-based VME master single board computer. The entire VME card-cage can be considered as a high-speed machine vision co-processor to the host workstation. This division of labor, according to the required processing power and I/O bandwidths, provides a robust means of achieving real-time digital signal processing for a wide range of algorithms.

Reconfigurability, as opposed to traditional micro-controlled programming, was employed extensively in the ASIC. This provided a high level of flexibility in a real-time system that would otherwise be entirely dedicated to a single mode of operation. Virtually every module within the ASIC can be programmed for different modes of operation.

Substantial partitioning of the algorithm was required within the ASIC to meet the video data rate. The computation of the projection bin addresses were "transposed" or "twisted" so that they would correspond to the incoming video data in a raster-scan mode. This saved a tremendous amount of high-speed memory in the system, which would have had to have been accessed in a random fashion.

20.5.1. Design tools for Printed Circuit Boards

The printed circuit boards and ASICs of this project employ the `lager` suite of CAD tools that use the OCT database. The SDL files for the Radon transform printed circuit boards are written in a modular, hierarchical, and parameterizable fashion. Modules are instantiated by higher-level files, while calling lower level modules or "leafcells". As the lowest level in the hierarchy, the leafcells are the representations of the actual physical components of the board, such as a

"74LS244" tri-state buffer for the DSP micro-computer's address bus, or one of the custom ASIC.

The custom printed circuit board is intentionally designed in a highly modular and hierarchical fashion. This approach greatly enhances the feasibility of re-use of these modules, both on this board and in other designs. The modular approach increases the ease with which different modules interconnect, and it facilitates the layout of the board by treating the modules as physical "blocks" that are each placed and moved as a whole.

Synthesis of the desired Radon transform printed circuit board proceeds as with other described applications. DMoct generates SMV and SIV from SDL, and behavioral-level simulation with THOR is performed. oct2rinf is used to convert to the required PCB place-and-route CAD tool format.

An option in oct2rinf provides automatic insertion of 100nF bypass capacitors for every digital component on the board. These capacitors are placed next to the "pin 1" of the digital component, and oriented in accordance with the digital component. These capacitors, and the ten tantalum capacitors of value 50 to 100uF, stabilize the five-volt power supply on the board and reduce system noise.

As is standard procedure in current printed circuit board designs of this size and complexity, one conductor plane of the board is dedicated to power, and another is dedicated to ground. This greatly reduces board-level noise by providing the shortest possible return path for any generated signal. This is because the cross-sectional area of the emanating signal line and its shortest return path through ground is made minimal by the ground plane, thus reducing the transmitting antenna effects as well as the receiving antenna operation of various signal lines on the board.

Employing hierarchy in the SDL description of the design allows modules to be repeatedly instantiated on the board, so that only a single placement offset parameter needs to be entered for each module. An example of this instantiation is shown in the section of the "board.sdl" file listed in Figure 20.6. The "x" and "y" coordinates of the four "dspasic8" modules are entered in the "position" argument list.

The "dspasic8" module contains a complete DSP micro-computer module, and a second module of eight ASICs with associated circuitry. The DSP micro--computer module contains two memory modules, I/O buffer modules, and control

```
;;;;;;;;;;;;;;;;;;;;;;;;;;;;;;;;;;;;;;;;;;;;;;;;;;;;;;;;;;;;;;;;;
; Name: board.sdl
; Function: Supports 32 custom ASICs
; and four DSP micro-computer modules
; with video interface and vme interface circuitry provided.
; Called by: None; Top level.
; Calls: four dspasic8 modules, one vme module, one vidcon module.
;;;;;;;;;;;;;;;;;;;;;;;;;;;;;;;;;;;;;;;;;;;;;;;;;;;;;;;;;;;;;;;;;
(parent-cell board)
(layout-generator NONE)
(parameters (ROTATION "0.0") (POSITION "0.0 0.0"))
(subcells
    (dspasic8 (
        (inst dspasic8_0 (ROTATION "0.0") (POSITION "0.0 2.1"))
        (inst dspasic8_1 (ROTATION "0.0") (POSITION "3.6 2.1"))
        (inst dspasic8_2 (ROTATION "0.0") (POSITION "7.2 2.1"))
        (inst dspasic8_3 (ROTATION "0.0") (POSITION "10.8 3.1"))
    ) )
    (vidcon vidcon ((ROTATION "0") (POSITION "0.2 14.6")))
    (vme vme ((ROTATION "0") (POSITION "0.0 0.0")))
)
```

Figure 20.6: Sample SDL file for a board.

circuitry modules. These levels of hierarchy continue until the leafcells containing the actual components are called.

The 300 ICs on this printed circuit board consume almost the entire area of this 400mm x 280mm VME card. Careful consideration of placement was made before routing so that the number of lines that needed to traverse large distances across the board was minimized. Extra routing room was given at the VME interface end of the board and at the video interface side, to accommodate wide address and data busses traversing long distances through densely packed areas of PGA sockets.

A liberal use of header pins was made for access to some of the PGAs' signals, and for reconfiguring the board for different modes of operation. Where room was available, "berg sticks" were employed to provide local connection to the ground plane for connecting ground clips of oscilloscope probes. LEDs (with pull-up resistors) were used on a couple of ports for debugging the board's operation at a glance; these are convenient "5-cent oscilloscopes". Due to area limitations, the metal-enclosed crystal oscillators for each DSP micro-computer were designed to be socketed above the surrounding socketed chips. This recognizes the use of designing in "three dimensions". Even the pull-up resistors were given consideration on the board in terms of the area that they consume; 10-pin SIPs are more efficient than 16-pin DIP resistor packages. The routing of this board took several days to execute on a SUN-4/280. This routing was based on a technology file that allows two printed circuit traces to pass between adjacent (100 mil spaced) pins of a component. Approximately 6000 wires had to be routed, on a board with six conductor planes.

20.5.2. Design tools for ASICs

As with the printed circuit boards, the ASICs of this project employ the lager suite of CAD tools that use the OCT database. For a class of architectures, higher levels of CAD tools exist that can take a "C-like" description of the desired algorithm and translate it into the appropriate hardware through the OCTOCT database. This applies primarily to those architectures that have many clock cycles available for every input data sample, and can thus re-use a single pipe-lined datapath for multiple operations. This is in contrast to the architecture presented in this chapter that requires many operations for every input data sample.

In contrast with the printed circuit board design, placement and routing of the ASICs continues in a hierarchical fashion. DMoct is used to translate the SIV to placement and routing information, depending on the type of module under construction.

The modules in the Radon transform ASIC that route by abutment are "tiled" together by the TimLager program. These are typically RAM, ROM, PLA, or I/O pad modules, that accept parameter values such as the number of words of memory in the module, or the number of bits per word. For example, the PC_RAM and LC_RAM modules are constructed as two blocks of 512 words by 26 bits, by specifying "ram_words = 512" and "ram_bits = 26" in a parameter value file. Similarly, the 108 I/O pad drivers of the ASIC are generated by four calls to TimLager to create 27 I/O pads in each module.

The standard-cell logic of the ASIC uses the Stdcell program, which calls the TimberWolf placement and routing program. Standard cells are employed as the local control circuitry for each of the address generation, back-projection ALU, projection collector, and region of interest modules. Standard cells are also used for global control circuitry for the ASIC, establishing appropriate clocking and timing throughout the various modules. Bit-sliced datapaths are assembled by the dpp datapath assembler.

The back-projection ALU module contains a 24-bit wide datapath, while the forward projection and line calculation modules employ datapaths of 17 and 9 bits each. The address generation module actually contains two 10-bit wide datapaths merged into one, and the region of interest module requires a 9-bit wide datapath. The configuration register uses a 16-bit wide datapath.

Placement and routing of the various modules on the ASIC, at each level of hierarchy in the design, is accomplished with the global place-and-route program Flint. Each of the modules that contain both datapaths and control logic standard cells, (the ROI, BPALU, PCALU, and AG) use Flint to place and route the appropriate nets between the respective datapath cell modules and standard cell modules. The resulting "controlled datapath" is considered a succinct module at the next highest level of hierarchy. Flint then places and routes these modules (the ROI, BPALU, PCALU, LCALU, AG, PCRAM, LCRAM, and CONFIG modules) into an ASIC "core" that contains all of the functional circuitry. Special attention is given to the routing of power and clock lines at every level of hierarchy. Placement of the ASIC's I/O pads, as well as routing of these pads to the central circuitry of the chip, is provided by the Padroute routine at the highest level of hierarchy.

The real-time constraints imposed on this ASIC necessitated the design or enhancement of many new scalable CMOS circuits for the CAD tool libraries. These new cells include the tiled read-modify-write dynamic RAM, and TTL compatible I/O pads; several bit-slice datapath cells such as adders, multiplexors, registers, latches, buffers, conditional zeroes, programmable masks, etc.;and several standard-cells.

20.6. SUMMARY

This chapter has discussed the use of the Radon transform to enable efficient computation in image processing/analysis applications. An algorithmic-specific architecture has been designed, fabricated, and tested which supports many different image processing algorithms. Custom ASICs are presented that implement the forward and inverse Radon transforms at real-time video rates. Custom printed circuit boards are described that contain DSP micro-computers to process the transformed data. Some of the image processing algorithms that benefit from operating in the Radon domain are also discussed.

REFERENCES:

[Brac79] R. Bracewell, "Image reconstruction in radio astronomy", *Image Reconstruction from Projections - Implementations and Applications*, G. Herman, Ed., Springer-Verlag, NY, 1979.

[Bres65] J.E. Bresenham, "Algorithm for computer control of a digital plotter", *IBM Systems Journal*, Vol. 4, No. 1, 1965, pp. 25-30.

[Herm80] G. Herman, *Image Reconstruction from Projections- The Fundamentals of Computerized Tomography.*, Academic Press, NY, 1980.

[Petk89] D.Petkovic, W. Niblack, M. Flickner, "Projection-based high accuracy measurements of straight line edges", *Machine Vision and Applications*, Vol. 1.

[Sanz88] J.L.C.Sanz, E.Hinkle, *Radon and Projection Transform Based Machine Vision: Algorithms, A Pipeline Architecture, and Industrial Applications*, Springer Verlag, 1988.

[Shap80] S.D. Shapiro, "Use of the Hough transform for image data compression", *Pattern Recognition*, Vol. 12, pp. 333-337, 1980.

21

Speech Recognition

Anton Stölzle

This chapter describes a full custom processing board for a real-time speech recognition system [Stö90] and how LAGER was used to design it. The task of this board is to perform the most time critical parts of hidden Markov model based recognition algorithms. Due to the large amount of computation (520 million operations per second) and a large memory bandwidth requirement (8Gbits per second), the processing board has a full custom architecture that uses 12 ASICs and 8MBytes of static memory.

Speech recognition technology has come to a stage where the underlying recognition algorithm (Hidden Markov model) has stabilized and impressive recognition accuracies for certain task domains are obtained. For example, speaker dependent systems from various DARPA-sponsored research facilities achieve a word accuracy of 95.5% up to 98.5% (June 1990 DARPA benchmark test). However, the underlying recognition algorithm is computationally intensive. Attempts to perform real time recognition for large vocabulary connected speech (1,000 words and more) using general purpose processors or general purpose multiprocessor systems are only successful if necessary computations are skipped, thus severely penalizing recognition accuracy. Even existing full custom processors like the Graph Search Machine [Glin87] or full custom multiprocessor systems like BEAM [Bis89] cannot perform in real time for a vocabulary larger than 1,000 words.

21.1. ALGORITHM AND ARCHITECTURE

Currently, the most successful method for modelling continuous speech is to use hidden Markov models [Rab89].

In this approach, speech is segmented into *frames* which are time intervals of typically 10ms. The characteristics of every frame is described with a set of *features* (o_i for the feature values at frame i) where one feature could, for example, be a vector describing the energy of the speech signal in different frequency bands. The assumption is that these features were produced by a HMM speech process consisting of a finite set of states and a set of transitions between these states (Figure 21.1). The states in the HMM correspond to generic speech sounds (e.g. a phoneme or a part of a phoneme) where every state has a probability distribution that gives the output probability, $P(o|s)$, that the state will output feature o. Speech can be considered as being generated by transitions between these states yielding a state sequence, $S_N = s_1 .. s_N$. The likelihood of these transition is described with a set of transition probabilities, $A(s,t)$, which reflect the probability that state t follows state s.

Using this framework, the task in speech recognition is to find the sequence of word models that most likely could have produced the speech to be recognized. This corresponds to finding the most probable state sequence (path) through a composite hidden Markov model that connects all the word models in the vocabulary. This problem can be formalized as finding the maximum a posteriori probability $P(S_N|O_N)$, the probability of the state sequence $S_N = s_1 .. s_N$ given a sequence of N observations, $O_N = o_1 .. o_N$. A computational very efficient solution to this problem is the Viterbi algorithm, a forward dynamic programming scheme.

Let us define the *state probability*, $P(O_i,s)$, as the probability of the most probable state sequence that ends in s and generates O_i, a sequence of i feature values $o_1 .. o_i$. This probability will be used to compute the desired state probabilities $P(O_N,s)$ for all the states in the HMM using the Viterbi algorithm [Rab89]:

$$P(O_1,s) = \pi(s) \cdot P(o_1 | s) \qquad (21.1)$$

Chapter 21 Speech Recognition 323

$$P(O_i, s) = MAX_{p \varepsilon pred}[P(O_{i-1}, p) \cdot A(p, s)] \cdot P(o_i | s) \qquad (21.2)$$

In these equations, $\pi(s)$ is a probability distribution that gives the probabilities at the beginning of a sentence (first frame). The states p in (21.2) denote all predecessors of state s, which are defined in the topology of the HMM. Given these equations, $P(O_i, s)$ can be computed for all states for O_1, then for all states for O_2 and so on until all probabilities $P(O_N, s)$ for all states are computed. The state with the highest state probability then is the final state of the most likely state sequence. In order to recover this desired state sequence, we use backtrack pointers that are generated along with (21.2) [Stö91].

The above equations allow us to compute the most probable state sequence for a HMM without hierarchy. However, to save memory the composite HMM describing a language is stored hierarchically using a small set of basic units called unique phones. These unique phones are then instantiated and concatenated to describe words and grammars as shown in Figure 21.2. The instances of unique phones are called *wordarcs*. To be able to perform (21.2) on the hierarchical HMM, we use grammar nodes with an assigned probability in which the end

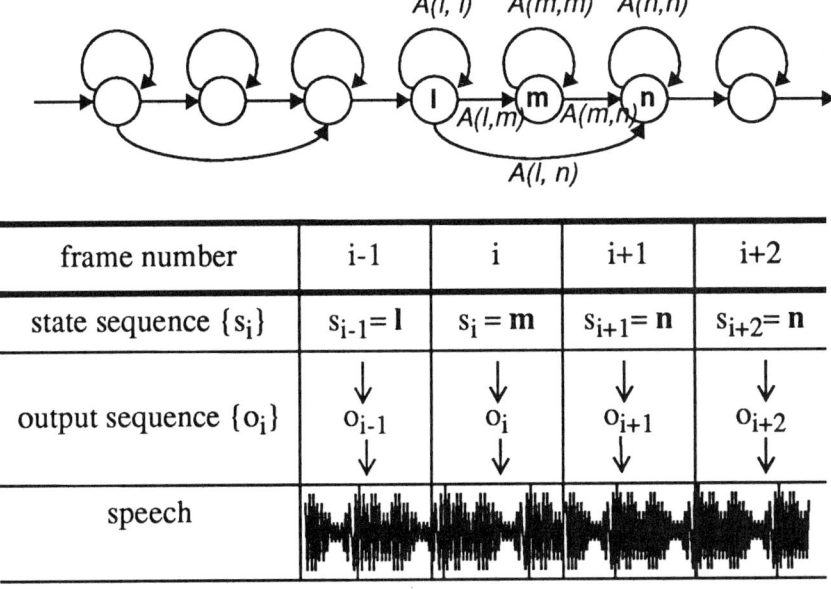

Figure 21.1: Hidden Markov Model for speech production.

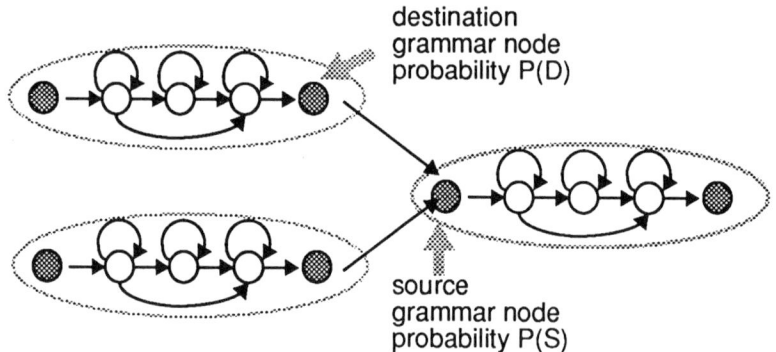

Figure 21.2: Concatenation of unique phones (inside shaded ellipses) using grammar nodes.

of a phone terminates the most probable state sequence (destination grammar node probability $P(D)$) or in which the beginning of a phone terminates the most probable state sequence (source grammar node probability, $P(S)$). Transitions between grammar nodes give the probability that a certain wordarc follows another. The grammar node probabilities are computed using

$$P(D^i) = max[P(O_i, s) \cdot A(s, D)] \qquad (21.3)$$

$$(P(S^i) = max[P(D^i) \cdot A(D, S)]) \sigma\sigma \qquad (21.4)$$

In these equations, i specifies the frame number for which the grammar node probabilities are computed. The transitions $A(s, D)$ from the states inside the wordarcs to the destination grammar node are called *null arcs* since they are not associated with a frame delay.

In order to minimize the amount of hardware needed to implement the recognition algorithm, we use the negative logarithm of probabilities. Thus, multiplications reduce to additions and the *MAX operations* are implemented using *MIN* operations. Also, we use a pruning algorithm which, in wordarc processing, discards wordarcs that have state and destination grammar node probabilities that are lower than a pruning threshold probability. Therefore, (21.2) is only performed for *active* wordarcs who's states that have a high likelihood of terminating the most likely path. Assuming that a state has an average of three predecessors, the

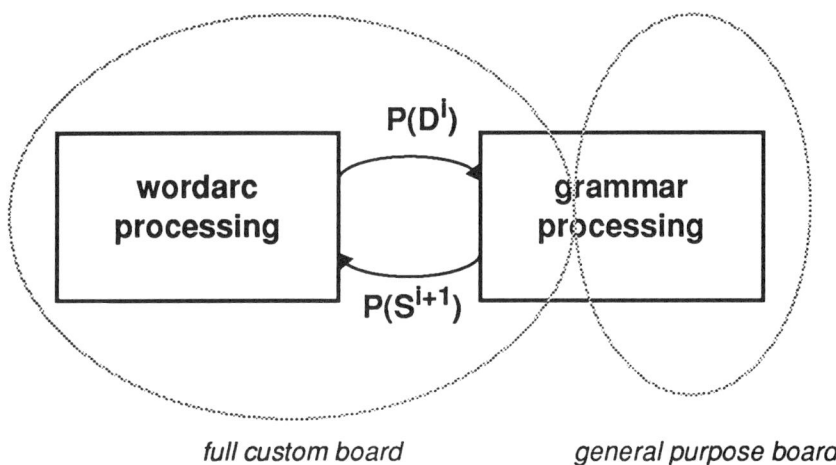

Figure 21.3: Hardware partitioning.

computation of the above equations involves 7 additions and 7 MIN operations per state (this includes normalization [Stö90]). Also, 9 address computations have to be performed to access 400 bits of data. Simulations showed, that a state has to be processed within 50ns for a real time implementation of a 60,000 word recognition system. This corresponds to 460 million (add and MIN) operations per second and a memory bandwidth of 8GBits per second.

21.2. SYSTEM ARCHITECTURE

The hierarchical representation of the HMM along with (EQ 25) and (EQ 26) make it possible to process the HMM in two levels: in the first level, the state probabilities in the wordarcs and the destination grammar node probabilities using (21.1), (21.2) and (21.3) are computed (*wordarc processing*). In the second level, called *grammar processing*, the source grammar nodes using (21.4) are computed using transitions between the phones. The communication between these two processes is formalized using the source and destination grammar nodes, as illustrated in Figure 21.3.

Besides the advantage of a potential computational speedup, it allows us to use different hardware approaches for the two levels. This is very important since the requirements for these two processes are fairly different. The computational throughput requirement for phone processing is very demanding while this requirement is more relaxed for wordarc processing. On the other side, the state transitions inside phones are local and strictly left-to-right, there is no predecessor state transitions that skips more than a few states, and there are only a few predecessor states for a given state. This is different for transitions between wordarcs: they are left-to-right inside a word, but at word boundaries they can theoretically go to any other word. Thus, they are also not local and a certain wordarc can have a very large number of predecessor wordarcs.

The actual partitioning of the subsystems into hardware entities is outlined with the shaded curves in Figure 21.3: wordarc processing along with the computation of the $MAX_{pred\ D}$ operation in (EQ 26) is performed on a full custom board with custom VLSI processors. The operation $P(D^i) \cdot A(S,D)$ in this equation, however, is performed on a set of general purpose signal processing boards using TI's TMS320C30 processors

This partitioning reduces the data rate between the subsystems since wordarcs that should stay active in the next frame but have a low destination grammar node probability do not have to be sent to the grammar processing system. Also, the multiplication in $P(D^i) \cdot A(S,D)$ can be implemented very efficiently on general purpose hardware while the *MAX* operation involves a sequence of operations that cannot be implemented on general purpose hardware in real time. Another benefit of this approach is that it is now possible to use a wide variety of grammar processing algorithms by reprogramming the grammar processing system.

The architecture of the full custom board is sketched in Figure 21.4. At any given frame two processes, each implemented with 3 custom VLSI circuits, are operating in parallel.

21.2.1. Viterbi Process

The Viterbi process sequentially computes the state probabilities and the destination grammar node probabilities ((21.2) and (21.3)) along with the backtrack pointers of active wordarcs that are listed in the ActiveWord memory. It also decides if a wordarc that has been processed should be active in the next frame and/or if the destination grammar node probability is high enough to send it to the general purpose grammar subsystem. All operations for the Viterbi process including address computations for the memories are done on a chip set consisting of 3 VLSI processors. To perform (21.2), the processors read external memory that contains the state probabilities of the active wordarcs from the previous frame, $P(O_{i-1}, s)$, the output probabilities $P(o_i|s)$ and the transition probabilities

Chapter 21 Speech Recognition

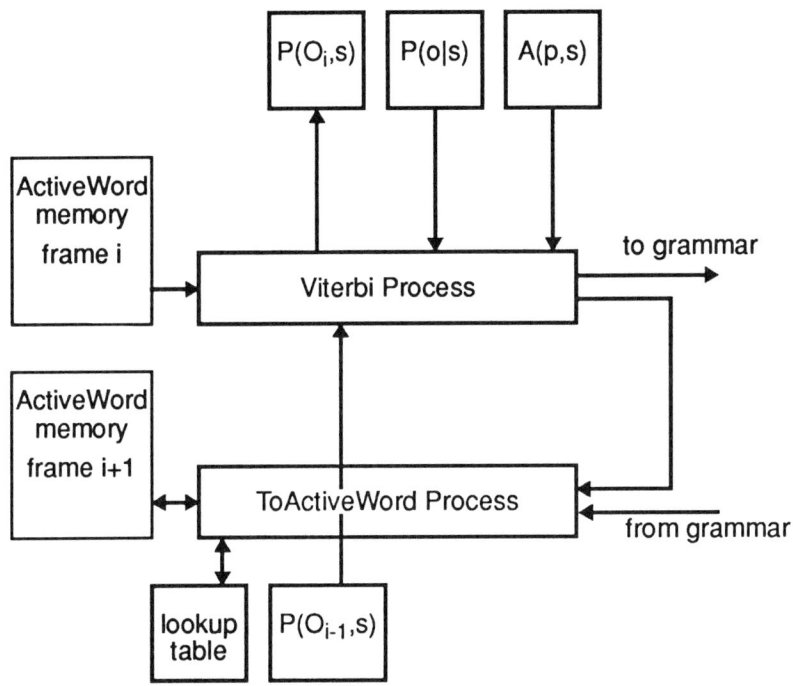

Figure 21.4: Basic block diagram of the full custom board.

$A(p,s)$ associated with transitions inside the wordarcs. To reduce the bandwidth between external memories and the VLSI processors, the topology of unique phones is stored using on-chip memories. The result of (21.2), is then stored in memory $P(O_i,s)$.

21.2.2. ToActiveWord Process

The ToActiveWord process, on the other side, generates a list of active wordarcs for the next frame. This list contains, among other data, the source grammar node probabilities that are computed by maximizing, for every source grammar node, over all the incoming contributions from destination grammar nodes of predecessor wordarcs. The source of a request to put a certain wordarc onto the active list can be the Viterbi process and the general purpose grammar susbsystem. To avoid replication after receiving multiple requests for the same wordarc, we use a lookup table that gives a pointer into the ActiveWord memory

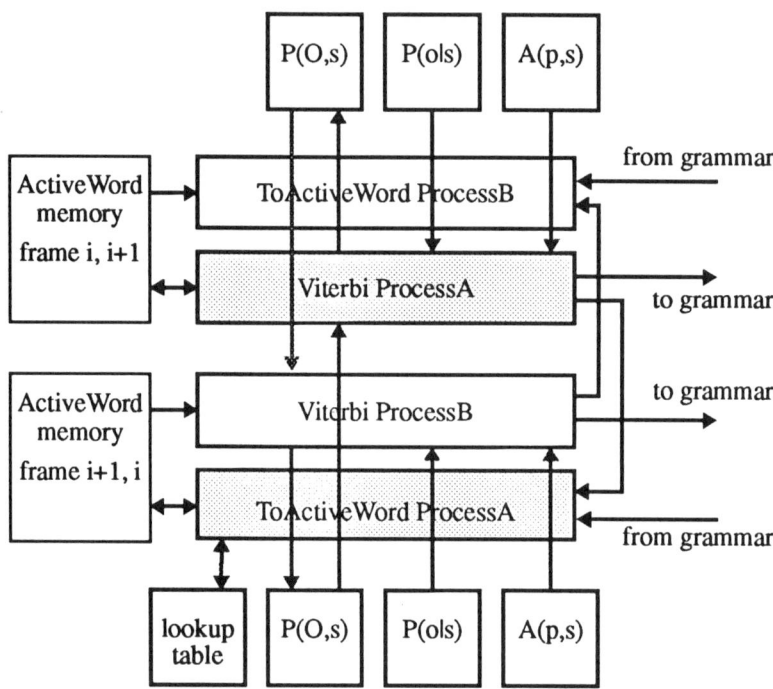

Figure 21.5: Switching processor architecture.

if the wordarc had been already processed. If it is the first request for that wordarc in a certain frame, the lookup table will yield a NIL pointer. In this case, the wordarc will be added to the ActiveList memory and the appropriate pointer written into the lookup table.

21.2.3. Switching Processors

Processing of a frame is finished after the Viterbi process updated all the active wordarcs in the ActiveWord memory and after the ToActiveWord process finished the generation of a new list of active wordarcs for the next frame. To proceed to the next frame, conceptually the memories have to be switched between processors. The ToActiveWord memory that contains the new list of active wordarcs and was written by the ToActiveWord process now has to be read by the Viterbi process. The lookup table that was used by the ToActiveWord Process has to be cleared and to save time, it is replaced by an already cleared version of the lookup table. In the Viterbi Process, the two memories containing the state probabilities of two consecutive frames have to be switched such that the memory that was written in the last frame is now read and vice versa.

To avoid having to switch all the address and data busses between the custom processors, we use a switching processor architecture. Instead of multiplexing the memories, we activate a second set of processors that is directly connected to the memories. Figure 21.5 sketches this principle: assume that during a certain frame the ToActiveWord process A is active and generated data for the ActiveWord memory A. During the same frame Viterbi B is active and processes the active wordarcs listed in the ActiveWord memory B. In the next frame, Viterbi B and ToActiveWord A are inactive while Viterbi A and ToActiveWord B are active. This way there are no multiplexors needed to switch memories, all that is required is to activate the right set of processors. This architecture also has the advantage that the resulting system is symmetric.

21.2.4. VMEbus Access to Memories

All memories on the system are accessible by the host CPU via the VMEbus. To reduce the number of discrete components on the system, the host CPU communicates only to the custom VLSI processors on the board. These processors have a small instruction set to read and write memories and internal status registers. Using this approach and the switching processor architecture, no address or data bus has to be multiplexed.

21.2.5. Caching Model Parameters

To decrease the amount of memory on the system board, we use a caching scheme for the output probabilities, the parameters with the biggest storage requirements: only a small subset of these parameters is loaded onto the board, the subset that corresponds to the output probabilities for a given speech segment. This loading operation is overlapped with the processing of the frame whose output probabilities had been downloaded in the previous frame. With this approach it is possible to use different modeling techniques for computing the output probability distributions. The current approach is to use up to 4 independent discrete probability distributions that are stored and combined on a separate board. Other modeling approaches such as continuous distributions and tied-mixtures are also possible, as long as the probabilities can be computed and loaded in real time.

21.3. CHIP ARCHITECTURES

21.3.1. Viterbi Process

To meet real time performance, the throughput requirement for the chip set that implements the Viterbi equations is to update 20 million states per second. We decided to use a clock cycle time of 50ns which is the minimum cycle time for the off-chip static memories, so the chip set has to keep up with that throughput of one state probability computation per clock cycle. To support that requirement, pipelined datapaths are used as shown in Figure 21.5.

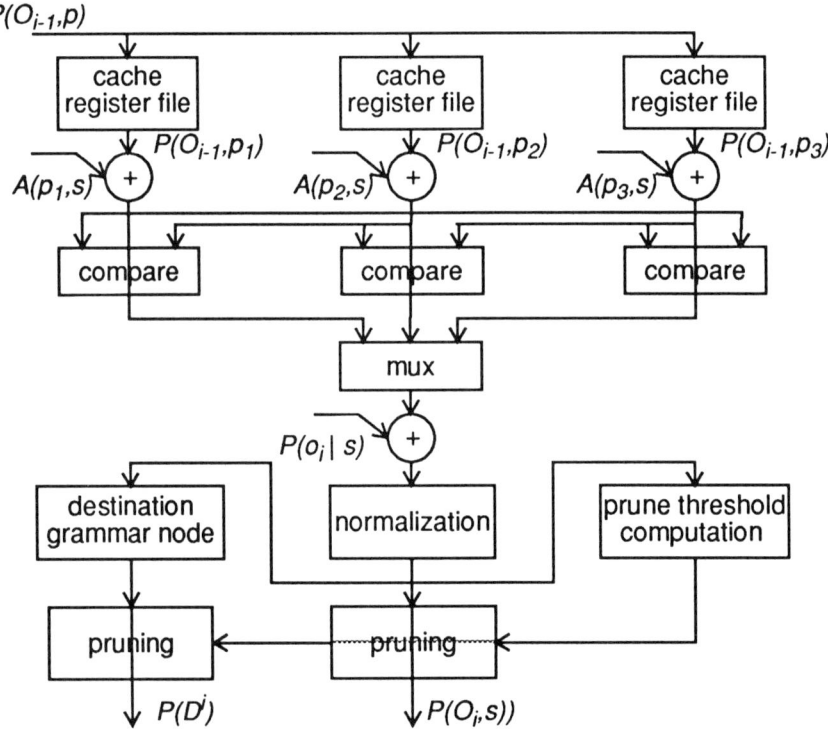

Figure 21.6: Architecture of the Viterbi chip.

For the computation of the inner loop, $(P(O_{i-1}, p) \cdot A(p,s))$, three identical pipelined datapaths are provided that work in parallel. Using this setup, the chips can process one state per processor cycle given there are not more than three predecessors for a state inside a wordarc.

To provide the necessary bandwidth to feed the multiple datapaths, we use three on-chip cache register files that keep three identical copies of a relevant subset of the off-chip memories containing the state probabilities $P(O_{i-1}, p)$. Thus, $P(O_{i-1}, p)$, can be accessed by the individual datapaths in parallel. This architecture is possible because transitions between states inside wordarcs are local and strictly left to right. Therefore, if the state probabilities of a wordarc are sequentially stored according to their occurrence in the HMM topology, all possible predecessors of a state are locally stored in the neighborhood of that state. The on chip cache memories can be updated in a sequential fashion since processing of the states is also done sequentially. Thus there is only one bus to external memory, but multiple busses to multiple internal memories to increase the bandwidth.

The processors are partitioned into VLSI chips according to their functionality. One chip computes the addresses for the external memories (add), another chip computes the backtrack pointers that are needed to recover the most probable state sequence after the termination of the Viterbi algorithm (back) and finally, the processor computes the state probabilities, performs normalization and pruning and controls the other two chips. Figure 21.7 shows a chip plot of that processor

The control structure chosen for the Viterbi chip set is data stationary control. This is possible because there are no local decisions at any pipeline step that globally change the control sequence for the entire pipeline. Using this setup, the size of the controller including delay registers can be minimized since the depth of the pipeline is fairly large (9 steps).

21.3.2. ToActiveWord Process

The time critical operation in processing a request to put a certain wordarc onto the list of active words is reading and writing memories: for every request the lookup table memory as well as ActiveWord memory have to be accessed twice (see 21.2.2.). Since one memory cycle corresponds to one processor cycle, the maximum throughput we can achieve is one request per two processor cycles. The ToActiveWord processors use a pipelined architecture to achieve this maximum throughput (Figure 21.8). If the pipeline is filled, three different requests are processed simultaneously.

If there is more than one request for the same wordarc in the pipeline of the processors, there might be a problem with memory coherence: decisions for the most recent request are based on data in the ActiveWord memory. However, these data might not yet be the up to date with respect to the former request in the pipe-

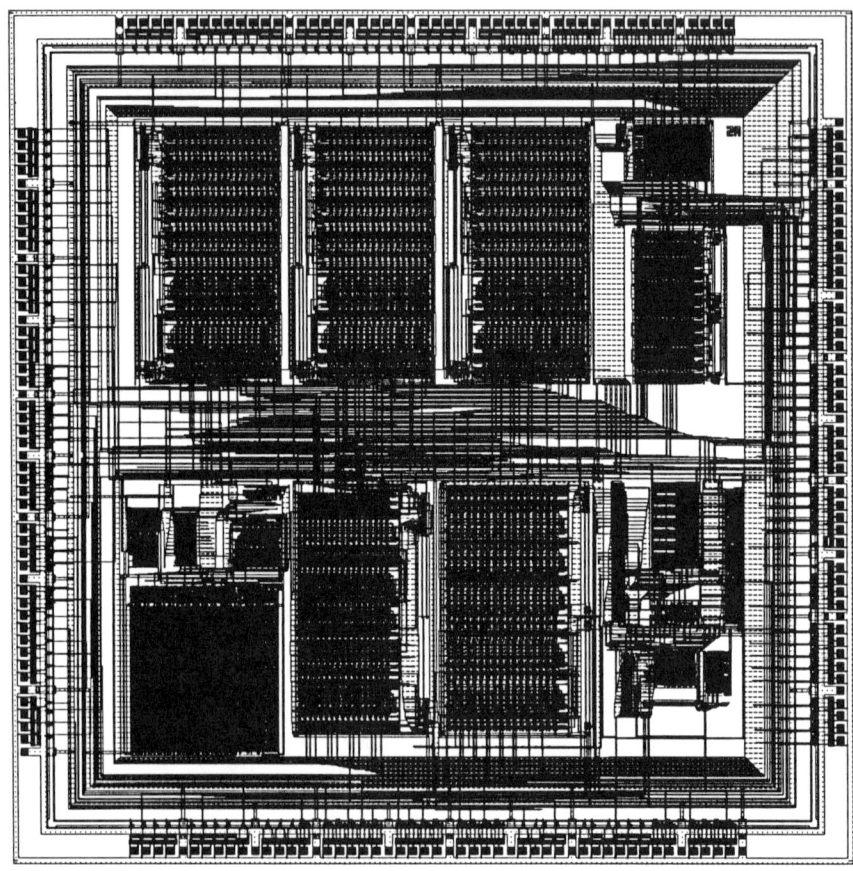

Figure 21.7: Chip plot of the Viterbi Processor.

line. To take care of that problem, we use 3 memory coherence registers that keep all the information associated with wordarcs that are currently processed in the pipeline. If there is a request for a wordarc that is already in one of the memory coherence registers, this request is only processed after processing the wordarc in the pipeline is finished.

The information associated with a particular wordarc request is 96 bits wide. It contains all data that are necessary to update that wordarc in the Viterbi process in the next frame. Since the ToActiveWord process has two input request busses and one 96 bit wide bus to the ToActiveWord memory, the implementation has to be partitioned into several chips. We used a bitsliced partitioning, where

Chapter 21 Speech Recognition 333

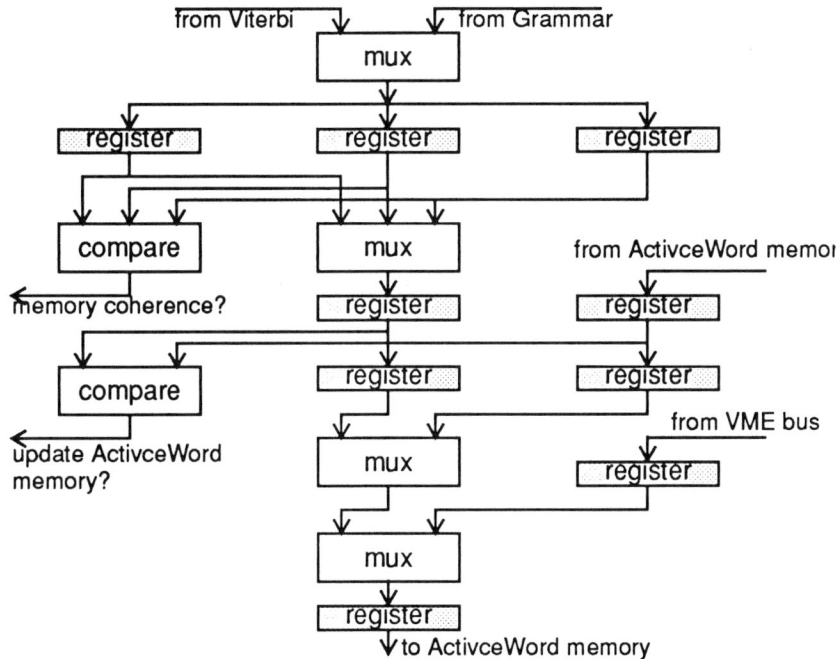

Figure 21.8: Architecture of the ToActiveWord Request Chip.

the process is implemented on three chips, each chip working on a small part of the 96 bit wide information.

The control structure for the ToActiveWord chipset is time stationary because decisions in one pipelinestep might influence the pipeline globally. Also, the depth of pipeline (6) is moderate, thus the size of the controller is not an issue.

21.3.3. Testing Strategy

The testing strategy for the custom processors is scanpath testing. In addition, they can also be tested on the board using the existing VME interface. A dedicated on-chip test controller supervises this VME test mode so that even the VME interface controller can be tested. This way, every state on the complete board (except the test controller itself) is observable and controllable without changing the hardware. Another provision for testability is that all memories are read and write accessible by the VME bus through the custom processors.

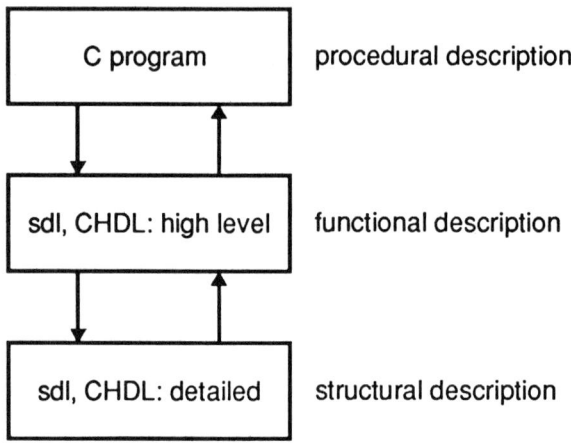

Figure 21.9: Top-down design procedure

21.4. DESIGN PROCEDURE

This section describes the individual steps and decisions that were made to design the hardware. The procedure is a series if partitions and refinements which repeatedly have to be verified. The partitioning was done in a top-down manner shown in Figure 21.9. The functional break down on the highest level used a few blocks, each still having a fairly high complexity. In the next level, these blocks are further partitioned and so on.

21.4.1. Procedural Description: Full Custom Hardware Specification

In the first step the parts of the algorithm that have to be mapped into custom hardware and the parts that can be performed with general purpose components have to be identified. For the Viterbi algorithm, the most critical operation in terms of computation and bandwidth is the performance of (21.2) and the MAX operation in (21.4). All other computation can be implemented using general purpose components.

In this partition we also took into consideration the fact that the multiplications (additions) in (21.4) ought to be flexible. There is still a lot of research being done to find the best way for deriving the transition probabilities between words:

these transition probabilities are not necessarily time stationary, they change depending on context (natural language processing). Thus, it is convenient to support this part with the flexibility of a general purpose solution even though it can only be implemented using complex and expensive hardware. Once the full custom partition is identified, its functionality has to be described and simulated to verify that the was done correctly. For that we used C to describe the procedure performed by the full custom hardware. We used C instead of a hardware description language like VHDL since this program was used as a subroutine in the DECIPHER speech recognition system [Murv89] for verification. The C code was then further refined to take into account finite word length effects. The result of these simulation was a compact, high level functional description of the full custom hardware.

21.4.2. Functional Description: Partition of the Full Custom Hardware

In this next level of refinement the function of the full custom hardware is further partitioned into smaller functions. The objective is to identify functional elements that have disjunct memory requirements to avoid memory contention. In this system, we identified two main functions, the Viterbi process and the ToActiveWord process (see above). There are only two memories that are shared between these processes, but it is easy to guarantee that they are never accessed simultaneously.

This description already defines the high level architecture of the custom board. Therefore it is important to take into account that the requirements for real time performance can be met. For example, we have to guarantee that all data needed to perform (21.2) can be accessed in only one memory cycle. This requirement defines a memory architecture that uses several data busses so that several memories can be accessed simultaneously. To support fast memory cycles, we fast static memories (SRAMs), but due to the memory requirements of a large vocabulary hidden Markov model, not all model parameters can be stored on SRAM. Also, we have to use full custom processors to perform the computations and these processors have an upper limit number of pins that can be used (204 pins in our packaging technology). Therefore, we designed a hierarchical memory structure that uses caching methods to achieve the required memory bandwidth.

Using this partition, we again described the system and verified it's correctness against the C code that described the full custom hardware. For this description we decided to use CHDL, the THOR netlist language, to describe memories and processing elements and to use SDL to connect the hardware blocks. However, this description did not take into account any pipelining artifacts, the hardware blocks (memory, Viterbi process, ToActiveWord process) were described procedurally.

This partition resulted in smaller functions that can directly be mapped into full custom processors or commodity components (memories, drivers). It also defined the connectivity between these hardware entities.

21.4.3. Detailed Structural Description

In the next step the functional blocks that were described in CHDL now had to be described as a structure of generic hardware elements using SDL. The blocks that are mapped into ASICs were described using LAGER library elements (datapath cells, finite state machines, standard cells, etc.), while the memory modules were implemented using an interconnection of discrete components (memories, drivers and FPGAs).

This structural specification results in a detailed description of the function where issues like timing (pipelining), parallelism and partitioning into several chips are implicitly contained. Also, architectural trade-offs to save hardware (switching processors, VME logic on ASICs) are to be described on this level.

The translation from the CHDL description to the detailed SDL structural description was done manually. On this level, the complete function was "flattened", mapped into library elements and then a new hierarchy was build based on structural entities (datapaths, standardcell blocks, controllers, chips). To avoid timing conflicts due to pipelining and to allocate hardware we used hardware allocation tables.

This structural hierarchy partitions the function into several chips and partitions the function of a certain chip into several datapaths. The objective in these portioning decisions was to minimize the amount of interconnects. On the level where chips are partitioned, we replicated hardware on several chips to minimize the interconnection. The chip partition for the Viterbi process was done based on functionality. That means, there is a chip that generates all addresses needed to access external data, a chip that generates the backtrack pointers and finally a chip that computes the state probabilities. The ToActiveWord process on the other side has a bitsliced partition: each chip performs the same function for a different block of data.

Chapter 21 Speech Recognition 337

The strategy we used to describe the individual chips was to build small functional entities like pipelined comparators or counters that are implemented using a few LAGER library elements. These entities were then used to describe larger datapaths. Using the structure processor Octflatten, the hierarchy that was built inside a datapath could be flattened so that the resulting layout was optimized. The objective in partitioning the function of an individual chip into several datapaths was, again, to minimize the number of interconnects and to optimize the floorplan. This resulted in datapaths that have approximately the same size and aspect ratio.

As a result we had a hierarchical structural description in SDL format for the full custom processing board. The lowest hierarchy levels described the individual chips of the board. This could be a single sdl file for a standard TTL part or a complex, hierarchical description of a full custom chip. The next sdl hierarchy described functional entities such as the Viterbi process as an interconnection of chips. This way entities like the Viterbi process that were described functionally in CHDL at an earlier stage, are now described structurally. Finally, these functional entities are connected to describe the complete full custom board. The hierarchical sdl files were then translated into an OCT structure_instance view that was used as a common database for verification and layout generation.

To verify the design against the higher level CHDL description, we used MakeThorSim to compile a THOR simulator of the initial hardware partitions and compared the patterns generated by the different levels. Once the individual partitions were verified, we checked the timing of critical paths between pipeline steps on the ASICs. For that, the physical design of the custom ASICs had to be generated and extracted to get a database for an IRSIM simulation.

We also compiled a THOR simulation for the complete full custom board and compared the patterns generated by this simulation against the results generated by the initial C code. This was necessary to verify the architectural decisions and partitions that were not considered in the C simulation. For example, the switching processor architecture uses busses that are shared between the subsystems. Thus the proper control of the tristate drivers connected to these busses had to be verified.

Finally, using the structure processor OCT2RINF, the OCT database describing the full custom board was used to drive RACAL to generate the printed circuit board layout.

21.5. SUMMARY

This chapter presented the architecture of a processing board for a real time large vocabulary speech recognition system and the procedure that was used to design it using the LAGER system. The fact that we used a unique description for both, simulation and layout generation, proved to be extremely valuable to handle the complexity of the system. It was also very important that we could generate a board level simulation. The printed circuit board uses 12 routing layers and any error in the structure of that board would be extremely difficult to correct. Due to the fact that errors could be detected through simulation, the board was operational after the first design cycle with a minimal amount of rework.

REFERENCES:

[Bis89] R. Bisiani, T. Anantharaman, L. Butcher: "BEAM: An Accelerator for Speech Recognition", *ICASSP 1989*, S15.3, pp782-784

[Glin87] S. Glinski, T. Lalumia, D. Cassiday, T. Koh, C. Gerveshi, G. Wilson, J. Kumar, "The Graph Search Machine (GSM): A VLSI Architecture for Connected Speech Recognition and Other Applications", *Proceedings of the IEEE*, Vol. 75, NO. 9, September 1987, pp1172-1184.

[Murv89] H. Murveit et al, "SRI's DECIPHER System", *Proc. of the Speech and Natural Language Workshop*, pp. 238 - 242, Feb. 1989

[Rab89] L. R. Rabiner, B. H. Juang, "A Tutorial on Hidden Markov Models and Selected Applications in Speech Recognition", *Proceedings of the IEEE,* pp. 257-2586, Feb. 1989.

[Stö90] A. Stölzle et al., "A Flexible VLSI 60,000 Word Real Time Continuous Speech Recognition System", *IEEE Press book: VLSI Signal Processing IV*, ISBN 0-87942-271-8, Chapter 27, pp 274 - 284, Nov. 1990

[Stö91] A. Stölzle et al, "Integrated Circuits for a Real-Time Large-Vocabulary Continuous Speech Recognition System", *IEEE Journal of Solid State Circuits*, vol. 26, no.1, pp. 2-11, Jan. 1990

22

Conclusions and Future Work

Robert W. Brodersen

The LAGER system met its primary requirement of reducing the time and effort required to implement an integrated circuit, and allowed the designers to focus on the chip architectures and the applications in which the chips were designed to be used rather than details of the physical implementation.

22.1. WHAT WORKED

The development of a clean interface to a set of tools which performed the *silicon assembly* task of taking a structural input and producing a complete physical description was found to have many benefits. A primary success was that it allowed a user to describe a chip at a sufficiently high level that the chip design task could be reduced to hours in simple cases and days for more complex circuits.

Another benefit was that it allowed the separate development of *silicon compiler* tools, which take a behavioral discription as an input and then produce the structural output. This was useful in that the high level tool development could focus on the complex issues involved in synthesizing structure from behavioral descriptions, with the knowledge that once this structural description was complete, other tools would provide the actual physical implementation. This separation comes with some disadvantage in that feedback of parameters from the actual physical implementation into the high level tools is complicated by the separation of function. More accurate feedback from the silicon assembly phase may be desirable through more direct estimation from the various physical design tools. The present strategy, however, has been to use the assembly process to develop accurate estimation formulas and tables to direct the design of the higher level tools.

The most important components of LAGER are the various circuit layout libraries. The reuse of these libraries made it possible for errors and limitations in the various circuits to be identified, and allowed most users to avoid the time consuming activity of physical design. A major limitation of libaries is that they must be redesigned when the underlying technology is changed. However, the use of scaleable design rules has allowed the libraries to stay essentially intact while the the technology has scaled from 3 micron to 1.2 micron.

The use of the object-oriented data manager OCT was found to be an excellent choice for our underlying database. Its efficiency was found to be more than adequate for our largest chip designs (100-200 thousand transistors) and its performace for recalling information was sufficient so that it rarely was a major time consumer. The simplicity and extensibility of the database worked well as we evolved the quantity and type of information that was stored well beyond that which was contemplated by the original developers.

22.2. WHAT DIDN'T WORK

The success of allowing designers to avoid the process of transistor level design, resulted in a considerable lack of interest in the further development of the libraries. This was due not only to the actual effort in the layout task but also to the documentation requirements if the new cells are to be made useful to other users. It was found to be a consistent effort to remind users of the necessity to move proven cells that were used in a design from a "private" library to the standard user library locations (with the required documentation). Support tools for this task would no doubt facilitate this transfer.

A related problem was the lack of interest in developing performance driven design tools. Once the capabilities to achieve design functionality was obtained, then architectural techniques were used to obtain the desired system performance. For the dedicated application for which the system was primarily used, this was found to be adequate. This approach has been found to make it pos-

Chapter 22   Conclusions and Future Work   341

sible to work up to frequencies well over 100 MHz (see Chapter 17), but if architectural techniques are not possible, it is then required to modify the underlying cell library (a task which has been undertaken a few times). This effort may have been avoided in some cases if performance driven tools had been developed.

The original goal was to remove the requirement that users (as opposed to developers) need to know about the physical design beyond that required for the structural description. This eventually was possible for the generation of the circuits, but not for the verification of functionality and performance. Though support was given to help set up the required files for final simulation, the user was required to exercise the simulator and interpret the results.

The verification phase had other limitations as well. It became the most time consuming portion of the design task. In addition, the use of a simulator for verification has the disadvantage of not being complete due to the difficulty (impossibility) of determining an adequate set of test vectors.

22.3. FUTURE

The future developments of LAGER are in a couple of directions. The basic tool set (particularly the control generation) is upgraded as new tools are developed and integrated into the underlying framework. Improvements in the verification phase include more sophisticated automatic test generation and a simulation strategy which can simultaneously simulate at various levels of description ranging from behavioral to structural.

The major thrust, however, is not in the enhancement of the basic chip generation capability, but rather to extend the capability of the compilation and assembly process to the system level. A system in this context contains a number of chips which may or may not be custom designs, and may include such components as programmable logic, software programmable processors, subsystems built from advanced packaging techniques such as multi-chip modules or even conventional printed circuit boards. This extension is being made following the same methodology developed at the chip level, but with extensions to handle the more heterogeneous nature of system design. In fact, the strategy for the system assembly task is to use basically the same design manager and structural description but with a number of physical target technologies. The libraries will contain not only leafcells for chip designs, but software libraries, board level module generators and programmable logic macrocells. The chip generation capability described here will then only be one of a number of possibilities for realization of user specifications.

APPENDIX A

Design Example

Brian C. Richards

A.1. RUNNING DMOCT TO GENERATE A DESIGN

To illustrate how DMoct uses a library cell to produce a customized layout, consider the following example of generating a latch. There are several command line options that control the design flow, allowing the user to generate the complete design with a single command, or to step through the design process incrementally, running DMoct several times.

To generate a latch in a single step, the following command can be issued:

```
DMoct -P latch MyLatch
```

In this case, the user will provide any necessary design parameters interactively, as needed. The option '-P' is thus given to prevent DMoct from asking for a parameter value file. The library template for the desired design is called 'latch', and the user wants the resulting circut to be named 'MyLatch'. The resulting diagnostics will be similar to Figure A.1:

After reading initialization files, DMoct begins by scanning the library for the file 'latch.sdl', and checking for the corresponding structure_master view (SMV). For a properly installed library circuit, the SMV should already exist, in which case the diagnostics show that it is up to date. If the SMV does not exist, then DMoct will attempt to create it in user's current working directory.

Given the parameterized structure_master view template for the latch, DMoct will then generate the structure_instance view in OCT (the SIV), with the name 'MyLatch'. The latch is parameterized, however, and DMoct determines that the parameter 'width' has not been defined. The user is asked to enter the

```
***************************************************************
*            LagerIV Silicon Assembly System                   *
*            Design Manager for the Oct Database               *
*               (email: lager@zion.berkeley.edu)               *
***************************************************************
Loading DMoct.ll ...
... Done loading DMoct.ll
DMoct>
DMoct>
DMoct> CREATING/READING STRUCTURE MASTER VIEWS (SMV's) ..
DMoct>
DMoct> (latch (reading ~lager/common/LagerIV/cellib/\
DMoct>    TimLager/blocks/latch/latch.sdl)
DMoct>    (latch:structure_master is up to date))
DMoct>
DMoct>
DMoct> CREATING STRUCTURE INSTANCE VIEWS (SIV's)..
DMoct>
DMoct> The value for the parameter 'width' is missing.
        Enter value for parameter ''width'': 4
DMoct> (MyLatch (creating the SIV 'MyLatch'))
DMoct>
DMoct>
DMoct> RUNNING LAYOUT GENERATORS
DMoct>
DMoct> (MyLatch (executing: home/zion4/lager/SUN4_2.0/\
DMoct>    LagerIV/bin/TimLager -L MyLatch.log -m\
DMoct>    MyLatch:structure_instance)
DMoct> )
```

Figure A.1: Sample of DMoct design generation feedback

value interactively ('4' in the above example). With all parameters entered, DMoct then proceeds to generate the SIV, containing the structure description of the latch (subcells, netlists), the CAD tool that is needed to generate the layout, and the parameters needed by the CAD tool.

During the layout generation stage, the SIV is passed to the CAD tool designated by the SIV, TimLager in this case, to produce the final layout. By default, both an OCT representation of the design (an OCT 'physical' view typically), and a MAGIC design file (for VLSI designs) are produced by the CAD tool.

In typical designs, it would be prohibitive to manually enter design parameters each time a design instance is generated. Most commonly, a parameter value

Appendix A   Design Example

file is given to DMoct. The parameter value file contains one or more pairs of parameter name - parameter value pairs, each enclosed in parentheses. For the above example, the file MyLatch.parval might be created, containing the following line:

```
(width 4)
```

DMoct can then be run as follows, to use the MyLatch.parval file:

```
DMoct -p MyLatch.parval latch MyLatch
```

The name of the parameter value file can have any suffix; .parval and .par are commonly used. Another alternative for a simple design with few parameters is to enter the parameters on the command line:

```
DMoct latch MyLatch -P -Dwidth=4
```

Note that options to DMoct can be before or after the design master and instance names. The only requirement is that the master name precede the instance name in the command list.

DMoct can also be run in several distinct steps, to control the design flow. For instance, only the SMV may be needed if the user is installing a personal library which will be used in several designs. Also, the user might want to postpone the layout generation until a functional simulation has verified that the system operates correctly. The design stages can be isolated as follows:

```
# SMV generation
DMoct -m latch

# SIV generation with parameters.
DMoct -s latch MyLatch -P

# Now a functional simluation using THOR can be run.

# Layout generation.
DMoct -l latch MyLatch
```

More than one of the options -m, -s, and -l can be given on the command line at the same time; often -m and -s are used concurrently, to prepare for a functional simulator after changing SDL files. If none of these three options are given, all three are implied by default.

A.2. DESIGN POST-PROCESSING WITH DMPOST

In the following, examples are shown how DMpost is used to do THOR and IRSIM simulations, design rule checking, CIF file generation, starting MAGIC in interactive mode and shipping a request to MOSIS. It is assumed that the design MyLatch, a simple latch shown in Figure A.2, has been generated using DMoct. The latch operates on a two phase clock where data at the input *in* is latched on *ld* and *phi2* and appears at the output *out* on *phi1* (see also Figure A.3). First, the *structure_instance* view is created to verify the behavior of the design, then the *physical* view is created to verify the logic level operation and to check for design rule errors. Next, the CIF file is generated. By splitting the simulation phase into functional and logic level simulation, time is saved since physical generation can be very time consuming especially on large designs.

A.2.1. Functional Simulation with THOR

First the structure instance view should be created from SDL by using DMoct with the -s option to prevent the creation of the physical view since this is not required for behavioral simulation. DMpost can now be run to generate the input files for the THOR simulator. This generates the THOR directory and stores the netlist in the THOR model files. An analyzer call containing all of the formal terminals is added to the netlist file, which allows observation of all signals that are external to the design being simulated.

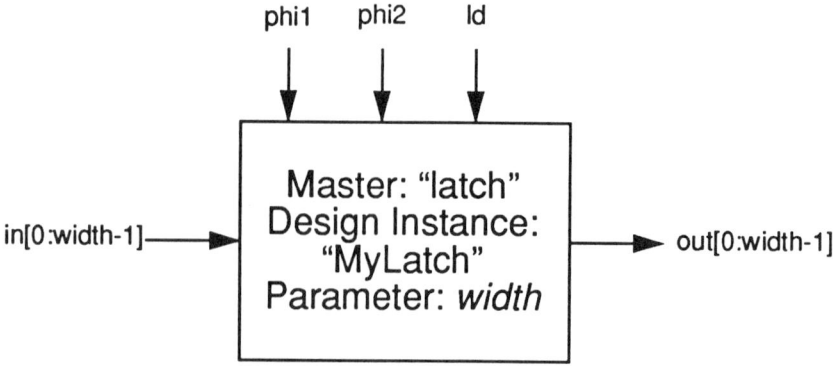

Figure A.2: DMpost example: MyLatch

Appendix A   Design Example   347

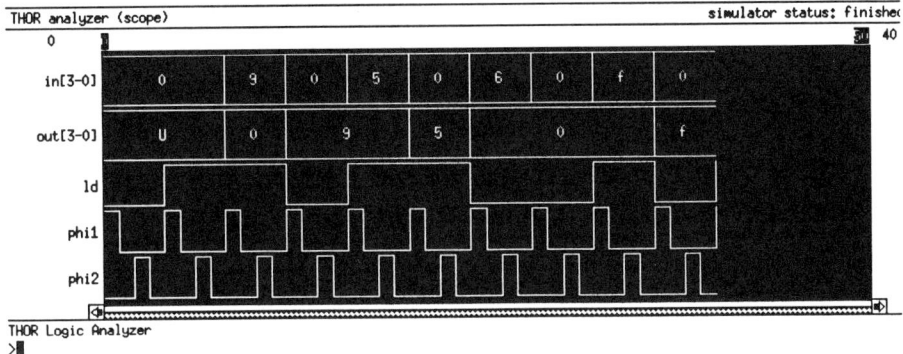

Figure A.3: Output from the THOR analyzer

At this point, the generators and any additional monitors have to be added by the user to the CSL file. THOR may then be started in two ways. The command *gensim* can produce an executable simulation file, which can then be run interactively, accepting both batch initialization files and user commands. Figure A.3 shows some time steps displayed by the THOR analyzer for this example, in which the values 9, 5, 6, and F_{16} are sequentially clocked through the latch. Alternatively, the *gensim* command can produce and run a simulation in batch mode, running for a specified number of time steps.

If the THOR simulation shows the desired functionality, the next step in the verification process can be taken, otherwise the design has to be modified and again simulated.

A.2.2. Simulation with IRSIM and Design Rule Checking

The next step in the verification process is to simulate the actual layout and check for design rule violations. Running DMoct -l on MyLatch will call all necessary layout generators and the physical view will be created in OCT and MAGIC representation since the -g option was not used. Then DMpost is able to extract the circuit data and check for design rule violations:

```
>> DMpost -l 0.8 -w pwell -sim -drc -t scmos MyLatch
```

The above command line specifies that the design should be checked and extracted for a 0.8 microns per lambda P-well process using the SCMOS technology of MOSIS and that the design should be read from Magic files. DMpost will generate the .ext files, the MyLatch.al and MyLatch.sim file and store them in the layout directory since the -sf option was not used. In addition IRSIM needs a technology file called scmos80.prm, which gives the parameters for the first-order timing checks, to which DMpost will establish a link in the layout directory.

When DMpost finishes running, the user will be instructed how to run the simulator:

```
DMpost>
DMpost> WELCOME to the LagerIV postprocessor
DMpost>
DMpost> Working on -sim -drc for MyLatch
DMpost> Reading input from magic files
DMpost> Using: technology=scmos, microns/lambda=0.8,
        well=pwell
DMpost>
phy2ext1.8> Reading names of cell instances
        from oct physical view
phy2ext1.8> Collecting magic files of all cells
        to "./layout" directory
DMpost> Running magic -Tscmos -dNULL MyLatch < magic.TmP4477
DMpost> Running ext2sim -L -R -c 10
        -a /class/tut1/layout/MyLatch.al
        -o /class/tut1/layout/MyLatch.sim MyLatch
DMpost>
DMpost>
DMpost> Options -sim -drc successfully completed
DMpost>
DMpost> drc: see file ./layout/MyLatch.drc
DMpost> sim: see file ./layout/MyLatch.sim
DMpost>
DMpost> Usage of irsim (in directory /class/tut1/layout):
DMpost>    irsim scmos80.prm MyLatch.sim -USER.COMMANDS
DMpost>
DMpost> See file DMpost.log in the current directory
```

The file -USER.COMMANDS contains all the startup commands to IRSIM provided by the designer. At the same time DMpost will generate the MyLatch.drc file in the layout directory containing information on design rule violations.

Appendix A   Design Example   349

A.2.3. CIF Generation

If simulation showed the correct behavior and if the design shows no design rule violation then the CIF file can be generated. For that purpose DMpost is used again:

```
>> DMpost -cif -w pwell -l 0.8 MyLatch
```

The technology parameters have to be specified again and should be the same as for the extraction and design rule checking (there is currently no mechanism to assure consistency). The MyLatch.cif file will be placed in the layout directory. The following lines show the first ten lines of the CIF file for the latch example:

```
DS 1 40 2;
9 MyLatch;
L CWP;
 B 4 24 146 800;
L CMS;
 B 16 16 32 988;
 B 16 16 64 988;
 B 16 16 96 988;
 B 40 48 284 972;
 B 40 48 372 972;
 . . .
```

The DS command defines the cell number 1 and the scale factor, followed by the name of the cell in the CIF extension command "9". The MOSIS layers are then referenced (the "L" command); in this example, the CWP layer defines the CMOS P-well mask, and the CMS layer defines the CMOS second metal layer. Boxes using those layers are defined with the "B" command, where the first two coordinates define the size of the box, and the latter two define the center of the box.

A.2.4. Interactive Magic

If it is desired to display a cell to investigate a design rule violation caused by an error in a cell or a CAD tool fault (which of course rarely happens!), print out a design, or use other possibilities supported by MAGIC but not by DMpost, MAGIC can be started in interactive display mode on a workstation that runs X-windows with all the necessary technology parameters:

```
>> DMpost -Xmag -w pwell -l 0.8 -t scmos MyLatch
```

350

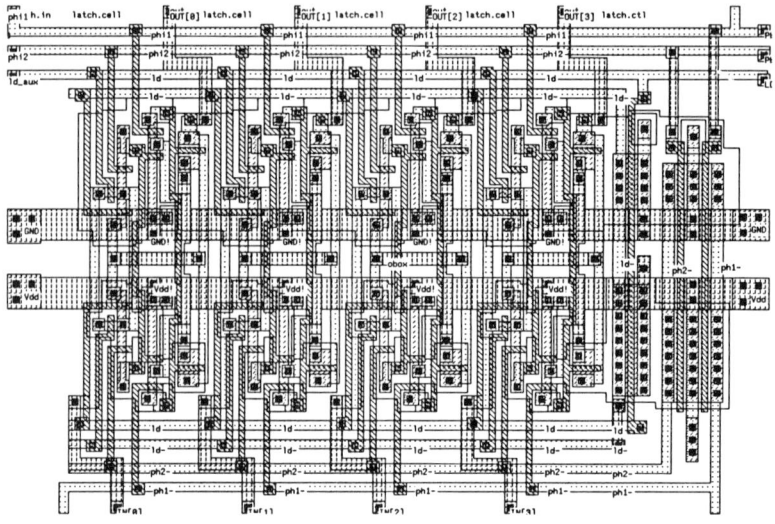

Figure A.4: Layout of MyLatch, from CIF

The final resulting layout of MyLatch, generated from a CIF file produced by MAGIC, is shown in Figure A.4. Notice that all of the named terminals are labelled in the plot.

A.2.5. Fabrication Request

Once the design exists as CIF file, a request for fabricating the design may be sent to MOSIS which must contain some data on the design. To ship the request to MOSIS, type:

```
>> DMpost -mosis MyLatch
```

The only thing DMpost will currently do is to ask for the request file and show you how to ship the request by e-mail to MOSIS. This is to avoid unintended shippment of requests to the MOSIS service.

A.2.6. Post Processing Run

The following output is displayed on the terminal when DMpost is run as follows:

Appendix A   Design Example   351

```
>> DMpost -THOR -sim -cif -drc -w pwell -l 0.8 MyLatch

DMpost>
DMpost> WELCOME to the LagerIV postprocessor
DMpost>
DMpost> Working on -THOR -sim -cif -drc for MyLatch
DMpost> Reading input from magic files
DMpost> Using: technology=scmos, microns/lambda=0.8,
 well=pwell
DMpost>
DMpost>
DMpost> Running MakeThorSim -L /class/tut1/DMpost.log -a
 MyLatch
phy2ext1.8> Reading names of cell instances from oct
physical view
phy2ext1.8> Collecting magic files of all cells to
 "./layout" directory
DMpost> Running magic -Tscmos -L/cad/lib/magic/sys -dNULL
 MyLatch < magic.TmP4735
DMpost> Running /cad/bin/ext2sim -L -R -c 10 -a
 /class/layout/MyLatch.al -o /class/tut1/layout/MyLatch.sim
 MyLatch
DMpost>
DMpost>
DMpost> Options -THOR -sim -cif -drc successfully completed
DMpost>
DMpost> cif:see file ./layout/MyLatch.cif
DMpost> drc:see file ./layout/MyLatch.drc
DMpost> sim:see file /class/tut1/layout/MyLatch.sim
DMpost> THOR:see file ./THOR/MyLatch.csl
DMpost>
DMpost> Usage of irsim (in directory /class/tut1/layout):
DMpost>    irsim scmos80.prm MyLatch.sim -USER.COMMANDS
DMpost>
DMpost> See file DMpost.log in the current directory
>>
```

APPENDIX B

Training and Distribution

Bob Reese

Training is a key issue when dealing with a complex system such as LAGER. A week-long training course and associated tutorial materials were developed to introduce new users.

B.1. TRAINING

The typical training takes four and a half days with each day consisting of two lecture periods and two laboratory periods (one lecture, one lab in the morning; one lecture, one lab in the afternoon). The schedule shown below covers the basic use of the toolset. After this training the user is able to produce designs which can use any of the cells in the three LAGER cell libraries (standard cell, datapath, macro-cell).

Day 1

The morning lecture covers the basic environment and SDL syntax. The Oct database is discussed and the process by which LAGER converts a SDL file to a silicon implementation is demonstrated. In the lab session the students are required to write a SDL file for a simple parameterized design. The afternoon lecture covers the functional and switch level simulators used in LAGER (THOR and IRSIM). The afternoon lab section has the students simulate the morning lab design with both IRSIM and THOR..

Day 2

The morning lecture covers the Standard cell design methodology. The laboratory session has the students create a small standard cell design. The afternoon lecture discusses the new SDL language features such as looping, conditional instances, conditional nets, and conditional terminals. The lab session has the students write the SDL file for a parameterized shifter using these new SDL features.

Day 3

The morning lecture discusses the logic synthesis methodology using the misII tool and BDS language. The lab session has the student synthesize the control for a small finite state machine. The afternoon session discusses the padframe creation tools. The lab session requires the student to put a padframe around a design created earlier in the week.

Day 4

The morning lecture covers the datapath library and Flint macro cell router. The lab session has the student create a simple datapath design. The afternoon session discusses the TimLager macro cell library concentrating on the multiplier and RAM macro cells. The afternoon session has the students instantiate and simulate one of the new macro cells in the TimLager library.

Day 5

This half day covered the TGS test generation tools. The lab session has the students generate test vectors for a sample design.

B.1.1. Training Materials

The primary training materials are overhead slides (approximately 200) which uses mixed text and graphics to convey LAGER concepts. All of the training slides are included in the LAGER distribution tape.

B.2. LAGER DISTRIBUTION

The LAGER distribution disk storage requirements vary from release to release but the total disk usage for the LAGER 3.0 release is approximately 200 Mbytes (includes Octtools, LAGER source on-line, all tools installed). Installation instructions also vary between releases and are shipped with the tape. Hints are given within the installation instructions on how to further reduce disk storage by removing source trees and not installing some optional systems.

All LAGER documentation is included on the tape in the form of "man" pages and postscript files. There are also seven bounds volumes of LAGER documentation which are available with the distribution. These are:

- Volume #1: LAGER Training Slides, LAGER Tutorials, Magic Tutorial
- Volume #2: LAGER Man pages; Thor, IRSIM, BDS User Guides
- Volume #3: Dpp, TimLager, Pad Library docs
- Volume #4: Stdcell Library docs
- Volume #5: GDT to LAGER Interface
- Volume #6: GE Bitserial Toolset Interface
- Volume #7: Viewlogic Interface

Volume #1 contains all of the training slides previously discussed. Volumes #5, #6 and #7 document LAGER interfaces to commercial tools.

INDEX

A
ACTUAL_PARAMETERS bag 28
Addright 72, 73, 78, 79, 81
   ALIAS 73, 80
   INDEX 73
   INDEX argument 80
   MX,MY 73
   OFFSETX,OFFSETY 73
   OVERLAP 73
   R90,R180,R270 73
   TD 73, 80
Addup 72, 73, 78, 79, 81
ASIC 267, 302, 319
automatic test pattern generation 195

B
bags
   ACTUAL_PARAMETERS 28
   CONNECTORS 23
   FORMAL_PARAMETERS 27, 32
   INSTANCES 24
   MAP 30
BDS 89
Bds2stdcell 38, 90
bdsyn 91
bipartitioning 110
bit-parallel compilers 270
bit-serial compilers 270
bit-slice datapath, see also dpp 127
BLIF 90
boundary scan 187, 189
   architecture 189

C
cables
   essential 117
canonical signed digit 274
cell 11
cell library 81
channel 114
   congestion 116
   ring-shaped 143
channel definition 104
channels
   undesired 116
chip under test 194

CIF 57
circumferential constraint graph 143, 149
C-like 251, 253
clock and power nets 146
Close_newcell 75, 78
computers
   very-long-instruction-word 266
CONDITIONAL property 29, 31
CONNECTORS bag 23
control flow graph 226
control terminals 129
control-flow 200
CORDIC 288
core 141
corner
   region 146
   terminal 147
corner pad 142
cost function 110
C-to-Silicon 285, 297
current array 72
customization 53
cycle 115
   removal 151

D
data flow graph 226
data terminals 129
data-flow 200
datapath
   automatic generation 243
   block 128
   signal flow 128
debugging 52
Denavit-Hartenberg 286
DEPENDS-ON property 38
design manager 45
design post-processing (DMpost) 57
design rule checking 57
DFT 187
Dijkstra 117
DMoct 45, 57
   design flow 49
   design flow strategy 46
   design management strategy 53
DMverify 161

dogleg 122, 151
dpp 127
   algorithms 134
   block stretching 136
   datapath layout restrictions 132
   feedthroughs 132
   floorplans 135
   global routing 136
   interaction with DMoct 128
   library organization 138
   paramaters 134
   placement 138
dppdotc 130
dynamic instruction count 292

E
examples
   channel equalizer 252
   dpp 132
   filter
      microcode 262
      program in RL 261
   FIR filter 279
   macrocells 83
   tic-tac-toe 91
ext2oct 161, 163

F
FACET
   PAD parameter 39
facet
   contents 12
   definition 12
   version 12
fdl 104, 107, 125
FIR filters
   high performance 269
Firgen 272
   architecture generation 274
   clock and data tree 277
   critical path 275
   Flint floorplan 277
   floorplan generation 277
   layout generation 278
Flint 103
   cable 104
   slicing structure 105
   use by Firgen 278
   use with dpp 127
floorplan description language 107
floorplanning
   interactive 104
FORMAL_PARAMETERS bag 27, 32
fplan parameter 144
FSM 81
functional simulation 48

G
gensim 177
Getpath 61, 143
GMedit 181
Graph Search Machine 321

H
hidden Markov models
   grammar node 324
   null arcs 324
   output probability 322
   pruning 324
   speech recognition
      hidden Markov models 322
   state probability 322
   transition probabilities 322
   unique phones 323
   Viterbi algorithm 322
   wordarcs 323
hierarchical datapaths 134
hierarchy 18, 227
   flatten 89
histogram
   gray-level 312
Hyper 223
   allocation 225, 241
   assignment 225, 240
   control path generation 244
   data path generation 243
   design space 235
   estimation 224, 232
   module selection 223, 229
   parallelism graphs 233
   partitioning 243
   retiming 239
   scheduling 225, 240
   simulation 225
   transformations 224, 238

I
image
   analysis 302
   projections 301
   reconstruction 301
   synthesis 303

image processing 301
INSTANCES bag 24
interactive floorplanning 104
inverse position-orientation problem 285
IPOP 285
IRSIM 58, 173, 182
   circuit file 182
   parameter file 182
   PUMA example 183
   with LAGER 183
isomorphism 160

J
JTAG 189

K
Kappa 252
kappa datapath 290

L
Lager 242
lager file 80
layout extractor 160
layout verification 158
layout-generator 48, 52, 87
LAYOUT-GENERATOR property 31
libraries 53
library 46
   blocks 81
   dpp 138
   generators 81
   leafcell directory 81
LightLisp 27, 54

M
machine description 251
Machine Vision 302
machine-dependent 80
Magic 48, 71, 106, 154, 164
   obox label 78
MakeThor 180
MakeThorSim 177
MAP bag 30
microcode assembly 264
   layout parameters 264
min-cut partitioning 108
misII 90
mkmod 177
mobile radio 252
Mocha Chip 72
Mosaico 127

MOSIS 58

N
natural language processing 335
netdiff 167, 169
netlist
   flattened 89
NETWIDTH property 30, 40

O
OCT 11
   change propagation 19
   Inconsistencies 19
   physical policy 20
   procedural interface 16
   symbolic policy 22
OCT physical view 154
OCT_BAG 15
OCT_FACET 14
OCT_INSTANCE 15
oct2rinf 29, 337
oct2tgs 187
oct2wir 37
octdiff 169
Open_newcell 74, 78
output-partitioned 304
overlap box 98
OVERLAP layer 79, 98

P
pad
   corner 142
   pad groups 142
   space 142, 145
Padroute 141
   placement algorithm 144
Parallel Pipeline Projection Engine 304
parameterization 46
parameters 50
   in Thor 178
Parsifal 270
partitioning
   in-place 111
phoneme 322
phy2ext 161
physical policy 25
physical view 12, 48
PLA 81, 264
placement 104, 109
   absolute 104
   relative 105

subblocks 113
PODEM 196
policy 13, 25
   LAGER 26
   physical 25
   symbolic 25
   versus mechanism 13
power
   net 146
pragma 257, 261
pragmas
   word_length 259
printed circuit board 302, 320
programmable processors 267
promoting terminals 80
property
   CONDITIONAL 29, 31
   DEPENDS-ON 38
   LAYOUT-GENERATOR 31
   NETWIDTH 30, 40
   SIVMASTER 38
   STRUCTURE-PROCESSOR 31
   TERM_EDGE 31, 39
   THOR_MODEL 178
   THOR_TEMPLATE 178
   WIDTH 30
   WOLFE-ROW 95
pruning 324
PUMA 252
   simulation
      fixed point 295
   Thor and IRSIM simulations 297
   wordlength 295
Puma 560 286
PUMA processor 285

R

radial constraint graph 143, 149
Radon transform 301
   discrete 304
   hardware subsystem 306
RAM
   dynamic read-modify-write 311
raster-mode 304
rational approximation 287
region-of-interest 312
register transfer notation
   microoperations 261
RL compiler
   back end 265
   code generation 266

front end 265
greedy scheduling 266
implementation 264
lazy data routing 267
local scheduling 266
network flow 267
spill path 267
straight-line segment 265
trace scheduling 266
RL language 251, 254
   floating-point operators 259
   functions
      abs() 259
      in() 259
      out() 259
   init() 260
   loop() 260
   machine description 262
   memory and register banks 258
   pragmas 257
      word_length 259
   profiling 263
      dynamic instruction count 264
      kprof 264
   register declarations 258
   register-transfer notation 261
   scaling by a power-of-two 295
   simulation 263
   type modifiers
      const 258
      volatile 258
   types
      boolean 258
      fix:0 295
      fix:n 259
      fixed-point 254
      register type modifiers 258
      type modifiers 257
      volatile bool 257
   user-defined operations 259
RL program 290
routing
   clockwise distance 147
   cycle removal 151
   detailed 104
   dogleg 122, 144, 151
   global 104, 105
   gridless 120, 144
   horizontal constraint graph 121
   left-right 121
   minimum radius 147, 153

power 105, 118
priority 122
switch-box 115
track assignment 152
tracks 121
vertical constraint graph 121
routing verification 161

S

scan path 187
   register cell 188
scanpath test 333
SCANTEST 194
schematic entry 35
schematic view 12
SDL 87, 264
   for trist_inverter 129
   instance declaration 28
   net declaration 28
   parameters declaration 27
   subcells declaration 28
selective generation 51
self-timed 311
shortest path 116
Silage 199
   block 207
   control variable 205
   expression 202, 203
   iteration 207
   pragma 218
   reduction operators 205
   sampling rate 207, 215
simulation
   batch mode 174
   behavioral 173, 229
   Hyper 225
   switch-level 173, 182
siv2vsm 37
SIVMASTER property 38
space pad 142
spanning
   forest 118
   tree 118
speech recognition
   BEAM multiprocessor system 322
   cache register file 331
   grammar processing 325
   graph search machine 321
   left-to right HMM 326
   memory coherence 331
   real-time 321

   recognition accuracies 321
   switching processor architecture 329
   ToActiveWord process 327
   Viterbi process 326
   wordarc processing 325
standard cell 87
   cell pitch 98
Steiner-tree 116
Structural Description Language (SDL) 46
structure_instance view 32, 46, 55, 142
structure_master view 26, 46, 54
structure-processor 47, 52, 80, 90, 128
STRUCTURE-PROCESSOR property 31
symbolic microcode 264
symbolic policy 25
symbolic view 12

T

TERM_EDGE property 31, 39, 88
   in dpp 129
test access port 189
Test Controller Board 187, 194
test generation system 187
testability
   design for 187
testing
   board level 189
   chip level 188
   Test Controller Board 193
      software 194
   test pattern generation 195
Thor 48, 58, 173
   CHDL 174
   CSL 174, 176
      generation from OCT 179
   delay handling 180, 181
   gensim 177
   MakeThorSim 177
   mkmod 177
   model 175
   parameters 178
THOR_MODEL property 178
THOR_TEMPLATE property 178
tiling
   bounding box 79
   box 78
   C versus SDL 76
   dpp blocks 130
   point 79
   SDL method 75
tiling procedure 72

TimberWolfSC 95
time-stamp checking 51
TimLager 67, 87, 141, 143
   constructs for dpp 130
   use in dpp 129

U
UNIX 11

V
vb2oct 37
verification
   layout 158
   routing 161
vertical constraint graph 143
video rates 304
view 11
   extract 161
   physical 12, 32
   schematic 12
   structure_instance 32
   structure_master 26
   symbolic 12
Viterbi algorithm 322
   backtrack pointers 323
VLIW 266
VME bus 306, 329

W
WIDTH property 30
wolfe 95
   ACCESSIBILITY property 95
   FEEDTHRU property 96
   WOLFE-CLASS bag 95
   WOLFE-ROW property 95

Y
YACR 95